Springer Tracts in Modern Physics

Volume 279

AF806226

Series editors

Yan Chen, Department of Physics, Fudan University, Shanghai, China
Atsushi Fujimori, Department of Physics, University of Tokyo, Tokyo, Japan
Thomas Müller, Institut für Experimentelle Kernphysik, Universität Karlsruhe, Karlsruhe, Germany
William C. Stwalley, Department of Physics, University of Connecticut, Storrs, USA
Jianke Yang, Department of Mathematics and Statistics, University of Vermont, Burlington, VT, USA

Springer Tracts in Modern Physics provides comprehensive and critical reviews of topics of current interest in physics. The following fields are emphasized:

- Elementary Particle Physics
- Condensed Matter Physics
- Light Matter Interaction
- Atomic and Molecular Physics
- Complex Systems
- Fundamental Astrophysics

Suitable reviews of other fields can also be accepted. The Editors encourage prospective authors to correspond with them in advance of submitting a manuscript. For reviews of topics belonging to the above mentioned fields, they should address the responsible Editor as listed in "Contact the Editors".

More information about this series at http://www.springer.com/series/426

Bernd Hönerlage · Ivan Pelant

Symmetry and Symmetry-Breaking in Semiconductors

Fine Structure of Exciton States

Springer

Bernd Hönerlage
Institut de Physique et Chimie des Matériaux de Strasbourg, UMR 7504, CNRS
Université de Strasbourg
Strasbourg, France

Ivan Pelant
Institute of Physics
Czech Academy of Sciences, v.v.i
Prague 6, Czech Republic

ISSN 0081-3869 ISSN 1615-0430 (electronic)
Springer Tracts in Modern Physics
ISBN 978-3-030-06817-2 ISBN 978-3-319-94235-3 (eBook)
https://doi.org/10.1007/978-3-319-94235-3

© Springer International Publishing AG, part of Springer Nature 2018
Softcover re-print of the Hardcover 1st edition 2018
This work is subject to copyright. All rights are reserved by the Publisher, whether the whole or part of the material is concerned, specifically the rights of translation, reprinting, reuse of illustrations, recitation, broadcasting, reproduction on microfilms or in any other physical way, and transmission or information storage and retrieval, electronic adaptation, computer software, or by similar or dissimilar methodology now known or hereafter developed.
The use of general descriptive names, registered names, trademarks, service marks, etc. in this publication does not imply, even in the absence of a specific statement, that such names are exempt from the relevant protective laws and regulations and therefore free for general use.
The publisher, the authors and the editors are safe to assume that the advice and information in this book are believed to be true and accurate at the date of publication. Neither the publisher nor the authors or the editors give a warranty, express or implied, with respect to the material contained herein or for any errors or omissions that may have been made. The publisher remains neutral with regard to jurisdictional claims in published maps and institutional affiliations.

Printed on acid-free paper

This Springer imprint is published by the registered company Springer Nature Switzerland AG
The registered company address is: Gewerbestrasse 11, 6330 Cham, Switzerland

To our families:

To Marja
To Alena, Irena, and Dita

Preface

Semiconductor physics is a very interesting field for both applied and fundamental research. The physical properties of materials are often studied by applying perturbations to a sample and studying the results of these perturbations. Then, if one wants to understand the observed effects in detail, one has to simulate the experimental conditions carefully. We are here especially interested in the understanding of experimental findings, reflecting the electronic or excitonic fine structure of crystalline semiconductors.

The detailed interpretation of perturbations applied to a material needs a theoretical modeling of the system. This task can be much simplified if the system is described through an effective Hamiltonian, which reproduces the physical properties of the system. It is important that this model Hamiltonian is invariant under the same symmetry operations as the system. It is the aim of this book to discuss the construction of such effective Hamiltonians describing the electronic elementary excitations of simple crystalline semiconductors (called "excitons"), give some examples of its parameterized form, and discuss the role of symmetry-breaking effects. In this book, we concentrate on excitons in direct-gap bulk semiconductors with zinc blende or wurtzite structure and on the fine structure of the exciton ground state.

We construct an effective Hamiltonian of a semiconductor system starting from angular momentum operators, acting on electron states of the conduction or valence bands. Their eigenfunctions are adapted to the crystal point-group symmetry. We now consider some symmetry-breaking interactions in this angular momentum subspace of electron states as an example. Then, the electron spin is considered through building the product space of the angular momentum eigenstates and the spin states. The resulting spin orbitals are eigenstates of the total angular momentum operator, and they were again adapted to the crystal symmetry. The effective Hamiltonian, defined in this product space, determines the multiplet structure of states in the presence of "spin–orbit" or "crystal–field" interaction. The full Hamiltonian may show (when compared to the former subspaces) new symmetry-breaking interaction terms, which have their origin in the spin–orbit coupling. They manifest themselves, e.g., in the electronic dispersion, in Stark or Zeeman effect, or in a dependence on applied strain.

Exciton states are then formulated in the product space of conduction- and valence-band states where the exciton-binding energy and the electron–hole exchange interaction show up in a parameterized form. Spin–orbit, crystal–field, and exchange interaction may lead to an energy shift of the multi-component exciton ground state and lift its degeneracy. These energy variations depend strongly on the symmetry of the considered exciton state and that of the interaction term. In addition, similar to spin–orbit coupling, exchange interaction can give rise to new, symmetry-breaking interaction terms in the effective Hamiltonian. Their relative importance can be estimated by comparing the strength of the spin–orbit coupling to that of the exchange interaction.

If exciton states are dipole active, they give rise to interesting quasiparticles that can be studied in optical experiments. Such "exciton polaritons" (the resulting coupled light–matter excitations) exhibit nicely in their energy dispersion the symmetry-breaking effects to which excitons are subject. The determination of exciton–polariton dispersion relations provides a powerful tool for understanding the influence of external or internal perturbations on the physical properties of semiconductors.

This textbook is written for graduate students or young scientists, who want to understand and simulate experimental findings reflecting the electronic or excitonic fine structure of crystalline semiconductors. The level of presentation throughout this book has been chosen to be intelligible to graduate university students. Of course, a basic knowledge of solid-state physics (crystalline structure, energy-band structure, reciprocal space, elements of group theory) is required, but principle concepts are recalled in the appendices of this book.

We would like to mention the intense and fruitful discussions that we had over the years of cooperation with our colleagues M. Gallart and P. Gilliot (Strasbourg) and J. Kočka, P. Malý, and J. Valenta (Prague), which we gratefully acknowledge. We also highly appreciate the creative atmosphere in the "Département d'Optique Ultrarapide et de Nanophotonique" at the "Institute de Physique et Chimie des Matériaux de Strasbourg (IPCMS)" and in the "Department of Thin Films and Nanostructures" of the "Institute of Physics," Academy of Sciences of the Czech Republic as well as the inspiring long-term tradition of solid-state physics research in the "Department of Chemical Physics and Optics" at the "Faculty of Mathematics and Physics" of Charles University in Prague, Czech Republic. Part of the results presented here was obtained within project P108/12/G108 of the "Czech Science Foundation." Support through the Ministry of Education, Youth and Sports of the Czech Republic via the research infrastructure "Laboratory of

Nanostructures and Nanomaterials," project LM2015087, is also greatly acknowledged. One of the authors (I. P.) thanks the IPCMS for kind hospitality during numerous visits in this institution. In addition, we are indebted to the IPCMS for the technical and instrumental assistance and support, which it has afforded to us.

Strasbourg, France
Prague, Czech Republic
December 2017

Bernd Hönerlage
Ivan Pelant

Contents

List of Tables

Chapter 1
Introduction

1.1 Influence of Symmetry Breaking on the Physical Properties of Semiconductors

This textbook deals with the electronic properties of bulk semiconductors. In the narrower sense of the word, it sets itself the task of explaining how symmetry of the crystal lattice and its breaking due to various external and internal perturbations are reflected in the physical (predominantly optical) properties of semiconductors. The level of presentation throughout the book is based on matrix formulation of quantum mechanics, on at least partial knowledge of group theory, and, of course, on solid state physics basic concepts such as energy-band structure and reciprocal space. All these disciplines might seem to be quite difficult, but one should keep in mind all the time that they have been developed in order to help us to make up proper experiments and to understand the replies we receive in the form of their results. To elucidate the motivation behind writing this book, let us now discuss a few selected experimental results from the field of semiconductor physics, in order to show where and why the concept of symmetry is helpful if not indispensable for understanding and/or designing the electronic properties of solids.

Non-contact optical probing belongs among the most frequent techniques being applied to gather information about electronic states in solids. Photoluminescence and reflectance measurements in the optical frequency range are usually very sensitive to optical excitation of electrons across the band gap. A typical photoluminescence set-up is shown in Fig. 1.1 as adapted from [1]. A c.w. laser beam, focused by the lens L_1, is used to excite a sample fixed in a liquid He-cryostat. The photoluminescence radiation is collected by the lens L_2 and sent to a monochromator. The dispersed luminescence radiation is detected by a photomultiplier tube (PMT). (Nowadays, the PMT is often replaced by a charge-coupled device (CCD) detector.) An electrical signal is then fed into a photon counter and displayed on a screen. Alternatively, the laser light can be switched on and off to reveal photoluminescence decay times at a fixed wavelength.

© Springer International Publishing AG, part of Springer Nature 2018

B. Hönerlage and I. Pelant, *Symmetry and Symmetry-Breaking in Semiconductors*, Springer Tracts in Modern Physics 279,
https://doi.org/10.1007/978-3-319-94235-3_1

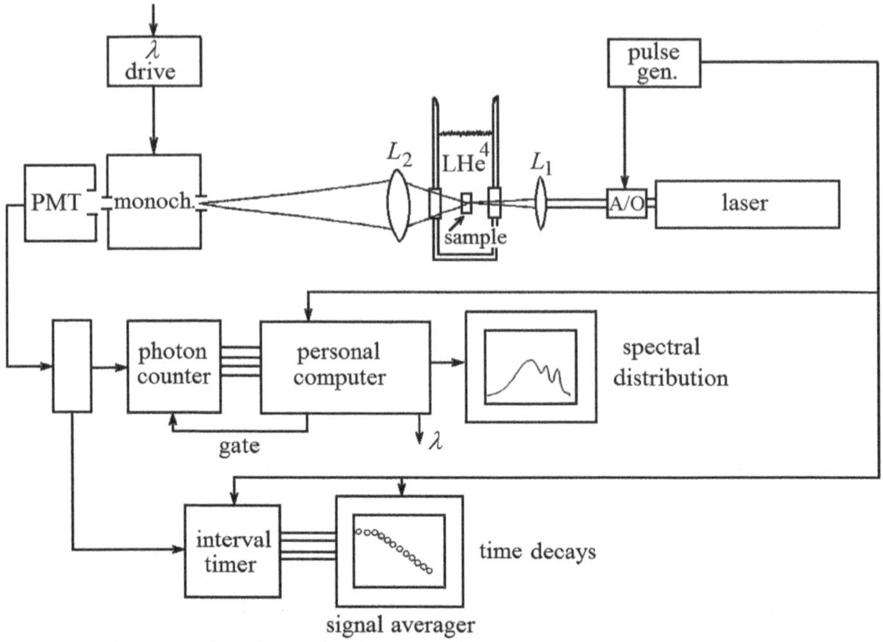

Fig. 1.1 Typical experimental arrangement for photoluminescence study. A c.w. laser beam, focused by the lens L_1, is used to excite a sample fixed in a liquid He-cryostat. The photoluminescence radiation is collected by the lens L_2 and sent to a monochromator. Dispersed luminescence radiation is detected by a photomultiplier tube (PMT). (Nowadays, the PMT is often replaced by a charge coupled device (CCD) detector.) Electrical signal is then fed into a photon counter and displayed on a screen. Alternatively, the laser light can be switched on and off to reveal photoluminescence decay times at a fixed wavelength. Adapted from Ref. [1]

A variety of experiments can be performed using photoluminescence spectroscopy, comprising both the spectral and temporal resolutions. An advantage of photoluminescence spectroscopy is that, in principle, samples of irregular shape and requiring (in most cases) no special care of their surface can be used for measurements. On the other hand, photoluminescence itself is extremely sensitive to the presence of unintentional perturbations as impurities, which can sometimes make the interpretation of experimental results ambiguous [2]. The same set-up as shown in Fig. 1.1 can be used, after proper modifications (consisting in replacing the excitation laser by e.g. a broadband lamp) to measure reflectance, i.e. to take spectra of radiation reflected from the sample surface. In this case, however, the surface should either be freshly cleaved or carefully polished, or high-quality thin films deposited on smooth substrates have to be used. Usually the spectra are acquired at very low temperatures ($T = 2$–10 K) since in this case the spectral features are neat and sharp, as they are got rid of the influence of crystal lattice vibrations.

An example of outputs from reflectance and photoluminescence measurements is shown in Fig. 1.2. At low temperatures, the principal response of a semiconductor

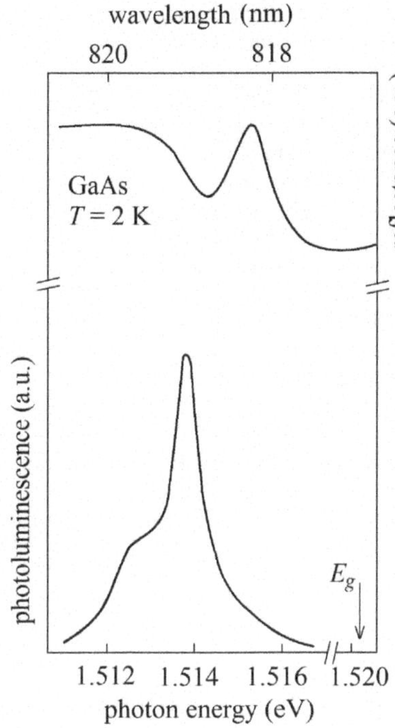

Fig. 1.2 Reflectance (upper curve) and photoluminescence (lower curve) spectrum of excitons in an epitaxial GaAs layer at $T = 2$ K. The photoluminescence was excited with a He-Ne laser at 633 nm (1.96 eV). E_g denotes the energy of the band gap. Adapted from Ref. [3]

to exciting radiation with photon energy close to the band gap is driven by the creation and annihilation of bound electron-hole pairs: the so-called excitons. These quasiparticles are treated in detail in the Chaps. 4, 5, and 7 of the present book. Their resonances manifest themselves, if explained in a simplified fashion, either as an "oscillation" or a "wobbling" in the reflectance spectrum (see Fig. 1.2, upper curve) or as a corresponding narrow line in the photoluminescence emission spectrum (see Fig. 1.2, lower curve) [3].

Now let us turn finally our attention to situations where a certain knowledge of symmetry principles is required if one wishes to interpret correctly all details of the experimental outputs. One of the most investigated semiconductors during the last decades seems to be gallium nitride GaN and its alloys (especially InGaN and AlGaN). This is since super-bright ultraviolet (UV), blue, green, and white light-emitting diodes and injection lasers based on these materials have been achieved. Their development was acknowledged by rewarding the Nobel Prize to three Japanese researchers (S. Nakamura, I. Akasaki, and H. Amano) in 2014. This means that predominantly the optical properties of GaN have found very successful applications in everyday life. Many prominent laboratories worldwide have been involved in research of GaN luminescence properties; in terms of the mission of this book it will be of interest to discuss and compare selected experimental results.

Figure 1.3 displays low-temperature photoluminescence (upper two curves) and reflectance spectra (lower two curves) of several GaN thin films in the exciton spectral

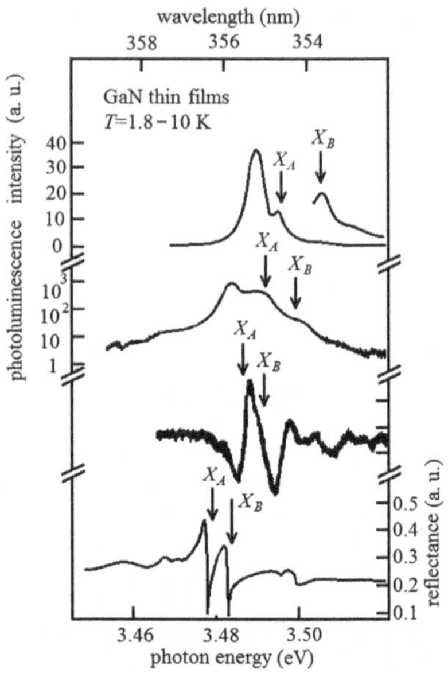

Fig. 1.3 Low-temperature photoluminescence (upper two curves) and reflectance (lower two curves) spectra of wurtzite GaN epitaxial films in the exciton region. X_A and X_B denote inter-band transitions associated with A (Γ_9^v - Γ_7^c) and B (Γ_7^v - Γ_7^c) excitons. The lowest curve characterizes a film grown on bulk GaN substrate, the remaining curves were acquired on films grown on sapphire substrates. Thicknesses of the films varied between 1.5 and 10 μm). Adapted after Ref. [4–7]

range, i.e. in the near UV region around 350 nm (photon energy ~ 3.5 eV). It should be noted here that GaN can crystallize in both, cubic zincblende (T_d point-group symmetry) and hexagonal wurtzite (C_{6v}) structure. By far the most common appears to be the hexagonal wurtzite structure and the spectra in Fig. 1.3 are obtained with this GaN modification. X_A and X_B denote inter-band transitions associated with A- and B- excitons, respectively. The lowest curve characterizes a film grown on bulk GaN substrate, the remaining curves were acquired on films grown on sapphire substrates. Thicknesses of the films vary between 1.5 and 10 μm. The curves are adapted after [4–7].

Figure 1.4 shows the band structure of wurtzite GaN around the center of the first Brillouin zone (i.e. around the electron wave-vector $\boldsymbol{Q} = 0$) according to Ref. [8]. The conduction band minimum (CBM) transforms like Γ_7. There are three valence bands with maxima (VBM) located also at $\boldsymbol{Q} = 0$. They are transforming like Γ_9, Γ_7, and Γ_7, respectively. The holes (and associated excitons) in these bands are called A-, B-, and C-type holes/excitons.

Free excitons manifest themselves in luminescence and reflectance (as outlined above) through distinct spectral maxima and wavy-like singularities, respectively. We focus our attention on the features denoted in Fig. 1.3 as X_A and X_B. As indicated in Fig. 1.4, the observed structures are associated with excitonic transitions from the two uppermost valence-band maxima (VBM) A and B to the conduction band minimum (CBM), having Γ_7 symmetry. The spectra in Fig. 1.3 were acquired in different laboratories but the spectral positions of the mentioned intrinsic exciton features are, of course, expected to be essentially (within the experimental accuracy) the same. However, a glimpse on Fig. 1.3 reveals that this is not the case. The experimental

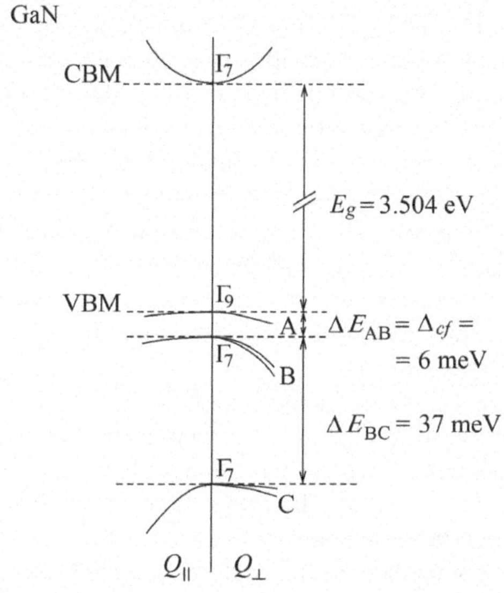

Fig. 1.4 Band structure of wurtzite GaN around the center of the first Brillouin zone $\boldsymbol{Q} = 0$. The conduction band minimum (CBM) transforms like Γ_7. There are three valence bands transforming like Γ_9, Γ_7, and Γ_7 with maxima (VBM) located also at $\boldsymbol{Q} = 0$. The holes (and associated excitons) in these bands are called A-, B-, and C-type holes/excitons. Δ_{cf} denotes the so-called "crystal-field splitting", E_g the band-gap energy, ΔE_{AB} and ΔE_{BC} the energy differences between A- and B- and B- and C-type holes, respectively. $Q_{\parallel}$ and $Q_{\perp}$ denote the wave-vector directions parallel and perpendicular to the crystallographic c-axis, respectively. According to Ref. [8]

accuracy of spectral devices in the near UV range is usually in the order of ~0.1 nm while the spectral positions of X_A and X_B vary over more than 2 nm.

What is, then, the origin of the obvious diversity in the experimental data? Disregarding from experimentalists' errors and temperature dependence of optical properties (which is known to be unimportant in GaN in the temperature range 1–10 K), one possibility remains only: mechanical strain inherently present in the thin films. Such films (with usual thickness of the order of 1 μm) must be grown on bulk substrates in order to assure their mechanical stability. Substrate materials with lattice constants close to that of epitaxially grown films are requested to make sure that the resulting films will be of good crystalline quality. Nevertheless, even a small lattice mismatch (a few percent) results in considerably strained films. Strain can be relaxed only in sufficiently thick films (>100 μm), usually by cracking, which deteriorates the film quality. The importance of strain thus varies with film thickness. Both compressive and tensile strain affect electronic and optical properties of solids due to symmetry breaking. The samples in Fig. 1.3, being grown on different substrates (bulk GaN and sapphire) with different thicknesses (1–10 μm), thus naturally exhibit different spectral manifestations of excitonic features. To evaluate properly the effect of strain (or mechanical deformation), detailed theoretical understanding of how the symmetry breaking modifies the physical properties of semiconductors is indispensable. This book is dedicated to a systematic exposition of such phenomena.

Here, a practical note for experimentalists is in order. Similar effects as above can enter the play when (before starting the experiment) fixing samples (as well in the case of thin films or bulk samples) to a holder. Excessive careless tightening of fixing screws/pads can stress the sample in an uncontrollable way and could make any measured results incorrect! The same problem can occurs if samples are fixed to a holder and temperature is varied, sample and holder having different dilatation coefficients. The list of possible sources of errors is long in experimental physics and (beside extremely careful operation) it is important to know the origin of errors in measurements that may show up.

The strain effects just discussed above can be called "unintentional". However, semiconductors can be strained intentionally to engineer their electronic band structure. For instance in silicon or germanium (indirect band gap materials) a properly applied mechanical stress modifies the many-valley band-structure, affecting the conduction- and valence-band degeneracy factors d_e and d_h, respectively.

Figure 1.5a shows a schematic of the well-known band structure of bulk silicon. In unstressed Si there are six equivalent ellipsoidal conduction-band minima along [100] directions (Δ-line), close to the Brillouin-zone boundary X. When neglecting the spin-orbit split-off band, the uppermost valence band Γ_8^+ is located at $\boldsymbol{Q} = 0$. It consists of two parabolic bands with different curvature. The structure is thus denoted as $[d_e, d_h]$ = [6,2]. Applying (via an external contact) stress along the crystal [100] axis, the number of equivalent conduction band minima decreases: only two of six are now the lowest ones. At the same time the valence band maximum splits into two bands ("heavy" and "light" holes). In many optical experiments only the upper band is involved and, consequently, this "stressed" structure is denoted as [2,1].

These stress-induced band structure modifications of silicon manifest themselves in a quite dramatic way in low-temperature photoluminescence spectra: In optically excited indirect gap semiconductors stability of electron-hole liquid (EHL) (a dense system of unbound free electrons and holes) with respect to free excitons (FE) is substantially enhanced by the multiplicity of the bands. This is because the multiple valleys delocalize the conduction-band electrons and thus tend to lower the kinetic energy of photo-carriers and thereby to increase the binding energy of EHL. Perturbing the crystal with an applied stress (i.e. removing the band degeneracy) makes the material less favorable for the existence of EHL, which is nicely seen in experiments as shown in Fig. 1.5b (adapted after [9]). At zero stress the recombination radiation from electron-hole pairs in droplets of EHL represents a broad band at 1.083 eV (1.145 μm) while a small peak at 1.098 eV (1.129 μm) is due to FE luminescence. Energy separation between these two lines gives the binding energy of EHL. This separation decreases with increasing stress, which indicates that the EHL binding energy relative to FE decreases, indeed. This observation correlates with an increased intensity of the FE line (meaning that more free excitons occur around the EHL droplets) while at the same time the EHL band gets narrower (meaning that equilibrium density of electron-hole pairs in the liquid state is reduced). Further increase of stress and a corresponding increase in density of the FE gas lead

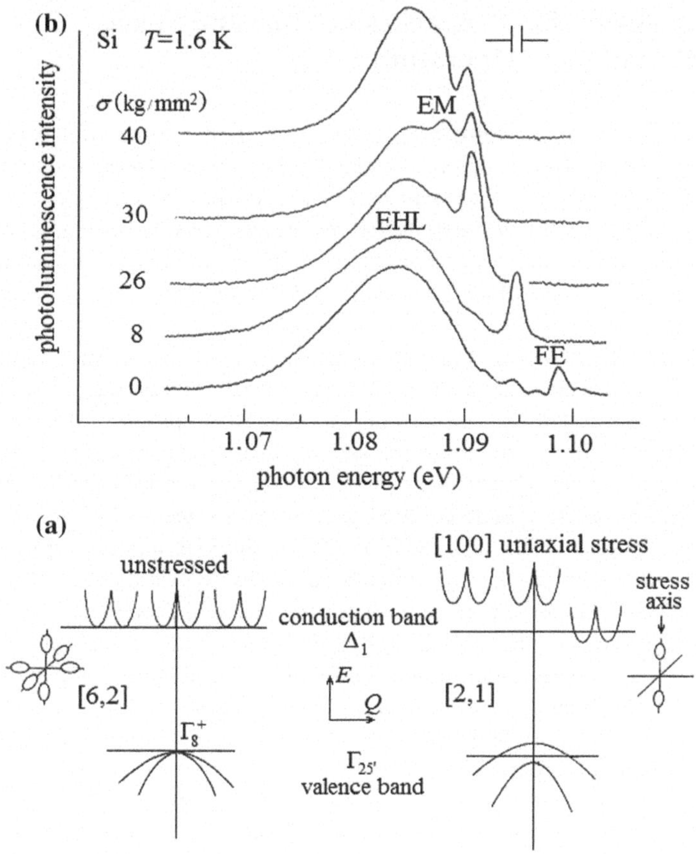

Fig. 1.5 (**a**) Schematic diagram of silicon conduction- and valence-band extrema and the effect of [100] stress. In unstrained Si there are six conduction-band minima and two degenerate valence-band maxima ([6,2] structure). With sufficient [100] stress a [2,1] structure is achieved. (**b**) Low-temperature photoluminescence spectra of bulk crystalline silicon under increasing [100] stress, the values of which are shown at each curve. FE, EHL, and EM denote emission lines of free exciton, electron-hole liquid, and excitonic molecule, respectively; see text. Adapted after Ref. [9]

then to an additional interaction: a substantial number of excitonic molecules (EM) or biexcitons can be created and, consequently, a new emission peak EM emerges, which naturally entails reduction in the FE gas density, as signalized by a reduced FE line intensity. Excitonic molecules are quasi-particles consisting of two electrons and two holes. They may originate via a fusion of two excitons when the density of the FE gas is sufficiently high.

It is obvious that in order to architect and interpret similar type of sophisticated experiments a deep understanding of symmetry-breaking influence on electronic states in semiconductors is essential.

1.2 Transformation Properties in Time and Space of Hamiltonian Operators

As we have seen in the preceding paragraph one is often confronted in experimental solid-state physics with the problem to understand the physical properties of a specific sample in detail. This is not always an easy task since the effects may largely depend on the intrinsic properties of the material under consideration (as e.g. its electronic band structure or phonon dispersion) or on extrinsic parameters (e.g. impurities, scattering processes, perturbation by fields, energy or phase relaxation processes, etc.).

To obtain a better understanding of the material properties, one often applies external perturbations (as electric or magnetic fields, strain or stress etc.) to the samples. This is for example necessary if one wants to modify or optimize different physical properties. Another important point is the detailed interpretation of the results, which depend also on the applied measuring process (linear or nonlinear optical processes, electrical, acoustical etc.). Such detailed interpretations need a thorough theoretical modeling of the system. This task can be much simplified if important system parameters can be extracted from the experimental results by describing the system through an effective Hamiltonian, which reproduces the physical properties of the system. It is important that this model Hamiltonian is invariant under the same symmetry operations as the system under consideration. This procedure is known as "Invariant Expansion of Effective Hamiltonians" [10–14]. Before starting to construct such effective Hamiltonians in Chap. 2, let us first discuss the general symmetry properties of a Hamiltonian in more detail.

Following Ref. [15] we consider a point-like free particle in the frame of non-relativistic classical mechanics. The particle is supposed to move without any interaction in an inertial system, i.e. a system, in which time is homogeneous and space isotropic and homogeneous. Since space and time are homogeneous, the Lagrange function L of the particle is independent of the position vector $\boldsymbol{r}$ and of time t. The Lagrange function can thus only depend on the particle velocity $\boldsymbol{v}$, which is defined as the variation of $\boldsymbol{r}$ with time:

- $\boldsymbol{v} = \mathrm{d}\boldsymbol{r}/\mathrm{d}t$.

Since the space is isotropic, however, L can only depend on the absolute value of $|\boldsymbol{v}|$, i.e.

- $L = L(v^2)$.

Since L is independent of $\boldsymbol{r}$, $(\dfrac{\partial L}{\partial \boldsymbol{r}}) = 0$ or

- $\mathrm{d}/\mathrm{d}t\ (\frac{\partial L}{\partial \boldsymbol{v}}) = 0$.

Integrating this with respect to time results in

- $\frac{\partial L}{\partial \boldsymbol{v}} = \mathrm{const}$

for the Lagrange function. Since $(\frac{\partial L}{\partial \boldsymbol{v}})$ is independent of the direction of the particle motion and since it depends only on the absolute value of $|\ \boldsymbol{v}\ |$, this results in $\boldsymbol{v} =$ const. This means that the motion of a free particle in an inertial system is described by a constant value of the velocity, which is in a fixed direction.

On the other hand, the Lagrange function determines the equations of motion of the particle, which have to have the same form in all possible inertial systems. Considering two different inertial systems, which move with the infinitesimal small velocity to each other, and transforming the coordinates of the mass point from one system to the other, one finds:

- $L = Av^2 = mv^2/2$

where the constant $A = m/2$ is positive and m denotes the mass of the particle. This defines the kinetic energy

- $T = mv^2/2$

of the particle. If we have a system containing many particles, their Lagrangians add up. If the particles are not interacting, we obtain

$$L = \sum_a \left(m_a \frac{v_a^2}{2} \right) \tag{1.2.1}$$

and $a = 1, \ldots, n$; where "a" indexes the particle and "n" is their total number. If the particles are interacting in-between themselves but not with other objects outside the system, and if the interaction is independent of time, the system is closed. Then, the Lagrange function can be given by adding a function $U(\boldsymbol{r}_1, \ldots, \boldsymbol{r}_n)$ depending only on the particle coordinates $\boldsymbol{r}_a$ to the Lagrange functions of the free particles

$$L = \sum_a \left(m_a \frac{v_a^2}{2} \right) - U(\boldsymbol{r}_1, \ldots, \boldsymbol{r}_n) \tag{1.2.2}$$

where $T = \sum_a m_a\, v_a^2/2$ and $U(\boldsymbol{r}_1, \ldots, \boldsymbol{r}_n)$ denote the total kinetic and potential energy of the system, respectively. The same form of the Lagrange function is obtained if a particle is moving in a time independent field. Since the time is homogeneous, the Lagrange function L does not explicitly depend on time. Then, the energy E

$$E = \sum_a \left(m_a \frac{v_a^2}{2} \right) + U(\boldsymbol{r}_1, \ldots, \boldsymbol{r}_n) \tag{1.2.3}$$

of the system is a constant of motion and is a conserved quantity. The total energy of the system is given by the sum of the kinetic energy, which is a function of the masses and velocities of the particles, and the potential energy, which is a function of the particle coordinates only. This equation is equivalent to the Hamilton function H (which is here expressed by coordinates and velocities of the mass points, and not through generalized coordinates and generalized momenta) describing a conservative system of particles. It is important to notice that energy and therefore the Hamilton function are scalar functions.

The form of the Lagrange function shows also that time is not only homogeneous but also isotropic: The Lagrange function or the Hamilton function are invariant if one changes $t \rightarrow (-t)$. This means that if a motion is possible, the reverse motion is also possible.

Energy conservation in a closed system or for a particle in a temporally constant field that is discussed above results from the fact that time is homogeneous and the interactions are independent of time (see Ref. [15]). (The Hamilton function does not explicitly depend on time, i.e. $\frac{\partial H}{\partial t} = 0$.) Such conservation laws are well known and are obtained in field theory from Noether's theorem. In our case it states that, if the Lagrange function is invariant with respect of infinitesimal shifts of its variables the corresponding conjugated generalized variables are conserved.

Besides energy conservation in systems, where the Lagrange function does not explicitly depend on time, conservation laws are also valid (Ref. [15]) for momentum components if the system is invariant against infinitesimal displacements along a coordinate axis. Let us consider an example taken from classical mechanics in order to discuss the result of an inhomogeneity or anisotropy in space. If space is homogeneous and isotropic and a particle is not accelerated, its movement is uniform and rectilinear, i.e. $\boldsymbol{v}$ = const. as discussed above. If, however, the space is inhomogeneous the movement is no longer uniform:

Consider a ball that is reflected by a flat wall orientated on the (z, y) plane. Then the momentum components parallel to the wall (p_z, p_y) are conserved during the reflection since the space is homogeneous in these two directions (z, y). The momentum component perpendicular to the wall p_x, however, is inverted after reflection: Since space is not homogeneous in the (x) direction, the momentum component p_x after the reflection is undetermined. It can be derived, however, from energy conservation. The equation

$$E = p^2/2m = p_x^2/(2m) + p_y^2/(2m) + p_z^2/(2m) = const.$$

allows two solutions for p_x after the reflection process: p_x and $-p_x$ such that the absolute value of the momentum $\mid \boldsymbol{p} \mid$ and therefore the energy are conserved. p_x = const. being excluded since the ball is reflected from the wall leads to the change $p_x \rightarrow -p_x$ after reflection in our example.

The same argumentation holds for the angular momentum components (l_x, l_y, l_z) if the system is invariant against infinitesimal rotations around any axis (x, y, z). Other examples concerning the Hamiltonian of a system are also given in [10].

We will mainly restrict here to electronic excitations in crystalline semiconductors and semiconductor structures, but the procedure can also be applied to other problems since it is quite general. As mentioned above, conservation laws are obtained in field theory from Noether's theorem, which is very useful whenever considering the effect of continuous transformation groups since it relates the symmetry properties of a physical system to these conservation laws [16].

In general, Lagrange and Hamilton functions depend explicitly on time. In classical mechanics the Hamilton function H determines entirely the mechanical state of the system. The total energy E then reads

$$E = H(q_1, q_2, \ldots, q_s; p_1, p_2, \ldots, p_s; t) \tag{1.2.4}$$

where $H(q_1, q_2, \ldots, q_s; p_1, p_2, \ldots, p_s; t)$ depends on the generalized coordinates q_i and the generalized momenta p_i in configuration space and on time. The number of degrees of freedom is indicated by "s". In addition, the equations of motion, which connect accelerations with the generalized coordinates q_i (which define the position vectors $\boldsymbol{r}_a$ with $a = (1, \ldots, n)$) and the generalized momenta $\boldsymbol{p}_a$, allow to calculate the evolution of the system. In quantum mechanics the state of the system can be determined solving the Schrödinger equation, which can be formally written as

$$i\hbar \frac{\partial \Psi}{\partial t} = H(q_1, q_2, \ldots, q_s; t)\Psi. \tag{1.2.5}$$

In Eq. (1.2.5) $\hbar = h/(2\,\pi)$, where h is Planck's constant. $\Psi(q_1, q_2, \ldots, q_s; t)$ denotes the wave function of the system, defined in configuration space and H is the Hamilton operator or Hamiltonian of the system. It can be derived from the classical Hamilton function using the correspondence principle. But, when doing this, care has to be taken [17] in order to ensure the validity of the Schrödinger equation under coordinate transformations:

First, one uses normally Cartesian coordinates where the generalized coordinates q_i are maintained and the generalized momenta p_i are replaced [17]. The correspondence principle involves the following substitutions:

- $E \Rightarrow i\hbar\, \partial/\partial t$

and

- $p_i \Rightarrow (\hbar/i\,)\, \partial/\partial q_i$

where $(i = 1, 2, \ldots, s)$. (Throughout the book we use the same font i both for the imaginary unit and also to denote consecutive subscripts/indexes. We believe, however, that confusion is excluded.) The use of Cartesian coordinates is not arbitrary but ensures the invariance of the Schrödinger equation under rotation of axes. One can remove this restriction and formulate the correspondence principle in a covariant form (see [17, 18]) but this generalization is beyond the scope of this book.

Second, a difficulty occurs if different generalized variables appear in an interaction term of the Hamilton function simultaneously, whose corresponding quantum mechanical operators do not commute (as e.g. p_i and q_i). In this case, the interaction term has to be symmetrized.

Let us consider an example [17]: in Cartesian coordinates the classical Hamilton function of interacting particles is assumed to be the sum of a quadratic expression in the generalized momenta p_i (which is completely independent of coordinates), a function, which depends only on the coordinates q_i of the particles, and a function containing the generalized momenta p_i in a linear form, typically

- $\sum_i p_i \, f_i(q_1, q_2, \ldots, q_s)$

where $f_i(q_1, q_2, \ldots, q_s)$ depends on the coordinate q_i. If this is the case, the forms

- $\sum_i p_i \, f_i(q_1, q_2, \ldots, q_s)$

and

- $\sum_i f_i(q_1, q_2, \ldots, q_s) \, p_i$

are no longer equivalent after the substitution according to the correspondence principle since after the operators have been applied to the wave function $\Psi(q_1, q_2, \ldots, q_s; t)$ the results are different. In such cases, the interaction term has to be replaced by its symmetric expression

- $(1/2)\sum_i [p_i \, f_i(q_1, q_2, \ldots, q_s) + f_i(q_1, q_2, \ldots, q_s) \, p_i]$

before the correspondence principle is applied to this symmetrized expression. As discussed in [17] this procedure removes the ambiguity resulting from the order of the quantum mechanical operators and secures that the statistical interpretation of the wave function remains consistent. The same symmetrization procedure has to be adopted in Chaps. 3–7 when interaction terms of effective Hamiltonians are constructed from matrix operators which do not commute.

Up to now, we have discussed how the Hamiltonian describing a quantum mechanical system is obtained from the classical Lagrange function

$$L = \sum_a \left(m_a \frac{v_a^2}{2} \right) - U(\boldsymbol{r}_1, \ldots, \boldsymbol{r}_n) \tag{1.2.6}$$

and that, if $U(\boldsymbol{r}_1, \ldots, \boldsymbol{r}_n)$ does not depend explicitly on time, the energy E is a constant of motion given by

$$E = \sum_a \left(m_a \frac{v_a^2}{2} \right) + U(\boldsymbol{r}_1, \ldots, \boldsymbol{r}_n). \tag{1.2.7}$$

Using the correspondence principle we obtain for the Hamiltonian:

$$H = \sum_i \left(\frac{p_i^2}{2m_i} \right) + U(\boldsymbol{r}_1, \ldots, \boldsymbol{r}_n) \tag{1.2.8}$$

with $i = (1, \ldots, s)$ denoting the degrees of freedom and $p_i = (\hbar/i)\,\partial/\partial q_i$. Equivalently, this reads

$$H = -\sum_a \hbar^2/(2m_a)\Delta_a + U(\boldsymbol{r}_1, \ldots, \boldsymbol{r}_n) \tag{1.2.9}$$

with ($a = 1, \ldots, n$) if n particles of mass m_a are considered, Δ_a being the Laplace operator acting on the ath particle. The Schrödinger equation reads in its time independent form:

$$H(q_1, q_2, \ldots, q_s; t)\Psi(q_1, q_2, \ldots, q_s) = E\Psi(q_1, q_2, \ldots, q_s), \tag{1.2.10}$$

the total kinetic and potential energies of the system being given by

- $\sum_a -(\hbar^2/2m_a)\Delta_a$

and

- $U(\boldsymbol{r}_1, \ldots, \boldsymbol{r}_n)$,

respectively.

It is important to mention at this point that the kinetic energy term has spherical symmetry and that the potential $U(\boldsymbol{r}_1, \ldots, \boldsymbol{r}_n)$ contains all the inhomogeneities and anisotropies of the system under consideration. It follows from group theory that, since the energy E is a real scalar quantity, the Hamiltonian H has to be a scalar operator. As we will discuss later, it then transforms according to the one dimensional identity representation Γ_1 of the symmetry group under consideration [11, 16] and in the case that is interesting for us it has to be invariant under time reversal as indicated by K^+.

1.3 Examples for Symmetry-Breaking Effects in Systems with Spherical Symmetry

Let us at this stage discuss some examples concerning possible terms, which can or cannot be present in a Hamiltonian. We consider for simplicity a single electron of mass m in a 3-dimensional attractive spherical potential $V(\boldsymbol{r})$. We denote by $\boldsymbol{p}$ and $\boldsymbol{r}$ the momentum and position vectors of the electron, respectively. The unperturbed Hamiltonian H_0 of the particle then reads

$$H_0 = p^2/(2m) + V(\boldsymbol{r}). \tag{1.3.1}$$

Let us now consider different perturbations H_i in the total Hamiltonian:

$$H = H_0 + H_i. \tag{1.3.2}$$

As discussed above, terms of the form

$$H_1 = a_1\boldsymbol{p} \tag{1.3.3}$$

or

$$H_2 = a_2\boldsymbol{r} \tag{1.3.4}$$

where the coefficients a_i are real numbers or real functions cannot occur in a Hamiltonian since H_1 and H_2 are no scalar expressions. A term of the form

$$H_3 = V_3(\boldsymbol{r}) \tag{1.3.5}$$

on the contrary, is possible if $V_3(\boldsymbol{r})$ is a scalar function.

Possible forms of H_i involving vectors would be an electric field $\boldsymbol{E}$, which interacts with an electric dipole moment $\boldsymbol{\mu}$ (describing the Stark effect)

$$H_4 = a_4 \boldsymbol{\mu} \cdot \boldsymbol{E} \tag{1.3.6}$$

where the dot designs the scalar product between $\boldsymbol{\mu}$ and $\boldsymbol{E}$, or

$$H_5 = a_5 \boldsymbol{m} \cdot \boldsymbol{B} \tag{1.3.7}$$

where $\boldsymbol{m}$ and $\boldsymbol{B}$ denote the magnetic moment and the magnetic field, respectively. H_5 is at the origin of the Zeeman effect or can be used to explain the spin-orbit coupling.

$\boldsymbol{E}$ ($\boldsymbol{B}$) are even (odd) functions under time reversal, their symmetry being indicated by K^+ (K^-), respectively. Products of two functions being both even or odd under time reversal are even functions; the product of an even and an odd function results into an odd function under time reversal. Therefore, $\boldsymbol{\mu}$ ($\boldsymbol{m}$) has to be even (odd), too, such that the product in H_4 (H_5) is invariant with respect to time reversal.

This is seen for example if we consider the form of the orbital magnetic moment $\boldsymbol{m}_{orb}$ in detail: We have $\boldsymbol{m}_{orb} = \gamma_e \boldsymbol{l}$ where γ_e is the gyromagnetic ratio of the electron and $\boldsymbol{l}$ the angular momentum operator. Using the definitions $\boldsymbol{l} = \boldsymbol{r} \times \boldsymbol{p}$ (where the vector product is indicated by "$\times$") and $\boldsymbol{p} = \mathrm{m}\ \boldsymbol{v}$ we find that $\boldsymbol{l}$ is also an odd function under time reversal. When multiplied with a magnetic field $\boldsymbol{B}$, the resulting perturbation operator H_5 is an even function under time reversal, which can be present in a Hamiltonian.

On the other hand, the dipole moment $\boldsymbol{\mu}$ scales linearly with the position vector $\boldsymbol{r}$ (i.e. $\boldsymbol{\mu} \propto \boldsymbol{r}$). Thus, $\boldsymbol{\mu}$ and $\boldsymbol{E}$ appearing in H_4 are even functions with respect to time reversal and can be present in a Hamiltonian when only their time-reversal symmetry is considered.

One can also notice that a term

$$H_6 = a_6 \boldsymbol{p} \cdot \boldsymbol{B} \tag{1.3.8}$$

is in principle possible when constructing a Hamiltonian (H_6 is a scalar operator and invariant under time reversal since $\boldsymbol{p}$ and $\boldsymbol{B}$ are odd functions with respect to time reversal) but it turns out that $a_6 \equiv 0$ since a charge moving parallel to the magnetic field is not accelerated, i.e. the Lorentz force is $\equiv 0$. Then, the velocity component parallel to the magnetic field is constant and for the other components the scalar product is equal to zero.

We see from these examples that the symmetry properties of the considered problem give information on whether an interaction term can be present in a Hamiltonian or not. It does not, however, give information about its absolute value, which has to be determined separately.

Effects of higher orders in $\boldsymbol{p}$, $\boldsymbol{r}$, $\boldsymbol{E}$, or $\boldsymbol{B}$ and terms, which involve products of these variables, can be constructed in the same way. Such perturbation terms are in the following called "Symmetry-Breaking Effects" [12–14] and it is the aim of this study to discuss their form and the consequences on the dispersion relation of electronic excitations, which are considered as quasi particles.

1.4 Some Considerations about Symmetry Properties of Crystalline Semiconductors and about Electron Band Structure

We are here mainly interested in semiconductor crystals. A crystal is characterized through a periodical arrangement of atoms or elementary building blocks of atoms, called "basis", in one, two, or three dimensions (dimension n= (1, 2, or 3)). These building blocks are attached to abstract "lattice points" that are arranged in such a way that they form a "Bravais lattice". The Bravais lattice is characterized by the fact that, if some well defined discrete spatial translations are applied to the lattice points, the environment of the point remains invariant under these translations. (Note that a periodic arrangement of points does not form necessarily a Bravais lattice.) Since the environment of a lattice point of the Bravais lattice has the same form for all points, the crystal structure can be decomposed into the (abstract) "lattice" and the (real) "basis". "Lattice" and "basis" fully describe together the crystal structure and determine the physical properties of the crystal. It is evident that such a perfect crystal takes into account only spatially periodic structures. Local (not periodic) structures as impurities or crystal imperfections are not considered here.

The shortest translation vectors, which fulfill this condition, are called "fundamental lattice vectors $\boldsymbol{a}_i$" (i = 1 to n) in direct space. These fundamental lattice vectors define the volume V of the unit cell, which is given by $V = \boldsymbol{a}_1 \cdot (\boldsymbol{a}_2 \times \boldsymbol{a}_3)$ in three dimensions.

The translation vectors $\boldsymbol{R}$ given by

$$\boldsymbol{R} = \sum_i m_i \boldsymbol{a}_i \tag{1.4.1}$$

span the Bravais lattice in direct space. In Eq. (1.4.1) the m_i are integer numbers ($m_i = 0, \pm 1, \pm 2, \pm 3, \ldots$) and the Bravais lattice has thus an infinite extension and is not limited by boundaries. In three dimensions there exist seven crystal systems which form 14 different kinds of Bravais lattices. These crystal classes and Bravais lattices are:

- Cubic: simple cubic, body-centered cubic, face-centered cubic
- Tetragonal: simple tetragonal, centered tetragonal

- Orthorhombic: simple orthorhombic, body-centered orthorhombic, face-centered orthorhombic, base-centered orthorhombic
- Monoclinic: simple monoclinic, centered monoclinic
- Triclinic: simple triclinic
- Trigonal: trigonal (or rhombohedral)
- Hexagonal: simple hexagonal.

Associated to the fundamental lattice vectors $\boldsymbol{a}_i$ are the "fundamental reciprocal lattice vectors $\boldsymbol{b}_i$", which are defined through $\boldsymbol{b}_1 = 2\pi(\boldsymbol{a}_2 \times \boldsymbol{a}_3)/V$ and cyclic permutations of the indexes "i" for $\boldsymbol{b}_2$ and $\boldsymbol{b}_3$. They generate the vectors

$$\boldsymbol{G} = \sum_i M_i \boldsymbol{b}_i \tag{1.4.2}$$

(with M_i integer numbers $M_i = 0, \pm 1, \pm 2, \pm 3, \ldots$), which span the reciprocal lattice. This reciprocal lattice is also a Bravais lattice.

The discrete translation vectors given in Eqs. (1.4.1) and (1.4.2) span Bravais lattices. When applied to a system the resulting operations are called "symmetry operations": A symmetry operation acting on an object leaves the object apparently unchanged. As mentioned above, this is fulfilled for example for the fundamental translation vectors of the crystal, but a crystal may be invariant under other, additional symmetry operations. When excluding translations, these symmetry operations define the "point-group symmetry" of the material. These operations leave at least one well defined point, line, or plane fixed. Together with the translations of the Bravais lattice they determine partly the space group of the crystal. For many physical properties of the crystal the point-group symmetry is very important and we will discuss it in detail in the following.

Let us first consider the consequences of translational invariance of crystals in solid state physics. As defined above, $\boldsymbol{b}$ and $\boldsymbol{G}$ are wave-vectors and $\hbar\boldsymbol{b}$ and $\hbar\boldsymbol{G}$ the corresponding quasi momenta. In this context, momentum conservation corresponds to the conservation of wave-vectors. As mentioned in connection with Noether's theorem, energy conservation remains valid in a crystal if the Hamiltonian is not explicitly time dependent. But, since the translations vectors $\boldsymbol{R}$ and the quasi momenta $\hbar\boldsymbol{G}$, under which the system is invariant, are discrete and not infinitesimal vectors, wave-vector is not strictly conserved in crystals. It is only conserved modulo a reciprocal lattice vector $\boldsymbol{G}$.

As discussed in more detail in [10], a similar problem arises in solid-state physics for the components of the angular momentum operator $\boldsymbol{l} = (l_x, l_y, l_z)$. If the Hamiltonian has spherical symmetry (as for example in simple atoms), it is invariant against infinitesimal rotations around any axis (x, y, z). Then, l^2 and one of the components of $\boldsymbol{l}$ are conserved under rotation. In crystals, depending on the point-group symmetry, only some rotations around discrete axes and for well defined angles are possible symmetry operations, i.e. they leave the system invariant. Taking also the spin $\boldsymbol{s}$ of the particle into account, these statements remain valid if the total angular momentum $\boldsymbol{j} = (j_x, j_y, j_z)$ (with $\boldsymbol{j} = \boldsymbol{l} \oplus \boldsymbol{s}$) is considered. Therefore, in general, angular momentum and total angular momenta are no longer conserved quantities in crystals. This also means that l and j are no longer good quantum numbers. We will see, however, that angular momentum can still be used in special cases in order to characterize crystal properties [10].

If we neglect translations, there are 5 other symmetry operations, which may leave an object invariant. To each operation belong symmetry elements, which specify the fixed points. The symmetry operations and symmetry elements are in general denoted by:

- E: The identity operation in which no action is applied to the object
- C_n: An n-fold rotation, i.e. the symmetry element is a rotation by $2\pi/n$ around an axis, which is called the "symmetry axis" of the system
- σ: A reflection on a mirror plane which has to be defined
- i: The inversion with respect to one point, which is called the "center of inversion". Taken as the origin, the coordinates $\boldsymbol{r}$ of a point are changed into $-\boldsymbol{r}$
- S_n: An n-fold improper rotation, i.e. the symmetry element is a rotation C_n by $2\pi/n$ around an axis, followed by a reflection on a plane perpendicular to this rotation axis

These symmetry operations are discussed in detail in Appendix D of this work.

Let us now consider the wave functions and energy levels in crystals in more detail. We have already stated that the crystal Hamiltonian remains invariant when applying the different translations of the Bravais lattice $\boldsymbol{R}$ to the system. Depending on the point group of the crystal under consideration, the Hamiltonian may also remain invariant under the other symmetry operations given above: e.g. rotations around axes or reflections on mirror planes etc. All axis, planes and points are well defined and are specific for a certain point group. Following the discussion in [10], we thus see that there exists a group of operators P_i, which transform the Hamiltonian H into itself. Accordingly, the Hamiltonian commutes with the operators P_i. Then, H and P_i have common eigenfunctions, corresponding to the same eigenvalues. The eigenvalues may be degenerate, i.e. they may be identical for different eigenfunctions.

Since all eigenvalues and eigenfunctions of H are supposed to be known, the eigenfunctions of P_i are linear combinations of those determined for H. Therefore, one may choose eigenfunctions, which are adapted to the crystal symmetry. They are said to transform according to a certain representation of the group. Such symmetry adapted eigenfunctions are very useful since they allow to make predictions about selection rules for transitions, level splittings, and degeneracies etc. when perturbations are applied to the system.

Before explaining the invariant expansion of a crystal Hamiltonian we will first consider the symmetry properties, which characterize in general a Hamilton operator. As discussed above, the Hamiltonian has to be a scalar operator with real eigenvalues. The symmetry of such an operator is labeled "Γ_1" in all crystal systems. If it does not explicitly depend on time, it has to be invariant under time reversal, i.e. it does not change if one considers $t \Rightarrow (-t)$. Then, energy is conserved and is a constant of motion. The operation of time reversal is called "Kramers' conjugation". In general, if an operator remains invariant under Kramers' conjugation its symmetry is denoted by K^{+}, if the operator changes its sign, it is denoted by K^{-}. In the language of group theory, one says that the Hamiltonian transforms as (Γ_1, K^{+}).

There is a further symmetry property: parity is a good quantum number if the system possesses a center of inversion. Even parity of an operator is indicated as Γ^{+}, odd parity by Γ^{-}. The full symmetry of a Hamiltonian is then denoted by (Γ_1^{+}, K^{+}) in crystal structures having a center of inversion. Starting from these symmetry properties, it is possible to give the general form of a Hamiltonian and the possible structure of the interactions.

As was stated above, we will construct an effective Hamiltonian, which shall model special properties of a semiconductor crystal in which we are interested. This effective Hamiltonian has to have the same symmetry properties as the system, which it approximates [11–13]. In the beginning, we are most interested in simple inorganic semiconductors (I–VII, II–VI, III–V compounds) with a direct energy band gap at the center of the Brillouin zone (the Γ-point). These semiconductors have mostly zincblende or wurtzite crystal structure, i.e. they have T_d or C_{6v} point-group symmetry, respectively. Diamond structure (O_h point-group symmetry) is similar to T_d structure. Both differ only by an additional center of inversion in O_h point-group symmetry such that parity is a conserved quantity. Therefore, a number of results obtained for T_d structure apply also for O_h symmetry.

To simplify the problem further, we will consider a certain subset of states we are interested in. Then, the Schrödinger equation can be written in matrix form with a finite dimension, corresponding to the number of states considered. These states are then coupled to each other due to different perturbations, which mix the states if the point-group symmetry is broken. The perturbations (as electric or magnetic fields, a finite wave-vector, strains, etc.) are then considered to different orders, which can have different symmetries. Taking into account the perturbations, the secular equation corresponding to the Schrödinger equation has to be diagonalized in order to find the eigenvalues and eigenvectors of the system. This approach is only reasonable if the perturbations are small and if the subset of states describes sufficiently well the considered physical properties of the semiconductor. In general, this is possible if the subset of states is energetically well separated from all other states. In that case, a small perturbation does not considerably mix the retained subset of states with the neglected ones.

In order to explain the technique of constructing the effective Hamiltonian in detail we will follow the procedure already described in Ref. [14]. It is, however, less general than that of Ref. [12] but perhaps easier to apply to a special problem. In both cases, we have first to specify the crystal structure and the subset of states, which we want to consider. It is essential to remind that the whole analysis, which follows now, has to be revised if a different crystal structure or subset of states is studied.

References

1. Wolfe, J.P., Jeffries, C.D.: In: Electron-hole droplets in semiconductors. Jeffries, C.D., Keldysh, L.V. (eds.) Modern Problems in Condensed Matter Sciences, vol. 6, p. 431. North-Holland, Amsterdam (1983)
2. Pelant, I., Valenta, J.: Luminescence Spectroscopy of Semiconductors. Oxford University Press, Oxford (2016)
3. Shay, J.L., Nahory, R.E.: Solid State Com. **7**, 945 (1969)
4. Kornitzer, K., Ebner, T., Thonke, K., Sauer, R., Kirchner, C., Schwegler, V., Kamp, M., Leszcynski, M., Grzegory, I., Porowski, S.: Phys. Rev. B **60**, 1471 (1999-I)
5. Chichibu, S., Okumura, H., Nakamura, S., Feuillet, G., Azuhata, T., Sota, T., Yoshida, S.: Jpn. J. Appl. Phys. **36**, 1976 (1997)
6. Monemar, B., Bergman, J.P., Buyanova, I.A., Amano, H., Akasaki, I., Detchprohm, T., Hiramatsu, K., Sawaki, N.: Solid-State Electron. **41**, 239 (1997)
7. Merz, C., Kunzer, M., Kaufmann, U., Akasaki, I., Amano, H.: Semicond. Sci. Technol. **11**, 712 (1996)
8. Shen, G.D., Smith, M., Lin, J.Y., Jiang, H.X., Wei, S.H., Khan, M.A., Sun, C.J.: Appl. Phys. Lett. **68**, 2784 (1996)
9. Gourley, P.L., Wolfe, J.P.: Phys. Rev. B **24**, 5970 (1981)
10. Klingshirn, C.: Semiconductor Optics, 3rd edn. Springer, Berlin, Heidelberg (2005)
11. Bir, G.L., Pikus, G.E.: Symmetry and Strain-Induced Effects in Semiconductors. Wiley, New York, Toronto (1974)
12. Cho, K.: Phys. Rev. B **14**, 4463 (1976)
13. Cho, K.: Excitons, Topics in Current Physics 14. Springer, Berlin, Heidelberg (1979)
14. Hönerlage, B., Lévy, R., Grun, J.B., Klingshirn, C., Bohnert, K.: Phys. Rep. **124**, 161 (1985)
15. Landau, L.D., Lifschitz, E.M.: Lehrbuch der theoretischen Physik I, Mechanik. Akademie Verlag, Berlin (1964)
16. Ludwig, W., Falter, C.: Symmetries in Physics: Group Theory Applied to Physical Problems, Springer Series in Solid State Sciences 64. Springer, Berlin, Heidelberg (1988)
17. Messiah, A.: Quantum Mechanics, vol. I. North Holland Publishing Company, Amsterdam (1967)
18. Brillouin, L.: Les Tenseurs en Mécanique et en Elasticité. Masson, Paris (1949)

Chapter 2
Symmetry-Breaking in Spin-Degenerate Conduction-Bands of Zincblende-Type Crystals

Most of simple binary crystalline semiconductors have either zincblende (T_d face-centered cubic point-group symmetry) or hexagonal closed packed wurtzite structure (C_{6v} point-group symmetry). They are derived from a cubic (fcc) or a hexagonal (hcp) structure. As discussed in detail in Refs. [1, 2] fcc and hcp structures are very similar to each other. This becomes apparent when regarding a simple model: We consider an ensemble of identical hard spheres that represent the unit cells, which will be used to build the crystal. The spheres can be arranged on a plane surface in a single closest-packed layer by surrounding each sphere by 6 other spheres. This gives rise to the hexagonal structure of the layer that is either the basal plane in the hcp structure or the (111) plane in the fcc structure. Let us note the positions of the spheres by "A". Then, the crystal is formed by placing the spheres of a second layer, identical to the first one, at the position of the holes of the first layer. Each sphere of the second layer touches thus 3 spheres of the first one. The positions of the spheres of the second layer are labeled "B". There are now two possibilities to position the third layer of spheres: either the new spheres are placed exactly above those of the first layer, i.e. at positions "A". When continuing the construction of a crystal in a periodic way, the positions of the spheres show the sequence "ABABAB…", which is the pile-up pattern of hcp crystals.

The second possibility to place the third layer is the position above the holes of the first layer, which are not occupied by the second layer. Calling this position "C", the pile-up sequence in the crystal growth becomes "ABCABC…". This structure is realized by (111) planes in fcc crystals. Both structures give rise to the same maximum volume-filling factor of the system. It is important to notice here that these structures (although similar) correspond, however, to different point-group symmetries of the crystals. The different positions in the pile-up procedure [3] are shown in Fig. 2.1 for the case of wurtzite (C_{6v} point-group symmetry, position "a") and zincblende structure (T_d point-group symmetry, position "b").

Let us first consider a direct semiconductor with zincblende structure, i.e. with T_d point-group symmetry. A typical one-electron band structure (i.e. the energy E of an electron state as a function of its wave-vector $\boldsymbol{Q}$) in the vicinity of the Γ-point is shown in Fig. 2.2. The semiconductor has an empty conduction band (index "c"),

© Springer International Publishing AG, part of Springer Nature 2018

B. Hönerlage and I. Pelant, *Symmetry and Symmetry-Breaking in Semiconductors*, Springer Tracts in Modern Physics 279,
https://doi.org/10.1007/978-3-319-94235-3_2

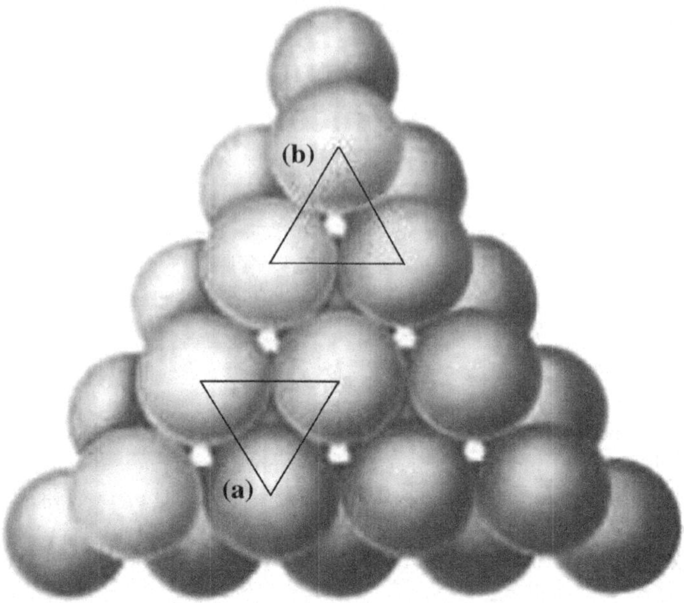

Fig. 2.1 The different positions in the pile-up procedure of wurtzite (C_{6v} point-group symmetry, position "a") and zincblende structure (T_d point-group symmetry, position "b"). Adapted from Ref. [3]

which, in most I-VII, II-VI or III-V compounds, is made up from spin-degenerate atomic s-orbitals. The symmetry adapted conduction-band states then transform as Γ_6 at the Γ-point [4, 5] when employing the double-group notation. We use in the following Cartesian coordinates, defined along the cubic axis (x, y, z).

The orbital part of the uppermost filled valence-band states (index "v") is mostly originating from atomic p-orbitals, which may have some admixture of atomic d-orbitals. These states are three-fold degenerate, described by an angular momentum $l = 1$. When including the electron spin, there are six degenerate valence-band states at the Γ-point. All other atomic states are supposed to be largely separated in energy from these states such that the considered subset of states describes the optical and electronic properties of the semiconductors close to the band gap.

Concerning the uppermost valence bands, as we will discuss in more detail, the degeneracy is partly lifted by spin-orbit interaction, giving rise to an energy splitting between a set of four-fold degenerate states, transforming as Γ_8, and the two-fold degenerate Γ_7 states [4, 5]. The band with Γ_7 symmetry is called the "split-off band" and the energy separation between the Γ_8 and the Γ_7 states is the "spin-orbit splitting" Δ^v_{so}. Usually, $\Delta^v_{so} > 0$, i.e. the energy of the electron states with Γ_8 symmetry is higher than that of the Γ_7 states. But, as for example in CuCl crystals, which have zincblende structure under normal conditions, the order can be reversed when compared to the other copper halides because of a different amount of admixture of atomic p- and d-orbitals in the crystal-wave functions.

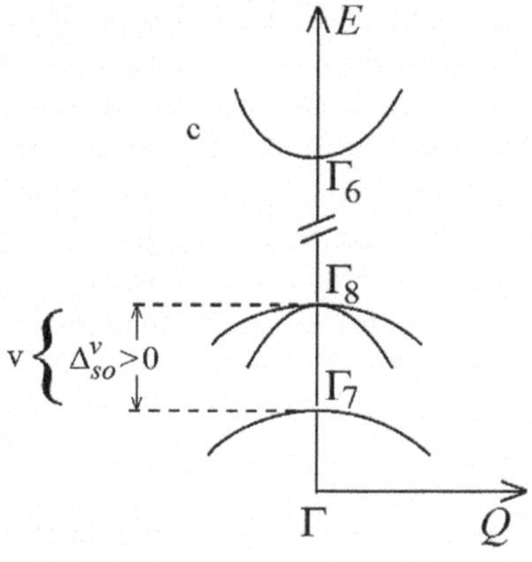

Fig. 2.2 Typical electron band structure (energy E as a function of wave-vector $\boldsymbol{Q}$) of a direct semiconductor with T_d point-group symmetry in the vicinity of the Γ-point. The direction of the wave-vector $\boldsymbol{Q}$ is not specified. Index "c": conduction band, index "v": valence bands, Δ_{so}^{v}: Energy splitting of the valence band due to spin-orbit coupling. Adapted from Ref. [9]

As we will see, a finite wave-vector $\boldsymbol{Q}$ breaks the T_d point-group symmetry. Then the band with Γ_8 symmetry at the center of the Brillouin zone is decomposed into a "heavy-hole" and a "light-hole band" at finite wave-vectors due to the dispersive terms varying as $\boldsymbol{Q}^2$ (see Fig. 2.2).

We will first consider a single, completely isolated electron in the lowest conduction band. All other electron states are neglected for the moment. The electron is described by an effective Hamiltonian H_e, acting only in the subspace of the lowest conduction-band electron states (index "e"). Due to the electron spin, this subspace is of dimension $p = 2$. Then, H_e can be written in matrix form as

$$H_e = \begin{pmatrix} a_{11} & a_{12} \\ a_{21} & a_{22} \end{pmatrix} \tag{2.0.1}$$

where the matrix-elements a_{ij} are functions of physical quantities, which have well defined symmetry properties with respect to translation, rotation, reflection etc. and time reversal.

2.1 Transformation Properties of Effective Hamiltonians in Crystalline Structures

As stated in the introduction, it is important to mention that, since the energy E is a real scalar quantity, the Hamiltonian H has to be a scalar operator. The Hamiltonian is compatible with the full point-group symmetry of the crystal, i.e. it is invariant under all symmetry operations. Therefore, it transforms as the one-dimensional identity operation. This operation corresponds to the irreducible representation, which is labeled "Γ_1" in all symmetry groups [6, 7].

In addition, the Hamiltonian in which we are interested is not explicitly depending on time, i.e. we look here only for stationary states whose energies are constants of motion. The electron wave-functions that are solutions of the Hamiltonian are Bloch functions, i.e. functions that are periodic on the crystal lattice multiplied by plane waves. The phase of the plane-wave part may fluctuate in time but the Bloch functions are time independent. Then, the squares of the electron eigenfunctions of the Hamiltonian are time-independent and the Hamiltonian itself is invariant under time reversal (that is labeled by K^+). This is a second necessary condition, which has to be fulfilled by any Hamiltonian considered in the following. These two symmetry conditions are the fundamental requirements of any invariant expansion by an effective Hamiltonian.

2.2 Effective Hamiltonians Involving Interaction Matrices Possessing the full Point-Group Symmetry

Because of the restrictions imposed onto the Hamiltonian, the different matrix-elements a_{ij} in Eq. (2.0.1) are not independent of each other but they are inter-connected and this interconnection depends on the sub-states considered. The eigen-functions of H_e are the pseudo-Bloch functions α_e and β_e where α and β denote the spin up- and down-states, respectively. The matrix $H_e = (a_{ij})$ can be decomposed into $n = p^2$ linearly independent basis matrices A_e^n:

$$H_e = \sum_n A_e^n. \tag{2.2.1}$$

This decomposition is arbitrary but the matrices A_e^n have to be linearly independent of each other in order to span the subspace for the total Hamiltonian. One could choose for example the following matrices A_e^n:

$$A_e^1 = a_{11}\begin{pmatrix}1 & 0\\ 0 & 0\end{pmatrix};\ A_e^2 = a_{12}\begin{pmatrix}0 & 1\\ 0 & 0\end{pmatrix};\ A_e^3 = a_{21}\begin{pmatrix}0 & 0\\ 1 & 0\end{pmatrix};\ A_e^4 = a_{22}\begin{pmatrix}0 & 0\\ 0 & 1\end{pmatrix}. \tag{2.2.2}$$

The advantage of another choice lies in the fact that one knows the transformation properties of angular-momentum operators in systems with a given point-group symmetry [5–9]. We therefore introduce an effective "pseudo-spin operator" $\boldsymbol{\sigma}_e$ with $\sigma_e = 1/2$, which only operates on the conduction-band spin-states. We choose for this two-dimensional problem the Pauli-spin matrices σ_e^i with $i = (x, y, z)$ as basis matrices:

$$\sigma_e^x = \begin{pmatrix}0 & 1\\ 1 & 0\end{pmatrix};\ \sigma_e^y = \begin{pmatrix}0 & -i\\ i & 0\end{pmatrix};\ \sigma_e^z = \begin{pmatrix}1 & 0\\ 0 & -1\end{pmatrix}. \tag{2.2.3}$$

These basis matrices transform like the components of an angular-momentum operator, i.e. like the irreducible representation Γ_4 in systems with T_d point-group symmetry [5–9]. In addition, since they describe components of an angular-momentum operator, they involve first-order time-derivatives and transform therefore as K^- under time reversal. When decomposing H_e into the Pauli-spin matrices we obtain

$$\begin{aligned} H_e =&(1/2)(a_{11} + a_{22})1_e + (1/2)(a_{12} + a_{21})\sigma_e^x \\ &+ (i/2)(a_{12} - a_{21})\sigma_e^y + (1/2)(a_{11} - a_{22})\sigma_e^z \end{aligned} \tag{2.2.4}$$

where the unit matrix 1_e is given by:

$$1_e = \begin{pmatrix} 1 & 0 \\ 0 & 1 \end{pmatrix} = (1/3)[(\sigma_e^x)^2 + (\sigma_e^y)^2 + (\sigma_e^z)^2] = (1/3)(\boldsymbol{\sigma}_e)^2. \tag{2.2.5}$$

If interactions with the physical system are considered, the first term of H_e of Eq. (2.2.4) may contain all interaction terms, which are diagonal in the considered states and have the same value for both states. This means that they do not couple the pseudo-Bloch functions α_e and β_e, which are eigenstates of the Hamiltonian H_e. The first term of Eq. (2.2.4) only shifts the energies of the states but does not split them, i.e. the states remain degenerate in energy. The corresponding interaction terms remain unchanged under all symmetry operations compatible with the point group of the crystal. In addition, since the matrix 1_e does not depend on time, these interaction terms have to be invariant under time reversal and their overall symmetry is (Γ_1, K^+).

As shown above, 1_e may also be constructed from the Pauli matrices by calculating $1/3(\boldsymbol{\sigma}_e)^2$. Concerning time-reversal symmetry one has to recall that operators transforming as K^- under time reversal (as e.g. $\boldsymbol{\sigma}_e$) give rise to operators having K^+ symmetry when taken to even orders, as it is the case for the operator 1_e.

We shall first consider the case that the coefficients a_{ij} are time independent real numbers or functions. Then, the term $\propto (1/2)(a_{11} + a_{22})1_e$ in Eq. (2.2.4) fulfills the above symmetry requirements and such interaction terms may be present in H_e, $(a_{11} + a_{22})$ being real. The interaction matrix 1_e possesses the full point-group symmetry, i.e. it does not mix the spin states and only shifts the eigenvalues. The corresponding interaction terms determine the energy of the spin-degenerate electron states in the conduction band at the center of the Brillouin zone.

Let us now discuss the three components in Eq. (2.2.4), which are directly proportional to Pauli's spin matrices. As stated above and given in Table 2.1, these matrices transform as Γ_4 in crystals with T_d point-group symmetry [4, 5] and are K^- under time reversal. Thus, when only multiplied by complex numbers or functions, they cannot be present in an effective Hamiltonian since they do not fulfill the above symmetry requirements. We thus find that

- $a_{11} = a_{22}$

and

- $a_{12} = a_{21} = 0$.

In conclusion, we have found that the matrix

$$H_e = a_e 1_e \tag{2.2.6}$$

(a_e being a real number or function) is the only form a Hamiltonian can take in its matrix representation if we restrict ourselves to a two-level system and preserve the full T_d point-group symmetry in the interaction matrix. Concerning the eigenvalues E_{α_e} and E_{β_e} of the eigenvectors α_e and β_e, Eq. (2.2.6) implies the relation $E_{\alpha_e} = E_{\beta_e}$. The special form of H_e is thus the reason why the conduction band in a simple semiconductor is always spin degenerate at the Γ point.

2.3 Construction of Effective Hamiltonians Containing Symmetry-Breaking Perturbations

Equation (2.2.4) shows that only symmetry-breaking perturbations can lead (when combined with the Pauli-spin matrices) to perturbation terms, which may lift the degeneracy of the states in the two-level system. In order to discuss the action of perturbations on electron states under T_d point-group symmetry we have to analyze their symmetry properties. As shown in [4, 5] there exist only 5 symmetry classes or irreducible representations in T_d point-group symmetry, into which all perturbations can (and have to be) decomposed. The irreducible representations are enumerated in Table 2.1 and labeled from Γ_1 to Γ_5. Table 2.1 gives also the denominations of the different components together with their transformation properties.

Let us denote by $\boldsymbol{l} = (l_x, l_y, l_z)$ an angular momentum operator and by $\boldsymbol{r} = (x, y, z)$ the position vector. With these operators we can work out the components belonging to the different irreducible representations, which are adapted to the symmetry of the crystal. The transformation properties of these objects (e.g. functions, operators, their matrix representations of the interacting states etc.) are indicated in square brackets in Table 2.1. As it is usual in group theory, Γ_1 denotes the symmetry of the identity operation, which is given in its matrix form by the unit matrix 1. Objects that have Γ_1 symmetry are spherically symmetric and transform as $[r^2 = x^2 + y^2 + z^2]$. They are labeled "S" in the following. Objects with Γ_2 symmetry are labeled "T". The irreducible representation Γ_2 is likewise Γ_1, an unidimensional representation.

Objects that transform as Γ_3 are labeled "(U, V)" and they transform as $[\sqrt{3}(x^2 - y^2), 2z^2 - x^2 - y^2]$, respectively. The elements of Γ_4 are labeled "(P, Q, R)" and transform as $[l_x, l_y, l_z]$, respectively. Γ_5 elements are called "(X, Y, Z)" and they transform as the components of the position vector $\boldsymbol{r}$, i.e. as $[x, y, z]$, respectively. It follows also from Table 2.1 that the components of Γ_2 (T) and those of Γ_4 (P, Q, R) transform as K^- under time reversal (Kramers' conjugation), the others have K^+ symmetry.

Table 2.2 gives the transformation properties of several perturbations under T_d point-group symmetry. Following [4, 9] we consider for example the transformation properties of a magnetic field $\boldsymbol{B}$, a finite wave-vector $\boldsymbol{Q}$, electric field effects (proportional to $\boldsymbol{E}$ or E^2), or strain components ϵ_{ij} (with $(i, j) = (x, y, z)$) and give their properties with respect to Kramers' conjugation (K^-, K^+).

Table 2.1 Notation of symmetry adapted operators or perturbation components in T_d point-group symmetry. Their transformation properties are given in square brackets. (K^-, K^+) give the transformation properties with respect to Kramers' conjugation. From Refs. [4, 8]

$\Gamma_1 : (S)$	$[\mathbf{1}]$ or $[\boldsymbol{r}^2 = x^2 + y^2 + z^2]$	$\lvert K^+\rvert$
$\Gamma_2 : (T)$	$[xl_x + yl_y + zl_z]$	$\lvert K^-\rvert$
$\Gamma_3 : (U, V)$	$[\sqrt{3}(x^2 - y^2), 2z^2 - x^2 - y^2]$	$\lvert K^+\rvert$
$\Gamma_4 : (P, Q, R)$	$[l_x, l_y, l_z]$ or $[\sigma_e^x, \sigma_e^y, \sigma_e^z]$	$\lvert K^-\rvert$
$\Gamma_5 : (X, Y, Z)$	$[x, y, z]$	$\lvert K^+\rvert$

Table 2.2 Transformation properties of perturbation components as magnetic field $\boldsymbol{B}$, wave vector $\boldsymbol{Q}$, electric field effects (proportional to $\boldsymbol{E}$ or E^2), and strain tensor components ϵ_{ij} (with $(i, j) = (x, y, z)$) in T_d point-group symmetry. From Refs. [4, 9]

	K^-	K^-	K^+	K^+	K^+
$\Gamma_1 : S$				E^2	$\epsilon_{xx} + \epsilon_{yy} + \epsilon_{zz}$
$\Gamma_2 : T$					
$\Gamma_3 : U$				$\sqrt{3}(E_x^2 - E_y^2)$	$\sqrt{3}(\epsilon_{xx} - \epsilon_{yy})$
$: V$				$3E_z^2 - E^2$	$2\epsilon_{zz} - \epsilon_{xx} - \epsilon_{yy}$
$\Gamma_4 : P$	B_x				
$: Q$	B_y				
$: R$	B_z				
$\Gamma_5 : X$		Q_x	E_x	$E_y E_z$	ϵ_{yz}
$: Y$		Q_y	E_y	$E_z E_x$	ϵ_{zx}
$: Z$		Q_z	E_z	$E_x E_y$	ϵ_{xy}

Very important information is contained in the multiplication scheme for the components of Table 2.1, which is given in Table 2.3 for T_d point-group symmetry. The Table follows directly from the multiplication Table 2.4 [4, 5] . When using Table 2.3 successively, perturbation components of higher order can be generated, which are adapted to the crystal symmetry. As we will see, when combined with the matrix components of the subspace of states one may then construct perturbation terms, which can be present in a Hamiltonian [4].

Let us discuss in detail the construction of interaction terms, which are present in a Hamiltonian, i.e. which have the full point-group symmetry and thus transform as Γ_1. Table 2.4 concerns the spatial part of the multiplication of objects in crystals with T_d point-group symmetry. Inspecting Table 2.4 one remarks that one can only form an object transformation as Γ_1 from two objects with transformation properties (Γ_i) and (Γ_j) if $(i = j)$. In addition, the combination of the different object components has to be carefully chosen in order to be adapted to the crystal symmetry.

Thus, the general strategy will be to construct from the components of symmetry-breaking interactions symmetry-adapted combinations that have the same symmetry as the matrix components of the subspace of states with which they are combined.

Table 2.3 Multiplication scheme for the components of Table 2.1 in T_d point-group symmetry. RC: Resulting component. From Refs. [4, 9]

RC	Product of components						
$\Gamma_1 : S$	SS'	TT'	$UU'+VV'$		$PP'+QQ'+RR'$		$XX'+YY'+ZZ'$
$\Gamma_2 : T$	ST'		$UV'-VU'$			$PX'+QY'+RZ'$	
$\Gamma_3 : U$	SU'	$-TV'$	$UV'+VU'$		$\sqrt{3}(PP'-QQ')$	$2RZ'-PX'-QY'$	$\sqrt{3}(XX'-YY')$
$: V$	SV'	TU'	$UU'-VV'$		$2RR'-PP'-QQ'$	$\sqrt{3}(QY'-PX')$	$2ZZ'-XX'-YY'$
$\Gamma_4 : P$	SP'	TX'	$(\sqrt{3}U-V)P'$	$-(\sqrt{3}V+U)X'$	$QR'-RQ'$	$QZ'+RY'$	$YZ'-ZY'$
$: Q$	SQ'	TY'	$-(\sqrt{3}U+V)Q'$	$(\sqrt{3}V-U)Y'$	$RP'-PR'$	$RX'+PZ'$	$ZX'-XZ'$
$: R$	SR'	TZ'	$2VR'$	$2UZ'$	$PQ'-QP'$	$PY'+QX'$	$XY'-YX'$
$\Gamma_5 : X$	SX'	TP'	$-(\sqrt{3}V+U)P'$	$(\sqrt{3}U-V)X'$	$QR'+RQ'$	$QZ'-RY'$	$YZ'+ZY'$
$: Y$	SY'	TQ'	$(\sqrt{3}V-U)Q'$	$-(\sqrt{3}U+V)Y'$	$RP'+PR'$	$RX'-PZ'$	$ZX'+XZ'$
$: Z$	SZ'	TR'	$2UR'$	$2VZ'$	$PQ'+QP'$	$PY'-QX'$	$XY'+YX'$

Table 2.4 Multiplication table for irreducible representations in crystals with T_d point-group symmetry

	Γ_1	Γ_2	Γ_3	Γ_4	Γ_5	Γ_6	Γ_7	Γ_8
Γ_1	Γ_1	Γ_2	Γ_3	Γ_4	Γ_5	Γ_6	Γ_7	Γ_8
Γ_2		Γ_1	Γ_3	Γ_5	Γ_4	Γ_7	Γ_6	Γ_8
Γ_3			$\Gamma_1+\Gamma_2+\Gamma_3$	$\Gamma_4+\Gamma_5$	$\Gamma_4+\Gamma_5$	Γ_8	Γ_8	$\Gamma_6+\Gamma_7+\Gamma_8$
Γ_4				$\Gamma_1+\Gamma_3+\Gamma_4+\Gamma_5$	$\Gamma_2+\Gamma_3+\Gamma_4+\Gamma_5$	$\Gamma_6+\Gamma_8$	$\Gamma_7+\Gamma_8$	$\Gamma_6+\Gamma_7+2\Gamma_8$
Γ_5					$\Gamma_1+\Gamma_3+\Gamma_4+\Gamma_5$	$\Gamma_7+\Gamma_8$	$\Gamma_6+\Gamma_8$	$\Gamma_6+\Gamma_7+2\Gamma_8$
Γ_6						$\Gamma_1+\Gamma_4$	$\Gamma_2+\Gamma_5$	$\Gamma_3+\Gamma_4+\Gamma_5$
Γ_7							$\Gamma_1+\Gamma_4$	$\Gamma_3+\Gamma_4+\Gamma_5$
Γ_8								$\Gamma_1+\Gamma_2+\Gamma_3+$ $2\Gamma_4+2\Gamma_5$

This condition is very helpful and will be often used in the following. It is important to notice at this point that this procedure enables one to reproduce all possible interaction terms that may exist for different types of symmetry-breaking perturbations.

Let us now discuss how perturbations may influence the eigenvalues of the conduction-band electron-states and eventually lift the spin degeneracy by symmetry-breaking effects. The interaction matrix then shows, which states are mixed and which type of perturbation may be involved.

In our example, in order to obtain a splitting of the spin-degenerate states of the conduction band, one needs a perturbation that transforms as the Pauli-spin matrices do, i.e. which transform according to Table 2.1 as (Γ_4, K^-). Then, following Table 2.3, the components of this perturbation can be combined with the corresponding Pauli-spin matrices to form interaction terms, which are compatible with the overall symmetry of a Hamiltonian.

This symmetry postulate defines in a general way the procedure of the invariant expansion by terms, which break the point-group symmetry of a system. The effect of the perturbations can be twofold:

- They may differently change the energies of the states and thus lead to a splitting of degenerate states of the unperturbed Hamiltonian. This is the case if the perturbation matrix is a diagonal matrix (in the basis of the unperturbed Hamiltonian).
- They may couple the different states of the unperturbed Hamiltonian and thus mix the eigenvectors. This implies that the perturbation matrix is a non-diagonal matrix (in the basis of the unperturbed Hamiltonian).

We see from Table 2.2 that the magnetic field $\boldsymbol{B}$ is a possible candidate to obtain such a coupling: It has the required symmetry properties and its components can be combined with the Pauli-spin matrices to give rise to a finite energy splitting and to a mixing of the unperturbed states: While σ_e^z is a diagonal matrix, which may lead to different energies of the spin up- and down states if a magnetic field is applied. σ_e^x and σ_e^y are non-diagonal matrices and if they are present in the interaction matrix these terms couple the spin up- and down states, leading in addition to a mixing of the eigenstates of the unperturbed Hamiltonian.

Or, in order to explain the construction of a perturbation interaction term in detail:

- identify from Table 2.1 the Pauli-spin matrices $[\sigma_e^x, \sigma_e^y, \sigma_e^z]$ with the expressions (P, Q, R)
- chose from Table 2.2 a perturbation that is (as the spin matrices) also an odd function K^- under time reversal. Let us consider the magnetic field $\boldsymbol{B} = (B_x, B_y, B_z)$ and label with (P', Q', R') the field components
- It then follows from Table 2.3 that the product $(PP' + QQ' + RR')$ has the required spatial symmetry (Γ_1) and is an even function (K^+) under time reversal.

Explicitly, following Table 2.3, a term $\propto \boldsymbol{B}$ leads to

$$H_{B1}^e = a_{B1}^e \boldsymbol{B} \cdot \boldsymbol{\sigma_e} = a_{B1}^e (B_x \sigma_e^x + B_y \sigma_e^y + B_z \sigma_e^z) \tag{2.3.1}$$

which corresponds to the linear Zeeman effect. Removing of the degeneracy can be seen immediately if we consider $\boldsymbol{B} = (0, 0, B_z)$. Then, Eq. (2.3.1) reduces to

$$H_{B1}^e = a_{B1}^e \boldsymbol{B} \cdot \boldsymbol{\sigma_e} = (a_{B1}^e B_z)\sigma_e^z = (a_{B1}^e B_z) \begin{pmatrix} 1 & 0 \\ 0 & -1 \end{pmatrix}$$

indicating two different energy eigenvalues of the diagonal matrix corresponding to H_{B1}^e while the eigenstates of the unperturbed Hamiltonian α_e and β_e remain unchanged.

For arbitrarily chosen field directions where B_x or $B_y \neq 0$ Eq. (2.3.1) leads to a splitting of the initially spin-degenerate conduction-band states α_e and β_e as

$$E_{\alpha_e} = -E_{\beta_e} = a_{B1}^e B. \tag{2.3.2}$$

The parameter a_{B1}^e of our development is given by $a_{B1}^e = -(1/2)\mu_B g^c$ where g^c is the "electron g-factor" (or "Landé factor") and μ_B represents the electron magneton of Bohr. In general, as we will see, α_e and β_e are now no longer eigenstates of the perturbed Hamiltonian, but the new eigenstates are given by a linear combination of the states α_e and β_e.

2.4 Shift and Splitting of States: Examples of Perturbations in Effective Hamiltonians

Let us consider another example of the application of external fields. As given in Table 2.2 the components (E_x, E_y, E_z) of an electric field $\boldsymbol{E}$ transform like (Γ_5, K^+). Because of its temporal symmetry properties, $\boldsymbol{E}^n$ ($\boldsymbol{E}$ taken to all possible orders n) have K^+ symmetry. Therefore, in order to contribute to the Hamiltonian and thus change the eigenvalues of the states, they can only be combined with the unit matrix

1_e. Because of the different time-reversal symmetry of the Pauli-spin matrices and the electric field components it follows immediately that an electric field cannot split the different spin states nor mix them.

In addition, a term

$$a_{E1}\boldsymbol{E}1_e = a_{E1}(E_x 1_e + E_y 1_e + E_z 1_e)$$

(where according to Eq. (2.2.5) 1_e is the unit matrix in the subspace of conduction-band electron-states) is not allowed (although it has the required K^+ temporal symmetry) since $a_{E1}\boldsymbol{E}1_e$ is not a scalar operator but is transforming like the position vector $\boldsymbol{r}$ (see Table 2.2).

We see that a linear Stark effect is forbidden at the Γ-point for the subspace considered here, and only higher orders in $\boldsymbol{E}^n$ can lead to a Stark shift of the states. An example is the quadratic Stark effect, which results in a perturbation of the form

$$H_{E2} = a_{E2}E^2 1_e \tag{2.4.1}$$

with

$$E^2 = E_x^2 + E_y^2 + E_z^2, \tag{2.4.2}$$

a_{E2} being a real constant. As it can be seen in Table 2.3, no other combinations of electric field components up to second order can exist in the Hamiltonian. One then finds for the energy shift due to an applied electric field in its lowest order the quadratic Stark shift

$$E_{\alpha_e} = E_{\beta_e} = a_{E2}E^2. \tag{2.4.3}$$

Since the electric field components (taken to arbitrary orders) can only be combined with the unit matrix in the subspace of conduction-band electron-states but not with the Pauli-spin matrices, an electric field can only lead to an energy shift of the states but not to their splitting.

One should remark here that we have only determined the symmetry properties of possible interaction terms but not the value of the matrix elements. They have to be calculated separately and it may turn out that they are equal to zero.

A result similar to that of Eq. (2.4.1) is found for strain in zincblende-type semiconductors. Let us denote by ϵ_{ii} $(i = x, y, z)$ the diagonal components of the strain tensor in the following. The combination $(\epsilon_{xx} + \epsilon_{yy} + \epsilon_{zz})$ transforms like the electric field $\boldsymbol{E}^2$ as (Γ_1, K^+) in crystals with T_d point-group symmetry and can be present in the Hamiltonian. All other combinations are excluded and no linear strain splitting but only a shift is obtained for the isolated conduction-band states at the Γ-point. Thus, neither electric fields nor mechanical strain can remove the degeneracy of the conduction-band states.

Let us now discuss several other examples of such symmetry-breaking effects: Similar to the electric field, all second order terms in $\boldsymbol{B}$ are even functions under time reversal. Combining them with the unit matrix, 1_e leads to the quadratic Zeeman effect, which takes here the form

$$H^e_{B2} = a^e_{B2} B^2 1_e \text{ with } B^2 = B_x^2 + B_y^2 + B_z^2. \tag{2.4.4}$$

This term is also compatible with the overall symmetry of a Hamiltonian. H^e_{B2} leads to an energy shift of the states but not to an additional splitting. Together with Eq. (2.3.2) we find for the spin-splitting in a magnetic field up to second order:

$$E_{\alpha_e} = a^e_{B1} B + a^e_{B2} B^2 \text{ and } E_{\beta_e} = -a^e_{B1} B + a^e_{B2} B^2. \tag{2.4.5}$$

As for electric fields, one can continue the development of interaction terms in a power series in higher orders of $\boldsymbol{B}$.

It is interesting to notice at this point that we were not obliged up to now to specify the direction of the magnetic field with respect to the cubic axes. In this special case the amount of shift and splitting of the states is independent of the field direction. This is due to the spherical symmetry of the s-type conduction-band wave-functions. Then, the direction of the quantization axis (the z-axis) can be arbitrarily chosen (i.e. ($z \parallel \boldsymbol{B}$) is a possible choice) and spin up and down states remain eigenstates of the perturbed Hamiltonian. This can be different when considering mixed perturbations due to different fields.

One easily checks with Table 2.3 that mixed terms of $\boldsymbol{E}$ and $\boldsymbol{B}$ combined with the Pauli-spin matrices can in principle fulfill the symmetry requirements of a Hamiltonian. This does not mean, however, that such terms are relevant. We mention here only for completeness such a term:

$$H_{BE} = a_{BE}[(B_y E_z - B_z E_y)\sigma^x_e + c.p.] \tag{2.4.6}$$

(where "c. p." stands for "cyclic permutations"). When inspecting the symmetry of this mixed perturbation one notices that the mutual orientations of electric and magnetic fields are important. Parallel components of both fields have no influence on energy corrections due to Eq. (2.4.6), while crossed fields may lead to such a "magnetic-field induced linear-Stark shift" discussed above. If one considers for example $\boldsymbol{B} = (0, B_y, 0)$ and $\boldsymbol{E} = (0, 0, E_z)$ the eigenvalues $E^{(1)}_{BE}$ and $E^{(2)}_{BE}$ of Eq. (2.4.6) are given by

$$E^{(1)}_{BE} = -E^{(2)}_{BE} = a_{BE}(B_y E_z),$$

which vary linearly with B_y and E_z. The related eigenfunctions $\Psi^{(1)}_{BE}$ and $\Psi^{(2)}_{BE}$ are given by a superposition of spin up and down states α_e and β_e, which are here given by

$$\Psi^{(1)}_{BE} = (\alpha_e + \beta_e)/\sqrt{2} \text{ and } \Psi^{(2)}_{BE} = (\alpha_e - \beta_e)/\sqrt{2}.$$

In this simple case these energy eigenvalues and the related eigenfunctions of the perturbed Hamiltonian can be determined easily "by hand" by solving Eq. (2.4.6) as we show now as an example:

To find the eigenvalues of H_{BE}, we first write down Eq. (2.4.6) in its explicit matrix form, which reads

$$H_{BE} = a_{BE}(B_y E_z)\sigma_e^x = a_{BE}(B_y E_z)\begin{pmatrix} 0 & 1 \\ 1 & 0 \end{pmatrix} = \begin{pmatrix} 0 & a_{BE}(B_y E_z) \\ a_{BE}(B_y E_z) & 0 \end{pmatrix},$$

taking into account that all field components are $\equiv 0$ except B_y and E_z. Determination of the eigenvalues consists in diagonalization of the above matrix, which can be accomplished using an equation analogous to Eq. (A.6) (see Appendix A). This equation reads in our particular case

$$D = \begin{vmatrix} -E_{BE} & a_{BE}(B_y E_z) \\ a_{BE}(B_y E_z) & -E_{BE} \end{vmatrix} = 0$$

or

$$E_{BE}^2 - (a_{BE} B_y E_z)^2 = 0,$$

which leads immediately to the requested eigenvalues:

$$E_{BE}^{(1,2)} = \pm a_{BE}(B_y E_z).$$

One now finds the corresponding eigenfunctions by applying an equation analogous to Eq. (A.6) into which one injects the calculated eigenvalues. In the case of the positive eigenvalue $E_{BE}^{(1)} = a_{BE}(B_y E_z)$ this equation then reads

$$\begin{aligned} -a_{BE}(B_y E_z)S_{11} + a_{BE}(B_y E_z)S_{12} &= 0 \\ a_{BE}(B_y E_z)S_{11} - a_{BE}(B_y E_z)S_{12} &= 0 \end{aligned}$$

or

$$S_{11} = S_{12} (= S).$$

Hence, if we take into consideration Eq. (A.1), the eigenfunction $\Psi_{BE}^{(1)}$ concerning the eigenvalue $E_{BE}^{(1)}$ is given by a linear combination of the basis functions α_e and β_e as

$$\Psi_{BE}^{(1)} = S(\alpha_e + \beta_e)$$

if we remember that the basis functions of the conduction band are the spin up and down states α_e and β_e, respectively.

The value of S follows from the normalization condition of wave functions. It can be obtained using an equation analogous to Eq. (A.13) to

$$\begin{aligned} S_{11}S_{11}^* + S_{12}S_{12}^* &= 1 \\ 2S^2 = 1 \text{ or } S &= 1/\sqrt{2} \end{aligned}$$

and thus

$$\Psi_{BE}^{(1)} = (\alpha_e + \beta_e)/\sqrt{2}.$$

For $\Psi_{BE}^{(2)}$ the expression

$$\Psi_{BE}^{(2)} = (\alpha_e - \beta_e)/\sqrt{2}$$

can be obtained in the same way.

The procedure discussed above to generate perturbation Hamiltonians is very powerful and versatile and may be used to analyze many different physical properties of samples. The method is, however, only easy to handle if the number of states considered in the subspace is small and if one can restrict oneself to low orders in the power expansion of the perturbation. Otherwise, the procedure may become very clumsy and cumbersome.

One sees already with these simple examples the importance of the orientations of the fields and of the order, up to which these fields have to be considered. In addition, if different fields have the same transformation properties they act in the same way on the physical system. The observed effects depend, however, also on the states, which are involved in the experiment since they determine the interaction matrices, which have to be considered.

2.5 Dispersion Relation of Spin-Degenerate Conduction-Band States

Applying electric or magnetic fields in well defined directions is an important tool in order to study physical properties of semiconductor crystals. We have discussed how these extrinsic symmetry-breaking effects manifest themselves in the eigenvalues of the conduction-band states. Besides external fields the finite wave-vector $\boldsymbol{Q}$ of a quasi particle is an additional but intrinsic symmetry-breaking perturbation that affects the dispersion relation $E(\boldsymbol{Q})$ of the states, which can often be determined experimentally.

Let us discuss the dispersion relation of the conduction-band states in more detail. On the contrary to external fields, for which often the axes of the fields can be chosen arbitrarily with respect to the cubic crystal axes, the quasi-particle propagation is defined with respect to the crystal axes. Thus the amount of shift and splitting of the states will depend on these directions for an anisotropic energy dispersion.

According to Table 2.2 the wave-vector components Q_i $(i = (x, y, z))$ transform like (Γ_5, K^-). Because of their temporal symmetry interaction terms proportional to $\boldsymbol{Q}^n$ (n taken to all odd orders) can appear in a Hamiltonian only in combination with the Pauli-spin matrices σ_e^i. On the contrary, as in Eq. (2.4.1), even order terms can only be combined with the unit matrix 1_e. Therefore, even orders n of $\boldsymbol{Q}^n$ may lead to an energy shift of the states (i.e. they may be responsible for the curvature of the conduction band outside of the Γ-point) and only odd orders may lead to their energy splitting: The pseudo-Bloch functions α_e and β_e are eigenstates of the Hamiltonian H_e given in Eq. (2.2.6) but no longer eigenstates of the full Hamiltonian if it contains symmetry-breaking effects. Taken to odd orders n, finite wave-vectors may thus lead to a coupling of the states and a mixing of the pseudo-Bloch functions α_e and β_e outside of the Γ-point.

Concerning terms linear in $\boldsymbol{Q}$, the multiplication scheme in Table 2.3 shows that no interaction term transforming as Γ_1 can be constructed and $\boldsymbol{Q}$-linear terms do not occur in the energy dispersion of the spin-degenerate conduction band. The band structure is then given in its lowest order by

$$H_{Q2} = a_{Q2} Q^2 1_e \tag{2.5.1}$$

with

$$Q^2 = Q_x^2 + Q_y^2 + Q_z^2 \tag{2.5.2}$$

where the constant value in Eq. (2.5.1) is given by $a_{Q2} = \hbar^2/(2m^*)$.

The effective-electron mass m^* is isotropic in this case. It is defined in terms of the curvature of the two-fold degenerate conduction band close to its minimum. The effective mass m^* is an important parameter since it influences the transport and the optical properties of semiconductors. As stated above, anisotropic Q^2-terms that show up in a Hamiltonian must have the same transformation properties as the Pauli matrices. Such terms are nevertheless forbidden in our case because of their wrong time-reversal symmetry of the resulting product. In addition, Table 2.3 shows that anisotropic terms of Γ_4 symmetry can in principle be constructed in second order from components of objects transforming as Γ_5. But such terms are all $\equiv 0$ since wave-vector components commute with each other, i.e. objects as $(YZ - ZY)$ appearing in Table 2.3 vanish if (Y, Z) denote wave-vector components.

It is interesting to notice, however, that higher than second order terms in $\boldsymbol{Q}$ can give rise to intrinsic perturbations transforming as (Γ_4, K^-). They can then be combined with the Pauli matrices σ_e^i, which have the same transformation properties, in order to construct an interaction term compatible with the requirements of a Hamiltonian. This leads, depending on the crystal direction, to a lifting of the conduction-band spin-degeneracy [10, 11]. This is important for problems concerning transport of electrons in different spin states and for spin-dephasing and spin-relaxation processes [12]. Since this wave-vector dependent perturbation has the transformation properties of a magnetic field $\boldsymbol{B}$ it can be looked upon as an intrinsic effective magnetic field. The effect is, however, a pure consequence of the symmetry breaking due to the wave-vector.

Let us consider as an example the construction of such a term in third order of $\boldsymbol{Q}$. Starting from the wave-vector components Q_i $(i = (x, y, z))$, which transform according to Table 2.2 as (x, y, z) and are labeled (X, Y, Z) we first construct combinations of them in second order, which transform also as (x, y, z). We see from Table 2.3 that

$$X' = (Q_y Q_z + Q_z Q_y) = 2(Q_y Q_z) \tag{2.5.3}$$

is the right combination. Similarly, we find

$$Y' = 2(Q_z Q_x) \tag{2.5.4}$$

and

$$Z' = 2(Q_x Q_y). \tag{2.5.5}$$

Using again Table 2.3, we construct combinations of third-order wave-vector components, which transform as (Γ_4, K^-) and are labeled (P', Q', R'). We obtain:

$$P' = 2Q_y(Q_x Q_y) - 2Q_z(Q_z Q_x) = 2Q_x(Q_y^2 - Q_z^2) \tag{2.5.6}$$

and similarly

$$Q' = 2Q_y(Q_z^2 - Q_x^2) \tag{2.5.7}$$

and

$$R' = 2Q_z(Q_x^2 - Q_y^2). \tag{2.5.8}$$

Combining (P', Q', R') with $(\sigma_e^x, \sigma_e^y, \sigma_e^z)$, which we denote as (P, Q, R), Table 2.3 leads to:

$$H_{BQ3} = a_{BQ3}[Q_x(Q_y^2 - Q_z^2)\sigma_e^x + c.p.] \tag{2.5.9}$$

where "c. p." stands for "cyclic permutation".

Table 2.3 shows that another combination to construct third-order components (P', Q', R') from the wave-vector components (X, Y, Z) is possible: One constructs first objects "U" and "V" from (X, Y, Z) in second order and then the (P', Q', R') objects by multiplying the (U, V) objects symmetry adapted with the corresponding (X, Y, Z) wave-vector components. This procedure results, however, in the same third-order objects as given in Eqs. (2.5.6)–(2.5.8).

When considering arbitrary low symmetry crystal directions one sees that this symmetry-breaking perturbation leads to an anisotropic dispersion (warping) and to a splitting of the very simple, spin-degenerate conduction band. Concerning the high symmetry directions [100] and [111] the perturbation H_{BQ3} of Eq. (2.5.9) does not contribute to the dispersion. This is more complicated in the (101)-plane. Taking for example $\boldsymbol{Q} = [Q_x, 0, Q_z]$ the perturbation takes the form

$$H'_{BQ3} = a_{BQ3}[Q_x(-Q_z^2)\sigma_e^x + Q_z(Q_x^2)\sigma_e^z]. \tag{2.5.10}$$

One sees that the high symmetry direction [101] of a cubic crystal with zincblende structure is influenced by the warping effect. Using $Q_z = Q_x$, Eq. (2.5.9) leads (after solving the eigenvalue problem in a way entirely parallel to that applied to the diagonalization of the Hamiltonian H_{BE} of Eq. (2.4.6)) to an energy difference with respect to the parabolic contribution of Eq. (2.5.1) given by

$$\gamma_{1,2} = \pm(\sqrt{2}a_{BQ3}Q_x^3). \tag{2.5.11}$$

Then, the spin states $\Psi_1^{[101]}$ and $\Psi_2^{[101]}$ that diagonalize the perturbed Hamiltonian in the [101]-direction are given as

$$\Psi_1^{[101]} = (-\sqrt{2}\alpha_e)/(2\sqrt{2-\sqrt{2}}) + (\sqrt{2-\sqrt{2}})\beta_e/2$$
$$\text{and} \quad (2.5.12)$$
$$\Psi_2^{[101]} = (\sqrt{2}\alpha_e)/(2\sqrt{2+\sqrt{2}}) + (\sqrt{2+\sqrt{2}})\beta_e/2$$

for the eigenvalues

$$\gamma_1 = (\sqrt{2}a_{BQ3}Q_x^3) \text{ and } \gamma_2 = -(\sqrt{2}a_{BQ3}Q_x^3),$$

respectively.

Figure 2.3 gives the dispersion (i.e. the energy E as a function of wave number Q) of conduction-band electron-states in the vicinity of the Γ-point, including terms up to third order in $\boldsymbol{Q}$ for the [111] and [101] directions. α and β denote different electron states, respectively. While the states remain degenerate in the [111] direction, Fig. 2.3 clearly shows how the splitting of the electron states $\Psi_1^{[101]}$ and $\Psi_2^{[101]}$ increases with increasing wave-vector.

Let us consider the splitting of the two spin states induced by the coupling given in Eq. (2.5.9) in more detail. The absolute value of the wave-vector $\boldsymbol{Q}$ is taken fixed and its direction varied in the (101)-plane, the component Q_x^2 of the wave-vector squared being taken as variable. The energy deviations $\gamma_{1,2}(Q_x)$ of the two electron states from the value obtained when considering dispersive terms up to second order only are given by

$$\gamma_{1,2}(Q_x) = \pm a_{BQ3}\lambda(Q_x)Q^3, \quad (2.5.13)$$

where $\lambda(Q_x)$ is illustrated in Fig. 2.4. $\lambda(Q_x)$ determines the anisotropic dispersion of the conduction-band electrons.

Similar to the anisotropic dispersion of conduction-band electrons the hybridization of the spin states is also induced by the coupling given in Eq. (2.5.9). Then, the electron-wave functions are defined as

$$\Psi_1^{(101)} = -a_1\alpha_e + b_1\beta_e$$
$$\text{and} \quad (2.5.14)$$
$$\Psi_2^{(101)} = b_1\alpha_e + a_1\beta_e.$$

Figure 2.5 shows the mixing coefficients a_1 and b_1 of the spin states α_e and β_e in the (101)-plane. The absolute value of the wave-vector $\boldsymbol{Q}$ is taken fixed and its direction varied in the (101)-plane, the wave-vector component squared $Q_x^2 \ / \ Q^2$ being taken as variable in the calculation of a_1 and b_1 as in the case of the parameter λ given in Fig. 2.4.

It is interesting to discuss also the coupling given in Eq. (2.5.9) in the (110)-plane, i.e. the directions $\perp$ to the z-axis. Proceeding as above, the same values as those

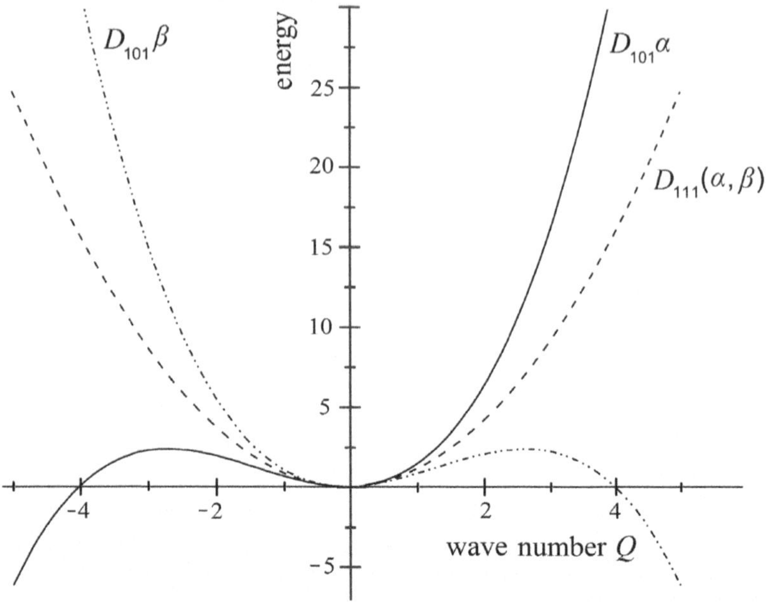

Fig. 2.3 Dispersion D_{ijk} (energy E as a function of wave number Q) of conduction-band electron-states of a semiconductor with T_d point-group symmetry in the vicinity of the Γ-point. $[ijk]$ are the Miller indexes, specifying the crystal direction. D includes terms up to third order in $\boldsymbol{Q}$ for the [111] and [101] directions. α and β denote the electron spin up and down states, respectively. The graphs are determined with arbitrarily chosen parameter values a_{Q2} and a_{BQ3} in Eqs. (2.5.1) and (2.5.10), respectively

Fig. 2.4 The anisotropic dispersion of conduction-band electrons of semiconductors with T_d point-group symmetry in the vicinity of the Γ-point is determined by $\lambda(Q_x)$. The absolute value of the wave-vector | $\boldsymbol{Q}$ | is taken fixed and its direction varied in the (101)-plane, the wave-vector component squared Q_x^2/Q^2 being taken as variable

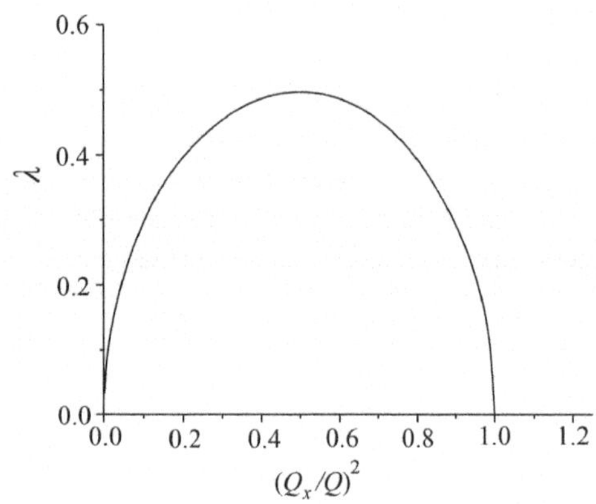

shown in Fig. 2.4 are obtained for $\lambda(Q_x)$ when $\boldsymbol{Q}$ is varied in the (110)-plane. The hybridization given by the mixing coefficients a_2 and b_2 of the spin states is, however, different from those obtained for the variation in the (101)-plane: the absolute values of a_2 and b_2 are equal to $1/\sqrt{2}$, independent of Q_x^2. One of the coefficients being real, the second is complex with an arbitrary phase.

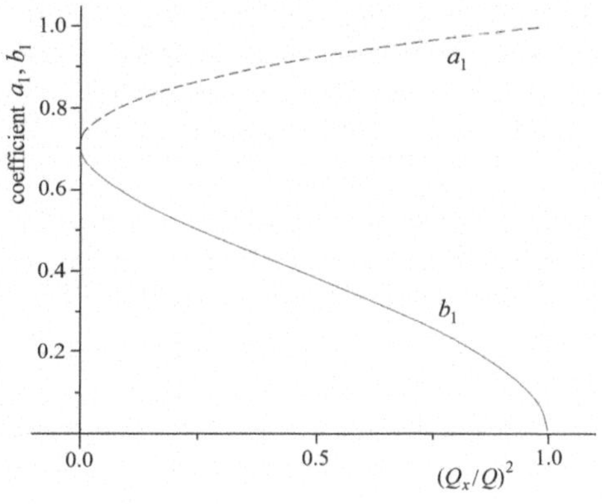

Fig. 2.5 Mixing coefficients a_1 and b_1 of the spin states α and β due to the anisotropic dispersion of conduction-band electrons of semiconductors with T_d point-group symmetry. The absolute value of the wave-vector $| \boldsymbol{Q} |$ is taken fixed and its direction varied in the (101)-plane, the wave-vector component squared Q_x^2/Q^2 being taken as variable

As pointed out by Dresselhaus [10] a H_{BQ3}-term may appear in crystals with zincblende structure. It is interesting to notice that such terms do not appear in diamond structure: In O_h symmetry the parity "+" (indicated by the superscript in (Γ_i^+)) has to be conserved and Hamiltonians transform according to (Γ_1^+, K^+). Here, the wave-vector $\boldsymbol{Q}$ transforms as (Γ_5^-, K^-) and the pseudo-spin $\boldsymbol{\sigma}_e$ as (Γ_4^+, K^-). Therefore, the transformation properties of the considered product (involving the third-order wave-vector component $\boldsymbol{Q}^3$) are (Γ_1^-, K^+), which is not compatible with an interaction term in a Hamiltonian.

Similar to the approach leading to Eq. (2.5.9) one could also construct invariant terms bilinear in $\boldsymbol{Q}$ and $\boldsymbol{E}$. It would have a similar form as Eq. (2.4.6):

$$H_{QE} = a_{QE}[(Q_y E_z - Q_z E_y)\sigma_e^x + c.p.] \tag{2.5.15}$$

(where "c. p." stands again for "cyclic permutations"). We thus notice that a linear Stark effect can occur when considering conduction-band states outside of the center the Brillouin zone. This can be important for the analysis of optical transitions in semiconductors, which may have a high density of states outside the Γ-point or in the study of indirect semiconductors.

Higher order terms in $(\boldsymbol{E}, \boldsymbol{Q})$ can also be constructed in the same way. One concludes that the spin degeneracy of a simple conduction band can be lifted at finite wave-vectors in the presence of an electric field.

References

1. Suzuki, M., Uenoyama, T., Yanase, A.: Phys. Rev. B **52**, 8132 (1995)
2. Hopfield, J.J.: J. Phys. Chem. Solids **15**, 97 (1960)
3. Ashcroft, N.W., Mermin, N.D.: Solid State Physics. Saunders College Publishing, New York, Orlando, Austin, San Diego (1976)
4. Cho, K.: Phys. Rev. B **14**, 4463 (1976)

5. Koster, J.F., Dimmock, J.O., Wheeler, R.G., Statz, H.: Properties of the Thirty-two Point Groups. MIT Press, Cambridge, Mass. (1963)
6. Ludwig, W., Falter, C.: Symmetries in Physics : Group Theory Applied to Physical Problems, Springer Series in Solid State Sciences 64. Springer, Berlin, Heidelberg (1988)
7. Bir, G.L., Pikus, G.E.: Symmetry and Strain-Induced Effects in Semiconductors. Wiley, New York, Toronto (1974)
8. Cho, K.: Excitons, Topics in Current Physics 14, Springer, Berlin, Heidelberg (1979)
9. Hönerlage, B., Lévy, R., Grun, J.B., Klingshirn, C., Bohnert, K.: Phys. Rep. **124**, 161 (1985)
10. Dresselhaus, G.: Phys. Rev **100**, 580 (1955)
11. D'yakonov, M.I., Perel, V.I.: Sov. Phys. JETP **33**, 1053 (1971); D'yakonovm, M.I., Kachorovskii, V.Y.: Sov. Phys. Semicond. **20**, 110 (1986)
12. Meier, F., Zakharchenya, B.P. (eds.): Optical Orientation. Modern Problems of Condensed Matter Sciences, vol. 8. North Holland, Amsterdam (1984)

Chapter 3
Symmetry-Breaking Effects in Valence Bands of Zincblende-Type Crystals

As we have mentioned above, the uppermost valence band of simple semiconductors with zincblende structure is mainly built from atomic p-orbitals that are three-fold degenerate. In addition, electrons are spin (1/2)-particles such that each p-orbital can accept two electrons with different spins. Therefore six spin-orbitals make up the uppermost valence band at the Γ-point in these simple semiconductors.

In the absence of spin-orbit coupling and spin-scattering processes, the space of valence-band electrons is decomposed into two subspaces of electron states, one for the spin-up states and the other containing only the spin-down states. These two subspaces are completely decoupled from each other and the interaction within each of the subspaces is identical. In a matrix formulation of the Hamiltonian this results into two identical diagonal block matrices of dimension three and all matrix elements coupling both blocks are $\equiv 0$. One has to solve therefore only a problem for threefold degenerate states with the same spin orientation. One then obtains the complete solution by adding at the end the solution where the spin is reversed. Here, one has only to pay attention if the action of perturbations on the electron spin is explicitly considered, since they may differ for the two spin states.

When analyzing the valence band we will proceed similar to the case of the conduction band: We will first establish (Sect. 3.1) a basis for the interaction matrix of the valence-band orbital-states. In this subspace some symmetry-breaking interaction terms are constructed. Then the basis is extended (Sect. 3.2) by including the valence-electron spin. Because of the multiplication procedure spin-orbit coupling shows directly up and new symmetry-breaking interaction terms can be constructed. Thus, when comparing to Sect. 3.1, terms that have their origin in spin-orbit coupling can be easily identified. In Sects. 3.3 and 3.4 we present developments of the interacting states in the reduced valence-band subspaces of Γ_7- and Γ_8-symmetry, respectively. Comparing to Sect. 3.2 we show that an important part of possible interactions within the complete valence band is already accounted for in the differ-

© Springer International Publishing AG, part of Springer Nature 2018

B. Hönerlage and I. Pelant, *Symmetry and Symmetry-Breaking in Semiconductors*, Springer Tracts in Modern Physics 279,
https://doi.org/10.1007/978-3-319-94235-3_3

ent subspaces. Therefore, a detailed knowledge of the coupling scheme between the different subspaces is often of minor importance.

3.1 Three-Fold Degenerate Valence Band without Spin

Let us for the moment neglect the electron-spin: then, the valence-band states are three-fold degenerate and are eigenfunctions of the z-component l_z of the angular-momentum operator $\boldsymbol{l}$ with $l = 1$. $|m_j\rangle$ denote the eigenfunctions of l_z obeying to

$$l_z|m_j\rangle = m_j|m_j\rangle \text{ with } m_j = (1, 0, -1). \tag{3.1.1}$$

Equivalently, one may also use the symmetry properties of p orbitals, which transform like the components of the position vector $\boldsymbol{r} = (x, y, z)$ and express $|m_j\rangle$ in this basis. One then obtains:

$$\begin{aligned} |1\rangle &= -(x+iy)/\sqrt{2} \\ |0\rangle &= z \\ |-1\rangle &= (x-iy)/\sqrt{2}. \end{aligned} \tag{3.1.2}$$

As discussed above, one knows the transformation properties of angular-momentum operators, which transform like the irreducible representation Γ_4 in systems with T_d point-group symmetry. As a starting point we now choose the angular-momentum matrices l_i with $i = (x, y, z)$ as three of the basis matrices for the construction of the valence-electron subspace:

$$l_x = (1/\sqrt{2})\begin{pmatrix} 0 & 1 & 0 \\ 1 & 0 & 1 \\ 0 & 1 & 0 \end{pmatrix}; \; l_y = (1/\sqrt{2})\begin{pmatrix} 0 & -i & 0 \\ i & 0 & -i \\ 0 & i & 0 \end{pmatrix}; \; l_z = \begin{pmatrix} 1 & 0 & 0 \\ 0 & 0 & 0 \\ 0 & 0 & -1 \end{pmatrix}. \tag{3.1.3}$$

From these matrices l_i we construct the other basis matrices, which are necessary to span the three-dimensional subspace of the valence-band states. Forming the product $\Gamma_4 \otimes \Gamma_4$ we obtain, according to the multiplication scheme given in Table 2.4:

$$\Gamma_4 \otimes \Gamma_4 = \Gamma_1 \oplus \Gamma_3 \oplus \Gamma_4 \oplus \Gamma_5. \tag{3.1.4}$$

The nine matrices resulting from Eq. (3.1.4) may be used to form a basis of our system since they are linearly independent. In addition, they will be adapted to the crystal symmetry. According to Table 2.3 we introduce the following components:

$$
\begin{aligned}
S = l_x^2 + l_y^2 + l_z^2 &\Leftrightarrow \text{transforming as} \Leftrightarrow (\Gamma_1, K^+) \\
(U, V) = (\sqrt{3}(l_x^2 - l_y^2), 2l_z^2 - l_x^2 - l_y^2) &\Leftrightarrow (\Gamma_3, K^+) \\
(P, Q, R) = (l_x, l_y, l_z) &\Leftrightarrow (\Gamma_4, K^-) \\
(X, Y, Z) = (\{l_y l_z\}, \{l_z l_x\}, \{l_x l_y\}) &\Leftrightarrow (\Gamma_5, K^+),
\end{aligned}
\tag{3.1.5}
$$

where

$$
\{l_y l_z\} = (1/2)(l_y l_z + l_z l_y). \tag{3.1.6}
$$

Inspecting Eq. (3.1.5), we see that all basis functions are even functions with respect to time reversal (indicated by K^+), except the matrices $(P, Q, R) = (l_x, l_y, l_z)$, which are given by the components of the angular-momentum operator and transform as K^-. The basis matrices spanning our subspace are thus obtained from Eq. (3.1.5). We obtain explicitly the following matrices together with their transformation properties:

$$
\begin{aligned}
S = 2\begin{pmatrix}1&0&0\\0&1&0\\0&0&1\end{pmatrix} &\Leftrightarrow (\Gamma_1, K^+) \\
U = \sqrt{3}\begin{pmatrix}0&0&1\\0&0&0\\1&0&0\end{pmatrix};\ V = \begin{pmatrix}1&0&0\\0&-2&0\\0&0&1\end{pmatrix} &\Leftrightarrow (\Gamma_3, K^+) \\
P = (1/\sqrt{2})\begin{pmatrix}0&1&0\\1&0&1\\0&1&0\end{pmatrix};\ Q = (1/\sqrt{2})\begin{pmatrix}0&-i&0\\i&0&-i\\0&i&0\end{pmatrix};\ R = \begin{pmatrix}1&0&0\\0&0&0\\0&0&-1\end{pmatrix} &\Leftrightarrow (\Gamma_4, K^-) \\
X = (\sqrt{2}/4)\begin{pmatrix}0&-i&0\\i&0&i\\0&-i&0\end{pmatrix};\ Y = (\sqrt{2}/4)\begin{pmatrix}0&1&0\\1&0&-1\\0&-1&0\end{pmatrix};\ Z = (1/2)\begin{pmatrix}0&0&-i\\0&0&0\\i&0&0\end{pmatrix} &\Leftrightarrow (\Gamma_5, K^+).
\end{aligned}
\tag{3.1.7}
$$

Since the matrices of Eq. (3.1.7) are a basis of the valence-band electron-subspace, all possible interaction terms in-between the valence-band electron-states can now be formulated in the frame of these matrices. The number of considered states being finite, one can work out all desired interaction terms up to a given order of the perturbation.

The matrices S, V, and R are diagonal matrices. S is proportional to the unit matrix and a perturbation, which involves this basis matrix transforming as Γ_1 leads only to a uniform energy shift of the states but does not lift their degeneracy. Perturbations involving the basis matrices V and R may lead in addition to a splitting of the states, but they will not mix the states in-between themselves. This is different for perturbations transforming like $(U) \in \Gamma_3$, $(P, Q) \in \Gamma_4$ or $(X, Y, Z) \in \Gamma_5$, which mix the different angular-momentum states $|m_j\rangle$. Thus, the states $|m_j\rangle$ will no longer be eigenstates of the complete Hamiltonian if it includes perturbations that have the above symmetries.

Concerning the multiplication scheme of Table 2.4, we recall that the product of two objects with transformation properties (Γ_i, K^k) and (Γ_j, K^l) with $(i, j) \in (1, ..., 8)$ and $(k, l) \in (+, -)$ can only form an object transformation as (Γ_1, K^+) if $(i = j)$ and $(k = l)$. This condition is very helpful when constructing interaction terms that may appear in a Hamiltonian.

Let us consider as an example the dispersion of the valence band, which is threefold degenerate at the Γ-point. We have seen that the wave-vector $\boldsymbol{Q}$ is a symmetry-breaking perturbation, which transforms as (Γ_5, K^-) in zincblende structure. Then, because of time-reversal symmetry, the only basis matrices that could be combined with a $\boldsymbol{Q}$-linear perturbation would be the matrices (P, Q, R), which also transform according to (K^-). All other basis matrices have (K^+) symmetry and their combination with a $\boldsymbol{Q}$-linear term is forbidden in a Hamiltonian. The basis matrices (P, Q, R) transform, however, as Γ_4 in zincblende structure. As shown in Table 2.3 no product of the basis matrices given in Eq. (3.1.5) with the wave-vector components $(Q_x, Q_y, Q_z) = (X, Y, Z)$ can be constructed such that the result transforms as Γ_1. Therefore, $\boldsymbol{Q}$-linear terms cannot exist in our example where the electron spin is neglected.

The situation becomes different when considering perturbations that vary in second order with wave-vector $\boldsymbol{Q}$. Such terms have K^+ symmetry under Kramers' conjugation. Possible perturbations must transform as the basis matrices Γ_1, Γ_3, and Γ_5 do, which have also K^+ symmetry. Therefore interaction terms varying as $\boldsymbol{Q}^2$ can be present in a Hamiltonian. Using the multiplication Table 2.3 we can construct three different interaction terms, which vary as $\boldsymbol{Q}^2$. The first term is given by

$$H^v_{1Q2} = a^v_{1Q2}\boldsymbol{Q}^2 S = 2a^v_{1Q2}\begin{pmatrix} Q_x^2+Q_y^2+Q_z^2 & 0 & 0 \\ 0 & Q_x^2+Q_y^2+Q_z^2 & 0 \\ 0 & 0 & Q_x^2+Q_y^2+Q_z^2 \end{pmatrix}. \tag{3.1.8}$$

Indeed, the factor a^v_{1Q2} corresponds to an (isotropic) effective mass of the valence band. Since the matrix labeled "S" in Eq. (3.1.7) has the full point-group symmetry, the angular-momentum states $|m_j\rangle$ are also eigenstates of Hamiltonians, including this perturbation.

A second term varying as Q^2 (and transforming according Γ_3) takes the form

$$\begin{aligned} H^v_{3Q2} &= a^v_{3Q2}\left(\sqrt{3}(Q_x^2-Q_y^2)U + (2Q_z^2-Q_x^2-Q_y^2)V\right) \\ &= a^v_{3Q2}\left(3\begin{pmatrix} 0 & 0 & Q_x^2-Q_y^2 \\ 0 & 0 & 0 \\ Q_x^2-Q_y^2 & 0 & 0 \end{pmatrix} + \begin{pmatrix} 2Q_z^2-Q_x^2-Q_y^2 & 0 & 0 \\ 0 & -2(2Q_z^2-Q_x^2-Q_y^2) & 0 \\ 0 & 0 & 2Q_z^2-Q_x^2-Q_y^2 \end{pmatrix}\right) \\ &= a^v_{3Q2}\begin{pmatrix} 2Q_z^2-Q_x^2-Q_y^2 & 0 & 3(Q_x^2-Q_y^2) \\ 0 & -4Q_z^2+2Q_x^2+2Q_y^2 & 0 \\ 3(Q_x^2-Q_y^2) & 0 & 2Q_z^2-Q_x^2-Q_y^2 \end{pmatrix}. \end{aligned} \tag{3.1.9}$$

Let us consider the Schrödinger equation corresponding to a perturbation "p", formulated in the basis of the wave functions of Eq. (3.1.2):

$$H_p^v|\Psi_p\rangle = E_p|\Psi_p\rangle. \tag{3.1.10}$$

Let further H_p^v be given by Eq. (3.1.9) and $E_{a3Q2}^v(i) = E_p$ with $(i) \in (1, 2, 3)$ denote the eigenvalues of Eq. (3.1.10). Then, Eq. (3.1.10) can be solved for $a_{3Q2}^v \neq 0$ and arbitrary values of $\boldsymbol{Q}^2 = (Q_x^2 + Q_y^2 + Q_z^2)$. This solution consists in diagonalization of the matrix (3.1.9), the wanted eigenvalues being then the diagonal terms. This procedure is described in detail in Appendix A. One then obtains the following eigenvalues and eigenfunctions $|\Psi_{a3Q2}(i)\rangle$:

$$\begin{aligned}
&E_{a3Q2}^v(1) = a_{3Q2}^v\left(-4Q_x^2 + 2Q_y^2 + 2Q_z^2\right)\\
&\qquad\text{with eigenfunction } |\Psi_{a3Q2}(1)\rangle = (1/\sqrt{2})\,(-|1\rangle + |-1\rangle)\\
&E_{a3Q2}^v(2) = a_{3Q2}^v\left(2Q_x^2 + 2Q_y^2 - 4Q_z^2\right)\\
&\qquad\text{with eigenfunction } |\Psi_{a3Q2}(2)\rangle = |0\rangle\\
&E_{a3Q2}^v(3) = a_{3Q2}^v\left(2Q_x^2 - 4Q_y^2 + 2Q_z^2\right)\\
&\qquad\text{with eigenfunction } |\Psi_{a3Q2}(3)\rangle = (1/\sqrt{2})\,(|1\rangle + |-1\rangle)\,.
\end{aligned} \tag{3.1.11}$$

We see that the valence-band states, which are degenerate at $\boldsymbol{Q} = 0$, remain degenerate for finite wave-vectors only in the [111]-direction. If only $(Q_x^2 = Q_y^2)$ the energy dispersion of $l_z = \pm 1$ states remain degenerate but are different from that of the $l_z = 0$ states. In arbitrary (i.e. not high-symmetry) directions the states are split in energy for finite $\boldsymbol{Q}$-values and the parabolic dispersion is no longer isotropic, i.e. the energy dispersion depends on the direction of the wave-vector and shows a "warping". Then, if $(Q_x^2 \neq Q_y^2)$ the $l_z = \pm 1$ states become mixed at finite $\boldsymbol{Q}$-values.

Inspecting Table 2.3 we see that combinations of wave-vector components in second order of $\boldsymbol{Q}$ with the basis matrices (X, Y, Z) can give rise to a third parabolic contribution to the dispersion of the valence band. Using again Table 2.3 and Eq. (3.1.7), we find for this contribution to the Hamiltonian

$$\begin{aligned}
H_{5Q2}^v &= 2a_{5Q2}^v\left(Q_yQ_zX + Q_zQ_xY + Q_xQ_yZ\right)\\
&= 2a_{5Q2}^v\left((\sqrt{2}/4)\begin{pmatrix}0 & -iQ_yQ_z & 0\\ iQ_yQ_z & 0 & iQ_yQ_z\\ 0 & -iQ_yQ_z & 0\end{pmatrix} + (\sqrt{2}/4)\begin{pmatrix}0 & Q_zQ_x & 0\\ Q_zQ_x & 0 & -Q_zQ_x\\ 0 & -Q_zQ_x & 0\end{pmatrix} + \right.\\
&\qquad\left. +(1/2)\begin{pmatrix}0 & 0 & -iQ_xQ_y\\ 0 & 0 & 0\\ iQ_xQ_y & 0 & 0\end{pmatrix}\right)\\
&= (\sqrt{2}/2)a_{5Q2}^v\begin{pmatrix}0 & Q_zQ_x - iQ_yQ_z & -i\sqrt{2}Q_xQ_y\\ Q_zQ_x + iQ_yQ_z & 0 & -Q_zQ_x + iQ_yQ_z\\ i\sqrt{2}Q_xQ_y & -Q_zQ_x - iQ_yQ_z & 0\end{pmatrix}.
\end{aligned} \tag{3.1.12}$$

Proceeding as above and treating H^{v}_{5Q2} as a symmetry-breaking perturbation, we introduce $E^{v}_{a5Q2}(i)$ and $|\Psi_{a5Q2}(i)\rangle$ as eigen-values and eigen-functions of the perturbed system, respectively. One sees that this interaction also mixes the different states, depending on the direction of the wave-vector. This mixing of states is, however, much more sensible than the one given in Eq. (3.1.11), since now all three states are mixed and the amount of mixing depends on the wave-vector components. Therefore, we will give here only results for three high-symmetry directions.

Starting with the [001]-direction one obtains that the perturbation is ≡0. For the [110]-direction we find

$$\begin{aligned}
E^{v}_{a5Q2}(1) = (1/2)a^{v}_{5Q2}Q^2 \text{ with eigenfunction } |\Psi_{a5Q2}(1)\rangle &= (1/\sqrt{2})\,(-i|1\rangle + |-1\rangle) \\
E^{v}_{a5Q2}(2) = 0 \text{ with eigenfunction } |\Psi_{a5Q2}(2)\rangle &= |0\rangle \\
E^{v}_{a5Q2}(3) = -(1/2)a^{v}_{5Q2}Q^2 \text{ with eigenfunction } |\Psi_{a5Q2}(3)\rangle &= (1/\sqrt{2})\,(i|1\rangle + |-1\rangle)\,.
\end{aligned} \tag{3.1.13}$$

The (at $\boldsymbol{Q} = 0$ degenerate) states split into three parabolic bands with different energy dispersion and the splitting increases with increasing wave-vector.

For the [111]-direction one obtains

$$\begin{aligned}
E^{v}_{a5Q2}(1) &= -(2/3)a^{v}_{5Q2}Q^2 \text{ with eigenfunction } |\Psi_{a5Q2}(1)\rangle \\
&= (1/\sqrt{3})\left(i|1\rangle + (1/\sqrt{2})(1-i)|0\rangle + |-1\rangle\right) \\
E^{v}_{a5Q2}(2) &= (1/3)a^{v}_{5Q2}Q^2 \text{ with eigenfunction } |\Psi_{a5Q2}(2)\rangle \\
&= (1/\sqrt{2})\,(-i|1\rangle + |-1\rangle) \\
E^{v}_{a5Q2}(3) &= (1/3)a^{v}_{5Q2}Q^2 \text{ with eigenfunction } |\Psi_{a5Q2}(3)\rangle \\
&= (1/\sqrt{2})\left((1/\sqrt{2})(1-i)|1\rangle + |0\rangle\right).
\end{aligned} \tag{3.1.14}$$

where the states $|\Psi_{a5Q2}(2)\rangle$ and $|\Psi_{a5Q2}(3)\rangle$ are degenerate and can be orthogonalized if needed. In this direction we obtain only two parabolic bands with different curvatures (effective masses).

Comparing the results for the different crystal directions we see that the crystal dispersion is not isotropic but shows a "warping". All combinations of wave-vector $\boldsymbol{Q}^2$ components with the basis matrices (P, Q, R) being forbidden in a Hamiltonian, Eqs. (3.1.8), (3.1.9), and (3.1.12) give all possible quadratic contributions to the dispersion of the valence band in our model.

Equations (3.1.11)–(3.1.14) show also that eigenvectors, which diagonalize the Hamiltonian are different, depending on the crystal direction (i.e. they are expressed by different combinations of the angular momentum eigen-functions). This mixing of states is a typical signature of symmetry-breaking effects.

The above discussion concerning symmetry-breaking effects varying in second order with the wave-vector components remains valid if one studies the effects of an electric field $\boldsymbol{E}$ considered in second order. The same statement holds for the strain tensor components ϵ_{ij} with $(i, j) = (x, y, z)$ when studied in first order. This is due to the fact that these perturbations have the same temporal and spatial transformation properties. But it is self-evident that the constant factors in the different interaction terms are completely free and not related to each other.

Similarly, our discussion concerning the third-order dispersive terms of the conduction band can be directly applied to the three-fold degenerate valence band. One has only to replace the matrices $(\sigma_e^x, \sigma_e^y, \sigma_e^z)$ in Eqs. (2.5.9) and (2.5.10) by the angular momentum matrices (l_x, l_y, l_z), which are basis matrices for the valence band as given in Eq. (3.1.3). This is again possible since both sets of matrices have the same transformation properties (Γ_4, K^-). In addition, the components of the angular momentum operator $\boldsymbol{l} = (l_x, l_y, l_z)$ form the only set of basis matrices for the valence band, which is odd under time reversal and can therefore be associated to combinations of wave-vector components in third order. If other sets of basis matrices with K^- symmetry would exist, other additional combinations should be considered.

Concerning other symmetry-breaking effects, the above argumentation applies for example also for perturbations linear in the magnetic field (linear Zeeman effect). Following Table 2.3 and Eq. (2.3.1), a term $(\propto \boldsymbol{B})$ leads to

$$H_{B1}^v = a_{B1}^v \boldsymbol{B} \cdot \boldsymbol{l} = a_{B1}^v (B_x l_x + B_y l_y + B_z l_z). \tag{3.1.15}$$

This interaction term has the required symmetry properties and leads to a splitting of the degenerate valence band states. It is the only possible term of the valence band varying linearly with $\boldsymbol{B}$.

Let us consider as in Eq. (3.1.10) the Schrödinger equation corresponding to an applied magnetic field $\boldsymbol{B}$ as a perturbation and $E_{aB1}^v(i) = E_p$ denote the eigenvalues. Different from the electron-dispersion problem, there is no direction privileged in this problem, i.e. the perturbation is expressed in components of the angular momentum operator. We can therefore choose the direction of the magnetic field as quantization axis and label it "z". Then, Eq. (3.1.10) can be solved and one obtains the following eigenvalues and eigenfunctions $|\Psi_{aB1}(i)\rangle$ characterizing the linear Zeeman effect:

$$\begin{aligned} E_{aB1}^v(1) &= a_{B1}^v B_z \text{ with eigenfunction } |\Psi_{aB1}(1)\rangle = |1\rangle \\ E_{aB1}^v(2) &= 0 \text{ with eigenfunction } |\Psi_{aB1}(2)\rangle = |0\rangle \\ E_{aB1}^v(3) &= -a_{B1}^v B_z \text{ with eigenfunction } |\Psi_{aB1}(3)\rangle = |-1\rangle. \end{aligned} \tag{3.1.16}$$

Similar to the wave-vector, all second order terms in $\boldsymbol{B}$ are even functions under time reversal (K^+ symmetry). Combining them with matrices S, (U, V) and (X, Y, Z) may lead to quadratic Zeeman shifts of the valence band. We find for example as in Eq. (2.4.4) and similar to Eq. (3.1.8) (which concerns the wave-vector $\boldsymbol{Q}$)

$$H_{1B2}^v = a_{1B2}^v B^2 S, \tag{3.1.17}$$

which is also compatible with the overall symmetry of a Hamiltonian. H_{1B2}^v leads to an energy shift of the states but not to an additional splitting.

Comparing in Table 2.3 the construction of perturbation terms from wave-vector components $\boldsymbol{Q} = (Q_x, Q_y, Q_z)$ and magnetic-field components $\boldsymbol{B} = (B_x, B_y, B_z)$ in second order, one sees that one has only to perform in Eqs. (3.1.9) and (3.1.12) the following substitutions:

$$\begin{aligned} Q_x &\to B_x \\ Q_y &\to B_y \\ Q_z &\to B_z \end{aligned} \tag{3.1.18}$$

in order to obtain functions, which have the same transformation properties. Therefore, one can directly determine from these relations the symmetry-breaking interactions quadratic in $\boldsymbol{B}$. In second order in $\boldsymbol{B}$ one thus obtains the additional terms

$$H^v_{3B2} = a^v_{3B2}\left(\sqrt{3}(B_x^2 - B_y^2)U + (2B_z^2 - B_x^2 - B_y^2)V\right)$$
$$\text{and} \tag{3.1.19}$$
$$H^v_{5B2} = 2a^v_{5B2}\left(B_y B_z X + B_z B_x Y + B_x B_y Z\right)$$

where a^v_{1B2}, a^v_{3B2} , and a^v_{5B2} are the three parameters, which determine the strengths of these symmetry-breaking effects. The corresponding interaction matrices are obtained from Eqs. (3.1.8), (3.1.9) and (3.1.12) after the substitutions Eq. (3.1.18) have been performed.

In the example discussed above and leading to Eq. (3.1.16) we have chosen only $B_z \neq 0$. Concerning the quadratic Zeeman effect, we see that the perturbation term H^v_{5B2} has no influence since it vanishes for $B_x = B_y = 0$. Then the solutions of the Schrödinger equation treating an applied magnetic field up to second order in $\boldsymbol{B}$ (linear and quadratic Zeeman effect) are given by

$$\begin{aligned} E^v_{aB12}(1) &= a^v_{B1}B + (a^v_{1B2} + 2a^v_{3B2})B^2 \text{ with eigenfunction } |\Psi_{aB12}(1)\rangle = |1\rangle \\ E^v_{aB12}(2) &= (a^v_{1B2} - 4a^v_{3B2})B^2 \text{ with eigenfunction } |\Psi_{aB12}(2)\rangle = |0\rangle \\ E^v_{aB12}(3) &= -a^v_{B1}B + (a^v_{1B2} + 2a^v_{3B2})B^2 \text{ with eigenfunction } |\Psi_{aB12}(3)\rangle = |-1\rangle. \end{aligned} \tag{3.1.20}$$

Figure 3.1 shows a typical Zeeman splitting as function of the magnetic field for $\boldsymbol{B} \parallel$ [001]-direction. If $\boldsymbol{B}$ cannot be chosen $\parallel$ to the quantization axis "z" the Schrödinger equation has to be solved using the full matrices given in Eq. (3.1.19).

It is interesting to discuss the linear Stark effect and to see, which states are coupled by a linear external electric field $\boldsymbol{E} = (E_x, E_y, E_z)$. In order to do this, it is only necessary to replace in Eq. (3.1.12) the wave-vector components $2Q_iQ_j$ $((i, j) = x, y, z)$ by the corresponding components of the electric field that are characterized by the same label (here: (X, Y, Z)). This is possible since the different functions have the same transformation properties (Γ_5, K^+).

According to Table 2.3 the product $(YZ' + ZY')$ transforms as Γ_5. In compliance with Table 2.2 this combination may be obtained by e.g. $Q_yQ_z + Q_zQ_y = 2Q_yQ_z$ (having K^+-symmetry with respect to time reversal). According to the first column

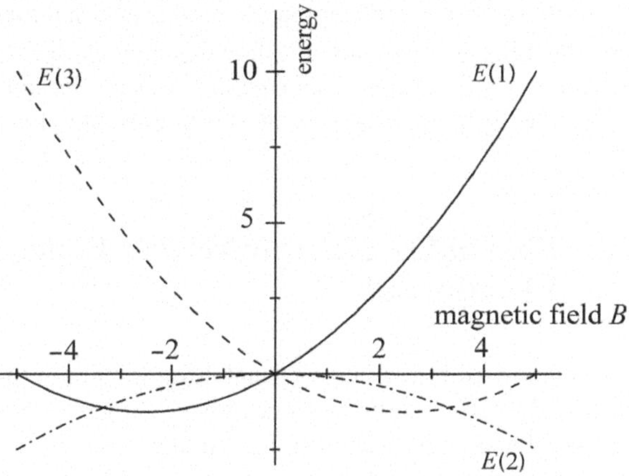

Fig. 3.1 Typical linear and quadratic Zeeman splitting as function of the magnetic field $\boldsymbol{B} \parallel [001]$-direction. $E(1)$, $E(2)$, and $E(3)$ denote the energies of the states $(|1\rangle, |0\rangle, |-1\rangle)$, respectively. The graphs are solutions of Eq. (3.1.20) with arbitrarily chosen parameter values a^v_{B1}, a^v_{1B2}, and a^v_{3B2}

of Table 2.3, the same symmetry reveals (apart from other terms) the product SX', or E_x. We thus obtain directly from Eq. (3.1.12) for the coupling of the valence-band states to the electric field the interaction Hamiltonian H^v_{E1}:

$$H^v_{E1} = (\sqrt{2}/4)a^v_{E1} \begin{pmatrix} 0 & E_y - iE_x & -i\sqrt{2}E_z \\ E_y + iE_x & 0 & -E_y + iE_x \\ i\sqrt{2}E_z & -E_y - iE_x & 0 \end{pmatrix}. \tag{3.1.21}$$

To find the corresponding energy eigenvalues $E^v_{E1}(i)$ with $i = (1 \text{ to } 3)$, the matrix (3.1.21) has to be diagonalized, which is achieved by establishing a characteristic determinant (see Appendix A). This leads to the polynomial

$$(E^v_{E1})^3 - (1/4)(E^v_{E1})(a^v_{E1})^2(E_x^2 + E_y^2 + E_z^2) + (1/4)(a^v_{E1})^3 E_x E_y E_z = 0 \tag{3.1.22}$$

the solution of which gives the energies of the valence-band states $E^v_{E1}(i)$.

Since the matrix form of the Hamiltonians given in Eq. (3.1.12) for wave-vector components and in Eq. (3.1.21) for the linear Stark effect are identical, the solutions of the respective Schrödinger equations (the resulting wave-functions and the eigenvalues with exception of the constant factors) are the same. Then, in the results for the eigenvalues given in Eq. (3.1.13) for the high symmetry directions [110] the energy factor $a^v_{5Q2}Q^2$ has only to be replaced by $a^v_{5E1}E$. Concerning the [111] direction the substitution consists in replacing in Eq. (3.1.14) $(2/3)a^v_{5Q2}Q^2$ by $(1/\sqrt{3})a^v_{5E1}E$.

The Stark effect is also $\neq 0$ if the electric field is orientated parallel to the [001] direction. Similar to Eq. (3.1.20) we obtain here:

$$\begin{aligned} E^v_{a5E1}(1) &= (1/2)a^v_{5E1}E \text{ with eigenfunction } |\Psi_{a5E1}(1)\rangle = (1/\sqrt{2})\,(-i|1\rangle + |-1\rangle) \\ E^v_{a5E1}(2) &= 0 \text{ with eigenfunction } |\Psi_{a5E1}(2)\rangle = |0\rangle \\ E^v_{a5E1}(3) &= -(1/2)a^v_{5E1}E \text{ with eigenfunction } |\Psi_{a5E1}(3)\rangle = (1/\sqrt{2})\,(i|1\rangle + |-1\rangle)\,. \end{aligned} \tag{3.1.23}$$

We have seen that the application of an external field or the finite wave-vector may mix the different states and lift their degeneracy. If these perturbations are in arbitrary directions, the amount of mixing depends on the material dependent factors, and if several perturbations are present it is important to know their respective orientations.

3.2 Six-Fold Degenerate Valence Band Including Electron Spin

We can now continue our invariant development of the Hamiltonian describing completely the semiconductor valence band. Since an electron is a spin 1/2 particle, the valence-band states with orbital angular momentum $l_v = 1$ are also two-fold spin degenerate. In the following, the orbital angular momentum $\boldsymbol{l_v}$ operates only on the orbital part of the valence-band states. Similar to the conduction band case, we consider spin degeneracy by introducing in addition an effective spin operator $\boldsymbol{\sigma_v}$ with $\sigma_v = 1/2$, which now only operates on the valence-band spin-states. The matrices describing this effective valence-electron spin transform again as (Γ_4, K^-) in zincblende-type material. We calculate now the direct product (or Kronecker product) of the angular momentum matrices (which span a subspace of dimension 3 and are given in Eq. (3.1.7)) with the spin matrices, i.e. we calculate the product $\boldsymbol{l_v} \otimes \boldsymbol{\sigma_v}$. Here, the spin matrices $\boldsymbol{\sigma_v} = (\sigma_v^x, \sigma_v^y, \sigma_v^z)$ together with the two-dimensional unit matrix 1_v describe the spin subspace of dimension 2. We thus build the product space of dimension six in which the valence-band states are defined. The valence-band eigenfunctions can be equally well determined using the Clebsch-Gordan coefficients for a system with angular momentum $l_v = 1$ coupled to a spin with $\sigma_v = 1/2$. The eigenstates of the system can be classified according to the total-angular momentum $\boldsymbol{j_v} = \boldsymbol{l_v} \oplus \boldsymbol{\sigma_v}$ and its projection component j_z onto the z-axis. Following this scheme, states with a total-angular momentum $j_v = 3/2$ and $j_v = 1/2$ can be constructed.

Using the above convention for the matrix products, valence-band electron-wave functions v_i^v are given in the new basis by

$$v_i^v = a_{i,1}|1\rangle\alpha + a_{i,2}|1\rangle\beta + a_{i,3}|0\rangle\alpha + a_{i,4}|0\rangle\beta + a_{i,5}|-1\rangle\alpha + a_{i,6}|-1\rangle\beta \quad (i = 1, ..., 6), \tag{3.2.1}$$

where $|m_j\rangle$ denote the eigenfunctions of the angular-momentum operator-component l_z as defined in Eq. (3.1.1) and α and β the electron spin up and down states, respectively.

Concerning the angular momentum, we start with the matrices $(S; U, V; P, Q, R; X, Y, Z)$, which are obtained from the components (l_x, l_y, l_z) of the angular momentum and are given in Eqs. (3.1.5) and (3.1.7). They transform as $(\Gamma_1; \Gamma_3; \Gamma_4; \Gamma_5)$, respectively. They are multiplied with the valence-electron spin-matrices, i.e. the two-dimensional unit matrix 1_v and the Pauli matrices $(\sigma_v^x, \sigma_v^y, \sigma_v^z)$, which transform as (Γ_1, K^+) and (Γ_4, K^-), respectively. These matrices are defined in the same way as in Eqs. (2.2.5) and (2.2.3) for the electrons in the conduction band. (For clearness, matrices defined in the angular-momentum subspace are denoted in the following discussion by capital Latin letters as in Eq. (3.1.7), those defined in the spin-subspace by a prime (′), and matrices in the product space by a double quotation mark (”). This convention will, however, not systematically be adopted throughout all this work.) We denote the spin matrices in the following by

$$\begin{aligned}(1/3)(\boldsymbol{\sigma}_v)^2 = 1_v &\rightarrow S' \\ \sigma_v^x &\rightarrow P' \\ \sigma_v^y &\rightarrow Q' \\ \sigma_v^z &\rightarrow R'.\end{aligned} \tag{3.2.2}$$

Since the transformation properties of the angular momentum and that of the spin matrices are known, we can now construct by successive use of Table 2.3 the symmetry adapted basis matrices giving the symmetries of all possible interactions, which can act within the valence-band states. We thus obtain the thirty-six symmetry adapted basis matrices:

$$\begin{aligned}(S; U, V; P, Q, R; X, Y, Z) \otimes S' &\Leftrightarrow \text{transforming as} \Leftrightarrow (\Gamma_1, K^+; \Gamma_3, K^+; \Gamma_4, K^-; \Gamma_5, K^+) \\ (P, Q, R) \otimes (P', Q', R') &\Leftrightarrow (\Gamma_1 \oplus \Gamma_3 \oplus \Gamma_4 \oplus \Gamma_5, K^+) \\ (X, Y, Z) \otimes (P', Q', R') &\Leftrightarrow (\Gamma_2 \oplus \Gamma_3 \oplus \Gamma_4 \oplus \Gamma_5, K^-) \\ (U, V) \otimes (P', Q', R') &\Leftrightarrow (\Gamma_4 \oplus \Gamma_5, K^-) \\ (S) \otimes (P', Q', R') &\Leftrightarrow (\Gamma_4, K^-).\end{aligned} \tag{3.2.3}$$

According to Table 2.4 we obtain in total fifteen sets of symmetry adapted matrices. 2 sets transform as (Γ_1, K^+), 1 as (Γ_2, K^-), 2 as (Γ_3, K^+), 1 as (Γ_3, K^-), 1 as (Γ_4, K^+), 4 as (Γ_4, K^-), 2 as (Γ_5, K^+), and 2 as (Γ_5, K^-). They fully account for all possible interactions in between the valence-band states.

It is convenient to enumerate these basis matrices explicitly since they can directly be used to establish (together with the perturbing field components) the different contributions to a Hamiltonian. The spatial transformation properties of the basis matrices are noted according to our convention in Eq. (3.1.5), they are classified according to their time reversal properties, and indicated by (”). Since different basis matrices may have the same spatial symmetry but arise from different components of the angular-momentum and spin operators, we indicate their origin by subscripts, where the first number indicates the irreducible representation to which the angular momentum components belong and the second that of the spin component. For example the matrix P_{34}” characterizes the matrix P (one of the components transforming as Γ_4 in the six-dimensional product space) obtained from the direct product of angular-momentum matrices transforming as Γ_3 with spin components transforming as Γ_4. As given in Eq. (3.2.3) it has K^- time-reversal’ symmetry. For shortness we write in the following (as in Table 2.3) the matrix product $S_{11}" = S \otimes S'$ in the form $S_{11}" = SS'$. We thus obtain the following matrices that transform

$$
\begin{aligned}
&\text{as } K^+ \text{ under time reversal :}\\
S_{11}" &= SS' \quad (\Gamma_1, K^+)\,;\\
U_{31}" &= US',\, V_{31}" = VS' \quad (\Gamma_3, K^+)\,;\\
X_{51}" &= XS',\, Y_{51}" = YS',\, Z_{51}" = ZS' \quad (\Gamma_5, K^+)\,;\\
S_{44}" &= PP' + QQ' + RR' \quad (\Gamma_1, K^+)\,;\\
U_{44}" &= \sqrt{3}(PP' - QQ'),\, V_{44}" = 2RR' - PP' - QQ' \quad (\Gamma_3, K^+)\,;\\
P_{44}" &= QR' - RQ',\, Q_{44}" = RP' - PR',\, R_{44}" = PQ' - QP' \quad (\Gamma_4, K^+)\,;\\
X_{44}" &= QR' + RQ',\, Y_{44}" = RP' + PR',\, Z_{44}" = PQ' + QP' \quad (\Gamma_5, K^+)\,;\\
&\text{and for those transforming as } K^- :\\
P_{41}" &= PS',\, Q_{41}" = QS',\, R_{41}" = RS' \quad (\Gamma_4, K^-)\,;\\
T_{54}" &= XP' + YQ' + ZR' \quad (\Gamma_2, K^-)\,;\\
U_{54}" &= 2ZR' - XP' - YQ',\, V_{54}" = \sqrt{3}(-XP' + YQ') \quad (\Gamma_3, K^-)\,;\\
P_{54}" &= YR' + ZQ',\, Q_{54}" = ZP' + XR',\, R_{54}" = XQ' + YP' \quad (\Gamma_4, K^-)\,;\\
X_{54}" &= -YR' + ZQ',\, Y_{54}" = -ZP' + XR',\, Z_{54}" = -XQ' + YP' \quad (\Gamma_5, K^-)\,;\\
P_{34}" &= (\sqrt{3}U - V)P',\, Q_{34}" = -(\sqrt{3}U + V)Q',\, R_{34}" = 2VR' \quad (\Gamma_4, K^-)\,;\\
X_{34}" &= -(\sqrt{3}V + U)P',\, Y_{34}" = (\sqrt{3}V - U)Q',\, Z_{34}" = 2UR' \quad (\Gamma_5, K^-)\,;\\
P_{14}" &= SP',\, Q_{14}" = SQ',\, R_{14}" = SR' \quad (\Gamma_4, K^-)\,.
\end{aligned}
\tag{3.2.4}
$$

Let us consider the construction of these matrices in more detail: Table 2.3 shows that e.g. the symmetry adapted basis matrices (U, V) and (P', Q', R') multiplied with each other can give rise again to symmetry adapted matrices in the product space if the correct combinations are chosen. Namely, one can construct from the direct product of (U, V) and (P', Q', R') six linearly independent, symmetry-adapted matrices which are:

$$(U, V) \otimes (P', Q', R') \rightarrow P_{34}" \oplus Q_{34}" \oplus R_{34}" \oplus X_{34}" \oplus Y_{34}" \oplus Z_{34}". \tag{3.2.5}$$

Since the final matrices ($P_{34}"$, $Q_{34}"$, $R_{34}"$) and ($X_{34}"$, $Y_{34}"$, $Z_{34}"$) are linearly independent they can form together with the other matrices of Eq. (3.2.3) a basis of the six-dimensional product space of the valence-band states.

Let us calculate as an example the matrix $P_{34}"$. As indicated in Table 2.3 it is given by

$$P_{34}" = (\sqrt{3}U - V)P'. \tag{3.2.6}$$

Using the definitions given in Eqs. (3.1.7) and (3.2.2) for U, V, and P'

$U = \sqrt{3}\begin{pmatrix} 0\,0\,1 \\ 0\,0\,0 \\ 1\,0\,0 \end{pmatrix}$, $V = \begin{pmatrix} 1 & 0 & 0 \\ 0 & -2 & 0 \\ 0 & 0 & 1 \end{pmatrix}$, and $P' = \sigma_v^x = \begin{pmatrix} 0\,1 \\ 1\,0 \end{pmatrix}$ one obtains:

$$P_{34}\text{''} = \left(3\begin{pmatrix} 0\,0\,1 \\ 0\,0\,0 \\ 1\,0\,0 \end{pmatrix} - \begin{pmatrix} 1 & 0 & 0 \\ 0 & -2 & 0 \\ 0 & 0 & 1 \end{pmatrix}\right) \otimes \begin{pmatrix} 0\,1 \\ 1\,0 \end{pmatrix} = \begin{pmatrix} 0 & -1 & 0 & 0 & 0 & 3 \\ -1 & 0 & 0 & 0 & 3 & 0 \\ 0 & 0 & 0 & 2 & 0 & 0 \\ 0 & 0 & 2 & 0 & 0 & 0 \\ 0 & 3 & 0 & 0 & 0 & -1 \\ 3 & 0 & 0 & 0 & -1 & 0 \end{pmatrix}. \tag{3.2.7}$$

The other matrices enumerated in Eq. (3.2.4) can be determined in the same way and thus a symmetry-adapted basis for the valence-band states is obtained. It is important to notice here that (according to the formation of the direct matrix product) the rows and columns of the matrix elements of P_{34}" (and of all other matrices in this section) are ordered according to the states $(|1\rangle\alpha, |1\rangle\beta, |0\rangle\alpha, |0\rangle\beta, |-1\rangle\alpha, |-1\rangle\beta)$. This allows one to analyze, which states are coupled within the matrices given in Eq. (3.2.4).

It is interesting to mention at this point that the first line of Eq. (3.2.3) describes all interaction terms, which are already present if only the angular momentum states of $l = 1$ are considered. For each state the spin variable is added, but the spin does not change the interaction between the states. This is expressed by the fact that the matrix S' is proportional to the unit matrix in the spin subspace. Similarly, the interaction terms of the last line of Eq. (3.2.3) act only on the spin part of the wave functions and the angular momentum of the actual product state is not influenced, S being proportional to the unit matrix of the angular-momentum subspace. All interactions within the pure spin and angular-momentum subspaces discussed in Chap. 2 and Sect. 3.1 can be described by these twelve matrices.

Interestingly, in the product space of spin and angular-momentum variables new matrices (allowing to construct new interaction terms) appear, which may have different symmetries (transformation properties) than the ones defined in the isolated spin and angular-momentum subspaces. They are thus due to spin-orbit coupling within the valence-band states and will be discussed in the following. It is very important to realize that in physical systems, which show no or only small spin-orbit coupling the interaction terms that are resulting from these matrices have also to vanish.

Let us now discuss the Hamiltonian of the valence band. We start with the terms, which have the full crystal symmetry at the Γ-point and define the Hamiltonian H^v. Using the valence-band electron-wave functions v_i^v given in Eq. (3.2.1) these terms can be formulated in form of a matrix. They are characterized by the fact that only scalar functions multiply the matrices of Eq. (3.2.3). Inspection of Eq. (3.2.4) shows immediately that only two products are compatible with interaction terms that may appear in a Hamiltonian, i.e. products transforming as (Γ_1, K^+). They are given by

$$S_{11}\text{''} = S \otimes S' \quad \text{and} \quad S_{44}\text{''} = P \otimes P' + Q \otimes Q' + R \otimes R'. \tag{3.2.8}$$

Therefore, the Hamiltonian H^v can be decomposed into

$$H^v = H_d^v + H_{so}^v = a_o^v S_{11}\text{''} + a_{so}^v S_{44}\text{''}. \tag{3.2.9}$$

The two terms in Eq. (3.2.9) are the only terms of a Hamiltonian that have the full point-group symmetry.

Following Eqs. (3.1.7) and (3.2.2) the matrix $S_{11}\text{''} = S \otimes S' = 2 \cdot 1\text{''}$ is proportional to the six-dimensional unit matrix 1". Then, the direct term $H_d^v = a_o^v S_{11}\text{''}$ is given by the product of a real scalar value $2a_o^v$ and the unit matrix 1". It determines the energy of the six-fold degenerate valence-band states, which can be taken in the following as the origin of the energy scale. H_d^v does not depend on the spin or angular-momentum structure of these states.

A second term, denoted $H_{so}^v = a_{so}^v S_{44}\text{''}$, describes the spin-orbit interaction in-between these states. After calculation of S_{44}" from Eq. (3.2.8) (by using explicitly the expressions for the matrices (P, Q, R) given in Eq. (3.1.7) and those for (P', Q', R') given in Eq. (2.2.3), H_{so}^v takes the matrix form

$$H_{so}^v = a_{so}^v S_{44}\text{''} = a_{so}^v \begin{pmatrix} 1 & 0 & 0 & 0 & 0 & 0 \\ 0 & -1 & \sqrt{2} & 0 & 0 & 0 \\ 0 & \sqrt{2} & 0 & 0 & 0 & 0 \\ 0 & 0 & 0 & 0 & \sqrt{2} & 0 \\ 0 & 0 & 0 & \sqrt{2} & -1 & 0 \\ 0 & 0 & 0 & 0 & 0 & 1 \end{pmatrix}, \tag{3.2.10}$$

where the constant a_{so}^v indicates the strength of the spin-orbit interaction. It depends on the strength of the electric potential, to which the electrons are exposed, and on the atomic orbitals, which give rise to the valence-band states. (Spin-orbit coupling is a relativistic effect and increases as discussed in atomic physics ($\propto Z^4$) with increasing atomic charge number Z. Therefore, spin-orbit interaction is more important in atoms with higher nucleus charges than in those with lower charges. This consideration remains important in semiconductor physics.).

The spin-orbit interaction H_{so}^v can be diagonalized and one finds in zincblende structure that the eigenvalues e_i (with $i = 1, ..., 6$) of Eq. (3.2.10) are given by

$$e_i = a_{so}^v \left(-2,\ -2,\ 1,\ 1,\ 1,\ 1\right). \tag{3.2.11}$$

The corresponding normalized eigenvectors of the valence band are given by the columns of v_i^v:

$$v_i^v = \begin{pmatrix} 0 & 0 & 0 & 0 & 0 & 1 \\ 0 & -\sqrt{6}/3 & 0 & 0 & \sqrt{3}/3 & 0 \\ 0 & \sqrt{3}/3 & 0 & 0 & \sqrt{6}/3 & 0 \\ -\sqrt{3}/3 & 0 & 0 & \sqrt{6}/3 & 0 & 0 \\ \sqrt{6}/3 & 0 & 0 & \sqrt{3}/3 & 0 & 0 \\ 0 & 0 & 1 & 0 & 0 & 0 \end{pmatrix}. \tag{3.2.12}$$

Let us in the following consider the total angular momentum $\boldsymbol{j}_v = \boldsymbol{l}_v \oplus \boldsymbol{\sigma}_v$ and its projection component j_z onto the z-axis. In the double-group representation, which has to be used if the spin and the orbital-angular momentum of the electron are considered, the valence-band states transform as Γ_7 (v_1^v and v_2^v; two-fold degenerate) and Γ_8 (v_3^v to v_6^v; four-fold degenerate), respectively [2, 4, 8]. We also use the notation $|j, m_j\rangle$ for indexing the valence-band states, where the quantum numbers (j, m_j) recall the total angular momentum j and its projection m_j onto the z-axis of the states. Because of the implicit use of the symmetry properties of the wave functions this notation is more convenient when discussing the matrices describing the effective Hamiltonians. Taking into account the ordering of the basis functions Eq. (3.2.1) the eigenvectors of the valence band are now explicitly given by

$$\begin{aligned} v_1^v &= (-|0\rangle\beta + \sqrt{2}|-1\rangle\alpha)/\sqrt{3} = |1/2, -1/2\rangle \\ v_2^v &= (|0\rangle\alpha - \sqrt{2}|1\rangle\beta)/\sqrt{3} = |1/2, 1/2\rangle \\ v_3^v &= |-1\rangle\beta = |3/2, -3/2\rangle \\ v_4^v &= (\sqrt{2}|0\rangle\beta + |-1\rangle\alpha)/\sqrt{3} = |3/2, -1/2\rangle \\ v_5^v &= (\sqrt{2}|0\rangle\alpha + |1\rangle\beta)/\sqrt{3} = |3/2, 1/2\rangle \\ v_6^v &= |1\rangle\alpha = |3/2, 3/2\rangle. \end{aligned} \tag{3.2.13}$$

Usually, one considers rather the transformation properties of the p orbitals in Cartesian coordinates, which are given in Eq. (3.1.2). The basis functions of the valence-band states then read:

$$\begin{aligned} v_1^v &= (-z\beta + (x - iy)\alpha)/\sqrt{3} = |\phi_6\rangle \\ v_2^v &= (z\alpha + (x + iy)\beta)/\sqrt{3} = |\phi_5\rangle \\ v_3^v &= (x - iy)\beta/\sqrt{2} = |\phi_4\rangle \\ v_4^v &= (2z\beta + (x - iy)\alpha)/\sqrt{6} = |\phi_3\rangle \\ v_5^v &= (2z\alpha - (x + iy)\beta)/\sqrt{6} = |\phi_2\rangle \\ v_6^v &= -(x + iy)\alpha/\sqrt{2} = |\phi_1\rangle. \end{aligned} \tag{3.2.14}$$

The notation of the states ($|\phi_i\rangle$) (with $i = 1, .., 6$) has been introduced in Ref. [1], which is recalled here. It is also used to define electron-hole pair states as will be discussed in the chapter concerning excitons.

As given in Eq. (3.2.11) and indicated schematically in Fig. 2.2 we see that the degenerate valence band is split by the spin-orbit interaction into two bands that are four- and two-fold degenerate at the Γ-point, respectively. The spin-orbit splitting of the valence-band states Δ_{so}^v is now obtained from Eq. (3.2.11) to be

$$\Delta_{so}^v = 3a_{so}^v. \tag{3.2.15}$$

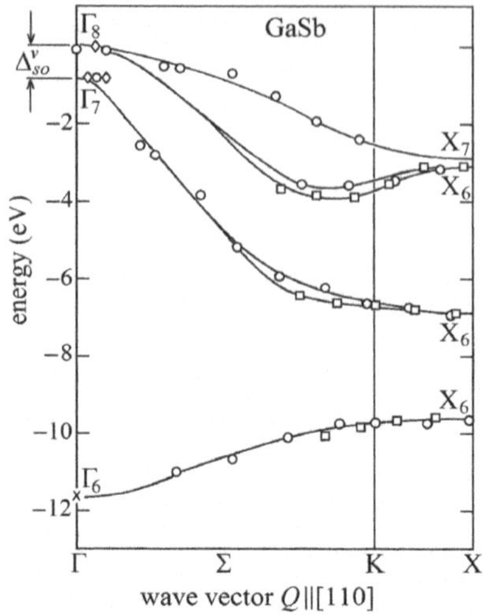

Fig. 3.2 Valence-band dispersion of zincblende GaSb single crystals along the $\Gamma - \Sigma - K$ direction of the first Brillouin zone. The spin-orbit splitting $\Delta^v_{so} = 0.8$ eV at the valence-band maximum (the Γ-point) is clearly seen. This splitting (being a relativistic effect) is particularly large in GaSb due to the large atomic number of Sb. Symbols indicate various experimental points, smooth curves represent theoretical band-dispersion calculations. According to Ref. [2]

The spin-orbit coupling of a valence band is illustrated in Fig. 3.2 that displays the valence bands of zincblende-type GaSb semiconductors (Ref. [2]). The spin-orbit splitting $\Delta^v_{so} = 0.8$ eV at the valence-band maximum (the Γ-point) is clearly seen. It is particularly large in GaSb because of the high atomic number of Sb.

The band having Γ_7 symmetry is also called the "split-off band". Usually, the spin-orbit interaction is positive in simple semiconductors and the valence band with Γ_8 symmetry has a higher energy than that with Γ_7 symmetry. But, as mentioned above, there are some exceptions (as for example CuCl). This unusual behavior is due to different admixtures of atomic d- and p-electron states to the uppermost valence-band states of the crystals. In addition, the contributions of atomic-electron states to the valence-band states may change with temperature and the spin-orbit splitting of the valence bands becomes temperature dependent.

As discussed above, symmetry-breaking interactions can now be considered in detail and it can be analyzed, in which way they modify the energy structure of the valence-band states and how they mix different states. The procedure is the same as the one, which we have adopted above: we develop the symmetry-breaking perturbation in a power series, use Table 2.3 to construct symmetry-adapted product-functions and, after being multiplied with the basis matrices of Eq. (3.2.4), search for interaction terms that are invariant under the symmetry operations of the crystal, i.e. that transform as (Γ_1, K^+). These terms may then appear in a Hamiltonian and contribute to the energy of the valence-band states.

Concerning the spatial symmetry properties of a system, we know from group theory (and it can also be seen in Table 2.4) that products of two irreducible representations of a group contain terms transforming as Γ_1 if, and only if, the two members of the product have the same transformation properties. Furthermore, in this case a term of Γ_1 symmetry appears only once. Or, to put it differently, only products of the form $\Gamma_i \otimes \Gamma_j$ contain a representation transforming as Γ_1 (and only once) if $i = j$. As we have seen in Eq. (3.2.3), the basis matrices can be decomposed into fifteen different irreducible representations. This means that, when applying an arbitrary perturbation to the valence band, it can give rise to at maximum fifteen symmetry breaking interactions.

Let us as a first example discuss the linear Stark effect and its relation to spin-orbit coupling in the semiconductor valence band. As discussed with Eqs. (3.1.21)–(3.1.23), when neglecting the electron spin, a linear Stark effect is already present in the valence band because of the orbital angular momentum. Including the spin and employing the matrices in Eq. (3.2.3), this interaction Hamiltonian H^v_{E1a} (corresponding to H^v_{E1} given in Eq. (3.1.21)) takes now the form

$$\begin{aligned} H^v_{E1a} &= a^v_{E1a}[XS'E_x + YS'E_y + ZS'E_z] = \\ &= a^v_{E1a}[X_{51}\text{''}E_x + Y_{51}\text{''}E_y + Z_{51}\text{''}E_z]. \end{aligned} \tag{3.2.16}$$

The inclusion of the spin leads, however, to an additional, second interaction term, that is also linear in the electric field $\boldsymbol{E} = (E_x, E_y, E_z)$. This term has its origin in the increase of the number of considered basis states, i.e. its influence on the valence-band electron-states is due to the spin-orbit coupling. In its matrix form the new interaction term H^v_{E1b} is proportional to a weight factor a^v_{E1b} that has to vanish if the spin-orbit coupling vanishes. We obtain for this additional term linear in the electric field explicitly:

$$\begin{aligned} H^v_{E1b} = a^v_{E1b}[(QR' + RQ')E_x + (RP' + PR')E_y + (PQ' + QP')E_z] = \\ = a^v_{E1b}[X_{44}\text{''}E_x + Y_{44}\text{''}E_y + Z_{44}\text{''}E_z]. \end{aligned} \tag{3.2.17}$$

Let us consider the matrix form of both terms in detail. We choose as an example the direction of the electric field $\boldsymbol{E}$ parallel to the z-direction, which is the quantization axis. Then, we obtain for H^v_{E1a} the coupling matrix

$$H^v_{E1a} = a^v_{E1a}[Z_{51}\text{''}E_z] = a^v_{E1a}(ZS')E_z = a^v_{E1a}E_z \begin{pmatrix} 0 & 0 & 0 & 0 & -i3/2 & 0 \\ 0 & 0 & 0 & 0 & 0 & -i3/2 \\ 0 & 0 & 0 & 0 & 0 & 0 \\ 0 & 0 & 0 & 0 & 0 & 0 \\ i3/2 & 0 & 0 & 0 & 0 & 0 \\ 0 & i3/2 & 0 & 0 & 0 & 0 \end{pmatrix}. \tag{3.2.18}$$

The order of eigenvectors of the valence band used to establish the matrix in Eq. (3.2.18) has been discussed in connection with Eq. (3.2.7). In a Hamiltonian non-diagonal matrix elements correspond to a mixing of the basis states in which the Hamiltonian is expanded (see e.g. Appendix A). We thus see from the matrix form of H^v_{E1a} that the electric field component E_z mixes the states $|1\rangle\alpha$ and $|-1\rangle\alpha$ as well as the states $|1\rangle\beta$ and $|-1\rangle\beta$, while the states $|0\rangle\alpha$ and $|0\rangle\beta$ states are not affected by this field component. This means that, when conserving the spin, E_z may act only on the orbital angular momentum and mix states with $\Delta l_z = \pm 2$.

H^v_{E1b} has a different structure than the linear Stark effect of H^v_{E1a}:

$$H^v_{E1b} = a^v_{E1b}[Z_{44}\text{''}E_z] = a^v_{E1b}(RR')E_z = a^v_{E1b}E_z \begin{pmatrix} 0 & 0 & 0 & -i\sqrt{2} & 0 & 0 \\ 0 & 0 & 0 & 0 & 0 & 0 \\ 0 & 0 & 0 & 0 & 0 & -i\sqrt{2} \\ i\sqrt{2} & 0 & 0 & 0 & 0 & 0 \\ 0 & 0 & 0 & 0 & 0 & 0 \\ 0 & 0 & i\sqrt{2} & 0 & 0 & 0 \end{pmatrix}. \tag{3.2.19}$$

When considering the matrix form of H^v_{E1b} given in Eq. (3.2.19) we see that the electric field component E_z mixes the states $|1\rangle\alpha$ and $|0\rangle\beta$ and the states $|0\rangle\alpha$ and $|-1\rangle\beta$. In this case the total angular momentum j is used to classify the states and one finds that states with $\Delta j_z = \pm 2$ become mixed (see Eq. (3.2.19). The value of the mixing is, however, different from that of H^v_{E1a} since the factors a^v_{E1a} and a^v_{E1b} are independent from each other. Similar, but more complicated mixing schemes occur if the field components E_x and E_y are $\neq 0$. In any case, an electric field of arbitrary strength and direction leads to a splitting of the sixfold degenerate valence-band states.

In view of the "Pseudo-Spin Development of the Valence Band" that we will discuss in the next sections, it is not so interesting to consider the thirty-six basis matrices given in Eq. (3.2.4), which are only adapted to the crystal symmetry. The rows and columns of these matrices are defined using the eigen-vectors of the valence band, ordered according to ($|1\rangle\alpha$, $|1\rangle\beta$, $|0\rangle\alpha$, $|0\rangle\beta$, $|-1\rangle\alpha$, $|-1\rangle\beta$). Instead, one better transforms the complete Hamiltonian to the set of valence-band states in which, in addition, the spin-orbit coupling is diagonal. This set of eigenstates v^v_1 to v^v_6 consists of a linear combination of the states ($|1\rangle\alpha$ to $|-1\rangle\beta$) and is given in Eq. (3.2.13) or Eq. (3.2.14). In order to obtain the new symmetry adapted matrices, we transform the matrices of Eq. (3.2.4) onto these normalized eigenvectors. We indicate the resulting matrices by the additional subscript "so". Then the matrix P_{34}" given in Eq. (3.2.7) becomes for example:

$$P_{34}\text{''} = \begin{pmatrix} 0 & -1 & 0 & 0 & 0 & 3 \\ -1 & 0 & 0 & 0 & 3 & 0 \\ 0 & 0 & 0 & 2 & 0 & 0 \\ 0 & 0 & 2 & 0 & 0 & 0 \\ 0 & 3 & 0 & 0 & 0 & -1 \\ 3 & 0 & 0 & 0 & -1 & 0 \end{pmatrix} \rightarrow \begin{pmatrix} 0 & -8/3 & -\sqrt{6}/3 & 0 & \sqrt{2}/3 & 0 \\ -8/3 & 0 & 0 & -\sqrt{2}/3 & 0 & \sqrt{6}/3 \\ -\sqrt{6}/3 & 0 & 0 & -\sqrt{3}/3 & 0 & 3 \\ 0 & -\sqrt{2}/3 & -\sqrt{3}/3 & 0 & 7/3 & 0 \\ \sqrt{2}/3 & 0 & 0 & 7/3 & 0 & -\sqrt{3}/3 \\ 0 & \sqrt{6}/3 & 3 & 0 & -\sqrt{3}/3 & 0 \end{pmatrix} = P_{34so}\text{''}. \tag{3.2.20}$$

(Concerning mathematical details, see Appendix B.)

For the above example of an electric field applied in the z-direction $\boldsymbol{E} = (0, 0, E_z)$, we obtain instead of Eq. (3.2.18) the coupling matrix

$$H^{v}_{E1a} = a^{v}_{E1a}[X_{51so}"E_x + Y_{51so}"E_y + Z_{51so}"E_z]$$

$$= a^{v}_{E1a}E_z \begin{pmatrix} 0 & 0 & 0 & 0 & 0 & i\sqrt{6}/2 \\ 0 & 0 & i\sqrt{6}/2 & 0 & 0 & 0 \\ 0 & -i\sqrt{6}/2 & 0 & 0 & i\sqrt{3}/2 & 0 \\ 0 & 0 & 0 & 0 & 0 & i\sqrt{3}/2 \\ 0 & 0 & -i\sqrt{3}/2 & 0 & 0 & 0 \\ -i\sqrt{6}/2 & 0 & 0 & -i\sqrt{3}/2 & 0 & 0 \end{pmatrix} \quad (3.2.21)$$

and instead of Eq. (3.2.19) for H^{v}_{E1b}:

$$H^{v}_{E1b} = a^{v}_{E1b}[X_{44so}"E_x + Y_{44so}"E_y + Z_{44so}"E_z]$$

$$= a^{v}_{E1b}E_z \begin{pmatrix} 0 & 0 & 0 & 0 & 0 & -i\sqrt{6}/3 \\ 0 & 0 & -i\sqrt{6}/3 & 0 & 0 & 0 \\ 0 & i\sqrt{6}/3 & 0 & 0 & i2\sqrt{3}/3 & 0 \\ 0 & 0 & 0 & 0 & 0 & i2\sqrt{3}/3 \\ 0 & 0 & -i2\sqrt{3}/3 & 0 & 0 & 0 \\ i\sqrt{6}/3 & 0 & 0 & -i2\sqrt{3}/3 & 0 & 0 \end{pmatrix}, \quad (3.2.22)$$

where the 2×2 block in the left, upper corner of the matrix concerns the interaction between the Γ_7-states and the 4×4 block in the right, lower corner the interacting Γ_8-states.

The coupling schemes of both interaction terms given in Eqs. (3.2.21) and (3.2.22) are very similar to each other: Within the blocks, states transforming as Γ_7 are not coupled directly with each other but the same states transforming as Γ_8 are coupled in both cases. These states are split into two doublets when applying an electric field.

Concerning the non-diagonal blocks, as shown in Eqs. (3.2.21) and (3.2.22), the same valence-band states with Γ_7 and Γ_8 symmetry are coupled through the E_z-linear terms. E_z thus mixes the irreducible representations of different symmetry. It is interesting to notice that the coupling strengths in these non-diagonal blocks and within the block of Γ_8-symmetry are different.

As we will discuss in the next sections, the mixing of the irreducible representations by the electric field can be neglected if the spin-orbit coupling is strong compared to the Stark effect. In order to show an intermediate situation and the competition between Stark effect and spin-orbit coupling, we have calculated the energy of the states resulting from the interaction Hamiltonian H^{v}_{E1a} of Eq. (3.2.18) and H^{v}_{E1b} of Eq. (3.2.19) as a function of E_z. It might be of importance to note that Figs. 3.3, 3.4, 3.5, 3.6, 3.7 and 3.8 display the expected phenomena in a qualitative way (i.e. in relative units) only; the symmetry-based approach itself is not able to predict the magnitude of the effects.

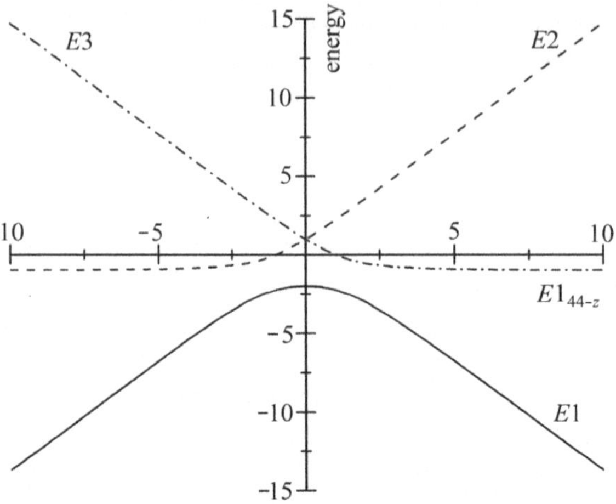

Fig. 3.3 Energy of the valence-band states at the Γ-point (given in units of $a^v_{so} = \Delta^v_{so}/3$) as a function of $E1_{44-z} = (a^v_{E1b}\, E_z\,)/a^v_{so}$, where the interaction energy $(a^v_{E1b}\, E_z)$ is measured in units of the spin-orbit coupling energy a^v_{so}. The electric field $\boldsymbol{E} \parallel [001]$-direction. $E1$, $E2$, and $E3$ denote the energies of the states. For $E_z = 0$ the energies of the Γ_8 and Γ_7 states are situated at $(+1a^v_{so})$ and $(-2a^v_{so})$ energy units, respectively

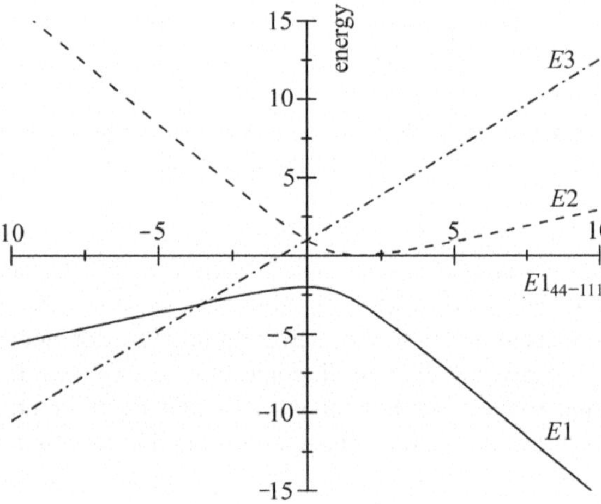

Fig. 3.4 Energy of the valence-band states at the Γ-point (given in units of $a^v_{so} = \Delta^v_{so}/3$) as a function of $E1_{44-111} = (a^v_{E1b}\, E_z\,)/a^v_{so}$, where the interaction energy $(a^v_{E1b}\, E_z)$ is measured in units of the spin-orbit coupling energy a^v_{so}. The electric field $\boldsymbol{E} \parallel [111]$-direction. $E1$, $E2$, and $E3$ denote the energies of the states. For E_z=0 the energies of the Γ_8 and Γ_7 states are situated at $(+1a^v_{so})$ and $(-2a^v_{so})$ energy units, respectively

Figure 3.3 shows the energy of the valence-band states at the Γ-point as function of a dimensionless energy $E1_{44-z} =(a^v_{E1b}\, E_z\,)/a^v_{so}$, i.e. the interaction energy $(a^v_{E1b}\, E_z)$ is measured in units of the spin-orbit coupling energy a^v_{so}. The notation $E1_{44-z} =(a^v_{E1b}\, E_z\,)/a^v_{so}$ reminds that the basis matrices with the indexes "44" are at the origin of the interaction and that the coupling is induced by the z-component of the electric field.

We observe in Fig. 3.3 three doublets; the branches are labeled $E1$, $E2$, and $E3$, respectively. The branches change their spectral position with increasing electric

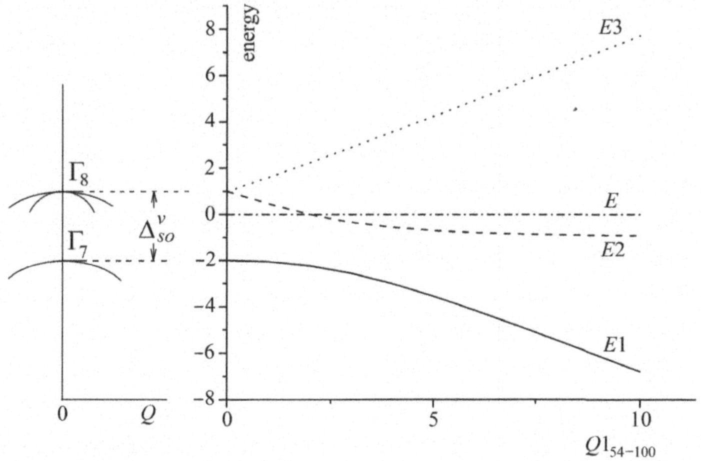

Fig. 3.5 Influence of the $\boldsymbol{Q}$-linear dispersion term: Left panel: Dispersion of the valence-band states in the "effective-mass approximation" including second-order dispersive terms and the valence-band spin-orbit splitting Δ_{so}^v. This dispersion has to be modified in the presence of $\boldsymbol{Q}$-linear terms. Right panel: Resulting energy corrections (given in units of $a_{so}^v = \Delta_{so}^v/3$) in the [100]-direction as a function of $Q1_{54-100} = (a_{Q1a}^v\ Q\ )\ /\ a_{so}^v$, where the interaction energy $(a_{Q1a}^v\ Q)$ and the energy corrections are measured in units of the spin-orbit coupling energy a_{so}^v. $E1$, $E2$, and $E3$ denote the energies of the twofold degenerate states, E gives the zero of energy. For $\boldsymbol{Q} = 0$ the energies of the Γ_8 and Γ_7 states are situated at $(+1a_{so}^v)$ and $(-2a_{so}^v)$ energy units, respectively. The twofold degenerate branches $E1$ and $E2$ are strongly coupled and show an anti-crossing behavior. Note the different scales in Figs. 3.5, 3.6 and 3.7

field and, in addition, the character of the states changes due to their increasing mixing. Since the branches are strongly coupled, the branches $E1$ and $E3$ show an anti-crossing behavior at positive values of $E1_{44-z}$. Similarly, the anti-crossing behavior between the branches $E1$ and $E2$ is important at negative values of $E1_{44-z}$. This is significant when the interaction energy $(a_{E1b}^v\ E_z)$ is comparable to the spin-orbit splitting $\Delta_{so}^v = 3a_{so}^v$. Similar to the discussion in Sect. 3.1 the same energy variation with increasing electric field of the branches is obtained when the field is orientated along the [110]-direction. However, as shown in Fig. 3.4 (i.e. the energy of the valence-band states as a function of $E1_{44-111} = (a_{E1b}^v\ E_z\ )/a_{so}^v)$ it differs for a [111]-orientation of the electric field.

These results remain qualitatively the same when the interaction Hamiltonian H_{E1a}^v of Eq. (3.2.18) as a function of $\boldsymbol{E}$ is considered. Similar to the interaction Hamiltonian H_{E1b}^v, the energy variation of an electric field $\parallel$ [110] is the same as that $\parallel$ [001]. Compared to the interaction Hamiltonian H_{E1b}^v the numerical values of the energies change, however, since the coupling coefficients in Eq. (3.2.22) are slightly smaller than in Eq. (3.2.21). In addition, the coupling strengths between the blocks with Γ_7- and Γ_8-symmetry are different in both interaction terms. This becomes evident when comparing the coupling coefficients in Eqs. (3.2.21) and

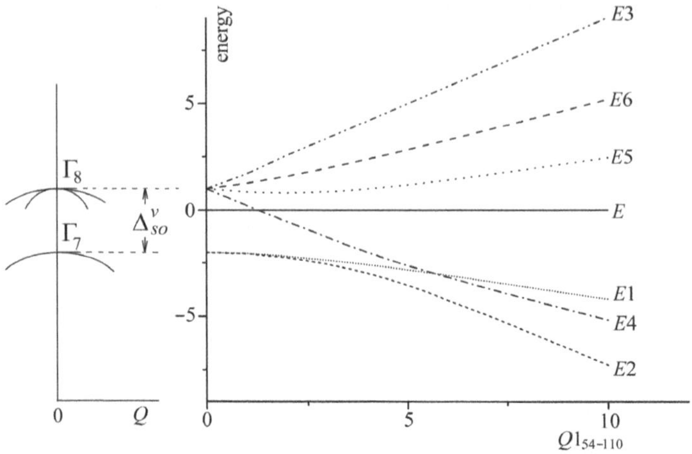

Fig. 3.6 Influence of the $\boldsymbol{Q}$-linear dispersion term: Left panel: Dispersion of the valence-band states in the "effective-mass approximation" including second-order dispersive terms and the valence-band spin-orbit splitting Δ^v_{so}. This dispersion has to be modified in the presence of $\boldsymbol{Q}$-linear terms. Right panel: Resulting energy corrections (given in units of $a^v_{so} = \Delta^v_{so}/3$) in the [110]-direction as a function of $Q1_{54-110} = (a^v_{Q1a}\ Q)\ /\ a^v_{so}$, where the interaction energy $(a^v_{Q1a}\ Q)$ and the energy corrections are measured in units of the spin-orbit coupling energy a^v_{so}. $E1$ to $E6$ denote the energies of the states, which are not degenerate, E gives the zero of energy. For $\boldsymbol{Q} = 0$ the energies of the Γ_8 and Γ_7 states are situated at $(+1a^v_{so})$ and $(-2a^v_{so})$ energy units, respectively. The branches ($E1$ and $E5$) as well as ($E2$ and $E4$) are mutually strongly coupled and show an anti-crossing behavior. Note the different scales in Figs. 3.5, 3.6 and 3.7

(3.2.22), which are given using the basis matrices in which the spin-orbit coupling is diagonal.

As shown in Figs. 3.3 and 3.4 the branches remain spin degenerate. This is due to the time reversal symmetry required for a Hamiltonian and the transformation properties of an electric field: Both quantities are time independent (symmetry K^+), while the spin state changes under time reversal. Therefore, when applying an electric field as a perturbation to the system, the different spin states must have the same energy. This is different when considering perturbations, which change their sign under time reversal as the wave-vector $\boldsymbol{Q}$ or a magnetic field $\boldsymbol{B}$.

An interesting example for a symmetry-breaking effect is the finite wave-vector $\boldsymbol{Q} = (Q_x, Q_y, Q_z)$, which transforms as (Γ_5, K^-) in zincblende structure. We have seen in Chap. 2 and Sect. 3.1 that $\boldsymbol{Q}$-linear terms cannot exist in a Hamiltonian describing only a spin-degenerate conduction band or a valence band, characterized by an angular momentum $l = 1$ but without spin. Therefore, neither the spin-degeneracy of the conduction band nor that of states with angular momentum $l = 1$ is lifted. Inspection of Table 2.3 shows, however, that combining $\boldsymbol{Q}$-linear terms with basis matrices that transform also as (Γ_5, K^-) can lead to terms possessing the transformation properties, required for a Hamiltonian. We see in Eq. (3.2.3) that (if the spin is included in the valence band) some basis matrices fulfill this condition. We can thus construct

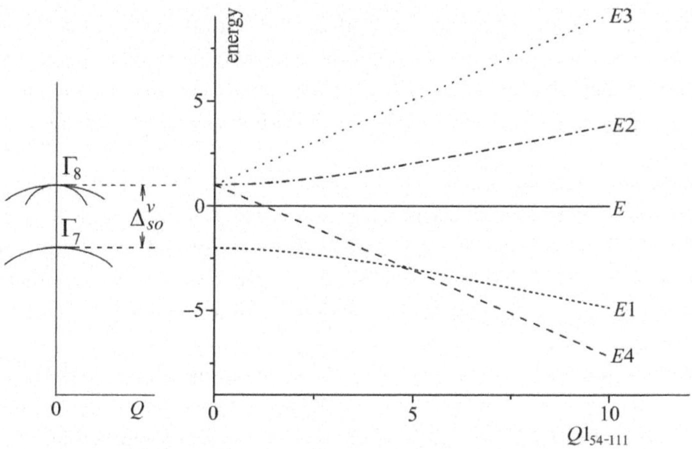

Fig. 3.7 Influence of the $\boldsymbol{Q}$-linear dispersion term: Left panel: Dispersion of the valence-band states in the "effective-mass approximation" including second-order dispersive terms and the valence-band spin-orbit splitting Δ^v_{so}. This dispersion has to be modified in the presence of $\boldsymbol{Q}$-linear terms. Right panel: Resulting energy corrections (given in units of $a^v_{so} = \Delta^v_{so}/3$) in the [111]-direction as a function of $Q1_{54-111} = (a^v_{Q1a}\ Q\ )/a^v_{so}$, where the interaction energy $(a^v_{Q1a}\ Q)$ and the energy corrections are measured in units of the spin-orbit coupling energy a^v_{so}. $E1$, $E2$, $E3$, and $E4$ denote the energies of the states, E gives the zero of energy. For $\boldsymbol{Q} = 0$ the energies of the Γ_8 and Γ_7 states are situated at $(+1a^v_{so})$ and $(-2a^v_{so})$ energy units, respectively. The branch $E1$, originating from the split-off band with Γ_7 symmetry at $\boldsymbol{Q} = 0$ remains (as well as the branch $E2$ originating from the states of the Γ_8 band) doubly degenerate, while the upper and lower Γ_8 bands ($E3$ and $E4$) are degenerate no more. The twofold degenerate branches $E1$ and $E2$ are strongly coupled and show an anti-crossing behavior. Note the different scales in Figs. 3.5, 3.6 and 3.7

two $\boldsymbol{Q}$-linear terms, coupling the valence-band states. They read explicitly in terms of the matrices defined in Eqs. (3.2.4) and (3.2.2):

$$\begin{aligned} H^v_{Q1a} &= a^v_{Q1a}[(ZQ' - YR')Q_x + (XR' - ZP')Q_y + (YP' - XQ')Q_z] = \\ &= a^v_{Q1a}[X_{54}"Q_x + Y_{54}"Q_y + Z_{54}"Q_z] \end{aligned} \tag{3.2.23}$$

and

$$\begin{aligned} H^v_{Q1b} &= a^v_{Q1b}[-(\sqrt{3}V + U)P'Q_x + (\sqrt{3}V - U)Q'Q_y + 2UR'Q_z] = \\ &= a^v_{Q1b}[X_{34}"Q_x + Y_{34}"Q_y + Z_{34}"Q_z]. \end{aligned} \tag{3.2.24}$$

It follows from the above discussion that these $\boldsymbol{Q}$-linear terms are only possible here because of the spin-orbit coupling, i.e. products of the type $(ZQ' - YR')$ etc. in Eq. (3.2.23) arise since the basis of the valence-band states has been increased by considering the electron spin. In the case of a vanishing spin-orbit coupling, the effective Hamiltonian would be given by the direct product of the unit matrix defined in the angular momentum subspace multiplied by the unit matrix defined in the spin subspace. Spin and orbital momentum would not see each other and the coefficients a^{v}_{Q1a}, which connect e.g. "Z" and "Q" in Eq. (3.2.23) would be $\equiv 0$. Then the matrix X_{54}" would not give rise to any symmetry breaking effect and the $\boldsymbol{Q}$-linear terms would vanish.

In general, such additional interaction terms discussed above may always appear if the number of considered states is increased corresponding to an increase of the dimension of the basis, on which the Hamiltonian is defined. For example, instead of considering the spin as additional variable, an increase of the basis is also achieved when respecting additional valence- or conduction-band states to which the former states are coupled.

In order to lift the spin degeneracy of states, the Hamiltonian matrix has to contain explicitly the spin-orbit coupling of Eq. (3.2.10) and the symmetry-breaking terms of Eq. (3.2.23) or Eq. (3.2.24). Both terms have to be diagonalized simultaneously if one wants to discuss remaining degeneracies. As discussed above, in the absence of spin-orbit coupling all angular momentum states would remain spin degenerate. Considering spin-orbit coupling leads in the presence of the dispersive terms to a symmetry-breaking interaction, which may lift the degeneracy of the states. As indicated in Eqs. (3.2.23) and (3.2.24) its value depends on the wave-vector components and can lead to an anisotropy of the dispersion.

In order to demonstrate possible degeneracies of the states and the anisotropy of dispersion due to the $\boldsymbol{Q}$-linear terms we have calculated the valence-band dispersion resulting from the term H^{v}_{Q1a} for the high-symmetry directions [100], [110], and [111], including spin-orbit coupling. The result is shown in Figs. 3.5, 3.6, and 3.7 where the interaction energy (a^{v}_{Q1a} Q) is measured again in units of the spin-orbit coupling-energy a^{v}_{so}. Concerning the [100]-direction, shown in Fig. 3.5, each of the three branches labeled ($E1$, $E2$, $E3$) is doubly degenerate. Again, hybridization is clearly observed when branch $E2$ gets close to branch $E1$ with increasing $Q1_{54-100}$ $= (a^{v}_{Q1a}\ Q\ )/a^{v}_{so}$.

As shown in Fig. 3.6, concerning the [110]-direction, the branches are only degenerate at $Q = 0$. Otherwise the degeneracy is completely lifted by the coupling of the states. A pronounced anti-crossing behavior is observed at finite wave-vectors between branches $E1$ and $E5$ as well as between $E2$ and $E4$.

In the [111]-direction (see Fig. 3.7) the split-off band (having Γ_7-symmetry at $Q = 0$ and giving rise to branch $E1$) and two of the states originating from the Γ_8 band (branch $E2$) remain doubly degenerate, while the upper and lower Γ_8 bands are not degenerate. The twofold degenerate branches $E1$ and $E2$ are strongly coupled and show an anti-crossing behavior.

Again, it is less interesting to discuss the dispersion in the frame of the thirty-six basis matrices, which are only adapted to the crystal symmetry, but rather to consider the set of basis matrices, in which the spin-orbit coupling is diagonal. When the matrices of Eq. (3.2.4) are transformed onto the normalized eigenvectors of the valence band given in Eq. (3.2.12) the $\boldsymbol{Q}$-linear term of the valence-band states given in Eq.(3.2.23) takes the matrix form:

$$H^v_{Q1a} = a^v_{Q1a}[X_{54so}"Q_x + Y_{54so}"Q_y + Z_{54so}"Q_z] =$$

$$= a^v_{Q1a}Q_x \begin{pmatrix} 0 & 0 & \sqrt{6}/12 & 0 & \sqrt{2}/4 & 0 \\ 0 & 0 & 0 & -\sqrt{2}/4 & 0 & -\sqrt{6}/12 \\ \sqrt{6}/12 & 0 & 0 & -\sqrt{3}/6 & 0 & -1/2 \\ 0 & -\sqrt{2}/4 & -\sqrt{3}/6 & 0 & 1/2 & 0 \\ \sqrt{2}/4 & 0 & 0 & 1/2 & 0 & -\sqrt{3}/6 \\ 0 & -\sqrt{6}/12 & -1/2 & 0 & -\sqrt{3}/6 & 0 \end{pmatrix} +$$

$$+a^v_{Q1a}Q_y \begin{pmatrix} 0 & 0 & i\sqrt{6}/12 & 0 & -i\sqrt{2}/4 & 0 \\ 0 & 0 & 0 & -i\sqrt{2}/4 & 0 & i\sqrt{6}/12 \\ -i\sqrt{6}/12 & 0 & 0 & i\sqrt{3}/6 & 0 & -i/2 \\ 0 & i\sqrt{2}/4 & -i\sqrt{3}/6 & 0 & -i/2 & 0 \\ i\sqrt{2}/4 & 0 & 0 & i/2 & 0 & i\sqrt{3}/6 \\ 0 & -i\sqrt{6}/12 & i/2 & 0 & -i\sqrt{3}/6 & 0 \end{pmatrix} +$$

$$+a^v_{Q1a}Q_z \begin{pmatrix} 0 & 0 & 0 & 0 & 0 & -\sqrt{6}/6 \\ 0 & 0 & -\sqrt{6}/6 & 0 & 0 & 0 \\ 0 & -\sqrt{6}/6 & 0 & 0 & -\sqrt{3}/3 & 0 \\ 0 & 0 & 0 & 0 & 0 & \sqrt{3}/3 \\ 0 & 0 & -\sqrt{3}/3 & 0 & 0 & 0 \\ -\sqrt{6}/6 & 0 & 0 & \sqrt{3}/3 & 0 & 0 \end{pmatrix}. \tag{3.2.25}$$

If we consider for simplicity the [100]-direction (i.e. $Q_y = Q_z = 0$), the second $\boldsymbol{Q}$-linear term given in Eq. (3.2.24) takes the matrix form

$$H^v_{Q1b} = a^v_{Q1b}[X_{34so}"Q_x + Y_{34so}"Q_y + Z_{34so}"Q_z]$$

$$= a^v_{Q1b}Q_x \begin{pmatrix} 0 & 0 & -\sqrt{2} & 0 & -\sqrt{6} & 0 \\ 0 & 0 & 0 & \sqrt{6} & 0 & \sqrt{2} \\ -\sqrt{2} & 0 & 0 & -1 & 0 & -\sqrt{3} \\ 0 & \sqrt{6} & -1 & 0 & \sqrt{3} & 0 \\ -\sqrt{6} & 0 & 0 & \sqrt{3} & 0 & -1 \\ 0 & \sqrt{2} & -\sqrt{3} & 0 & -1 & 0 \end{pmatrix}. \tag{3.2.26}$$

For both interaction terms one remarks that the first two states transforming as Γ_7 are not coupled directly with each other by the $\boldsymbol{Q}$-linear terms. Only the states transforming as Γ_8 experience such a direct coupling. Their coupling scheme is—with the exception of an unimportant phase factor—the same for both $\boldsymbol{Q}$-linear terms. This statement remains valid when considering the Q_y and Q_z components of the wave-vector $\boldsymbol{Q}$.

As shown in Eqs. (3.2.25) and (3.2.26) the valence-band states with Γ_7 and Γ_8 symmetry are (as in the case of the Stark effect) coupled with each other in the same way through the two different $\boldsymbol{Q}$-linear terms, which thus mix the irreducible representation at finite wave-vectors. The coupling strength between the states are again different in both terms, leading to different energy variations with wave-vector. Both $\boldsymbol{Q}$-linear terms lead, however, to the same mixing scheme of the states. In order to determine the entire anisotropic dispersion of the states Eqs. (3.2.25) and (3.2.26) have to be solved numerically.

We also obtain (quite similar to the $\boldsymbol{Q}$-linear terms) the additional linear terms for the magnetic field $\boldsymbol{B}$:

$$H^{v}_{B1c} = a^{v}_{B1c}[(ZQ' + YR')B_x + (XR' + ZP')B_y + (YP' + XQ')B_z] = \\ = a^{v}_{B1c}[P_{54}\text{''}B_x + Q_{54}\text{''}B_y + R_{54}\text{''}B_z] \tag{3.2.27}$$

and

$$H^{v}_{B1d} = a^{v}_{B1d}[(\sqrt{3}U - V)P'B_x - (\sqrt{3}U + V)Q'B_y + 2VR'B_z] = \\ = a^{v}_{B1d}[P_{34}\text{''}B_x + Q_{34}\text{''}B_y + R_{34}\text{''}B_z]\,, \tag{3.2.28}$$

which are also different from the linear Zeeman terms discussed in Eq. (3.1.15). The terms obtained from Eq. (3.1.15) take now the forms:

$$H^{v}_{B1a} = a^{v}_{B1a}[PS'B_x + QS'B_y + RS'B_z] = \\ = a^{v}_{B1a}[P_{41}\text{''}B_x + Q_{41}\text{''}B_y + R_{41}\text{''}B_z] \tag{3.2.29}$$

and

$$H^{v}_{B1b} = a^{v}_{B1b}[SP'B_x + SQ'B_y + SR'B_z] = \\ = a^{v}_{B1b}[P_{14}\text{''}B_x + Q_{14}\text{''}B_y + R_{14}\text{''}B_z]\,, \tag{3.2.30}$$

respectively. The degeneracy of the valence-band states is completely lifted in the presence of a magnetic field and their mixing and energy depends on the field direction with respect to the cubic axes. The simplest case is found if $\boldsymbol{B} \parallel [001]$. In this case, the wave functions remain similar to those given in Eq. (3.2.13), which have the form $|j, m_j\rangle$. Only the factors multiplying the spin orbitals depend on the strength of the magnetic field but the mixing scheme remains the same.

Let us as an example demonstrate the action of the $\boldsymbol{B}$-linear terms H^{v}_{B1a} to H^{v}_{B1d} (given in Eqs. (3.2.27)–(3.2.30)) on the energy of the valence-band states for $\boldsymbol{B} \parallel [001]$ and including spin-orbit coupling. If the field has other components the coupling scheme is much more complex. Figure 3.8 shows the result where the interaction energies $a^{v}_{B1a}B$ are measured again in units of the spin-orbit coupling-energy a^{v}_{so}. As mentioned above the notation

$$B1_{41-001} = a^{v}_{B1a}B_z/a^{v}_{so} \tag{3.2.31}$$

indicates for example that the basis matrices with the indexes "41" are at the origin of the interaction. Furthermore, the interaction term is linear in the magnetic field (indicated by "$B1$") with $\boldsymbol{B} \parallel [001]$. The branches resulting from this perturbation when varying the magnetic field $\boldsymbol{B}$ are labeled $E1$ to $E6$. The numbers i =(1–6)

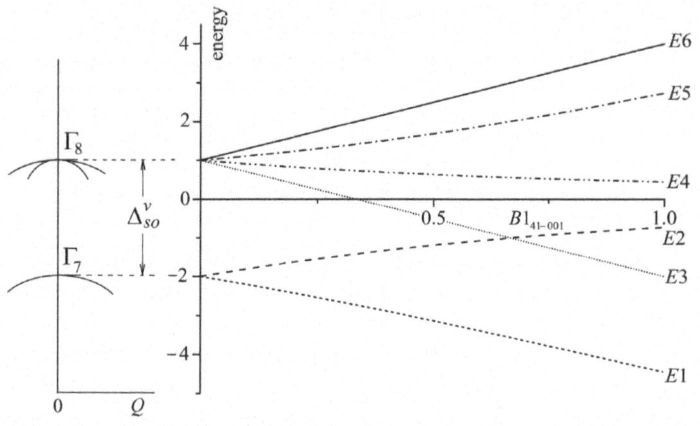

Fig. 3.8 Energy of the valence-band states (given in units of $a^v_{so} = \Delta^v_{so}/3$) as a function of $B1_{41-001}$ = $(a^v_{B1a}\ B\ )/a^v_{so}$ if the magnetic field $\boldsymbol{B}$ is applied in the [001]-direction. The interaction energy $(a^v_{B1a}\ B)$ is measured in units of the spin-orbit coupling energy a^v_{so}, Δ^v_{so} being the spin-orbit splitting of the valence band. $E1$ to $E6$ denote the energies of the valence-band states, including spin-orbit interaction. Hybridization of states is observed when branches ($E1$ and $E4$) or ($E2$ and $E5$)) are getting close to each other in energy when varying the magnetic field strength. For B=0 the energies of the Γ_8 and Γ_7 states are situated at $(+1a^v_{so})$ and $(-2a^v_{so})$ energy units, respectively

correspond to the eigenfunctions v^v_i given in Eqs. (3.2.13) and (3.2.14), which diagonalize the spin-orbit Hamiltonian in the absence of an external magnetic field. In this case the valence-band states are split into six branches, whose splitting increases with increasing magnetic field strength. The branches (2 and 5) and (1 and 4) are coupled, leading to hybridization of the states while branches (2 and 3) are not coupled and cross with increasing field strength. Hybridization of states is clearly observed when branches with $m_j = \pm 1/2$ (i.e. branches ($E1$ and $E4$) and ($E2$ and $E5$)) are getting close to each other in energy when varying the magnetic field strength.

It is interesting to notice that the perturbation terms H^v_{B1a} and H^v_{B1b} describe the direct action of a magnetic field on the angular momentum and the spin, respectively. On the other hand, the terms H^v_{B1c} and H^v_{B1d} describe the modification of the spin-orbit coupling, induced by the magnetic field. In perturbation theory, this would be a higher-order process and it should be less important than the direct terms if the spin-orbit coupling is small.

According to Table 2.4, symmetry-adapted perturbations transforming as (Γ_1, Γ_3, Γ_4, and Γ_5) can be constructed from perturbations ($\propto \boldsymbol{B}$, $\propto \boldsymbol{E}$, or $\propto \boldsymbol{Q}$) when they are taken to second order. The resulting symmetry-breaking interactions are then all transforming as K^+. It follows from Eq. (3.2.3) and the foregoing discussion that the perturbation Hamiltonians H^v_{A2} (where $\boldsymbol{A}$ denotes $\boldsymbol{E}$, $\boldsymbol{Q}$, or $\boldsymbol{B}$, respectively) have the form

$$
\begin{aligned}
H^{v}_{A2} = a^{v}_{A2a}[S_{11}"(A_x^2 + A_y^2 + A_z^2)] + a^{v}_{A2b}[S_{44}"(A_x^2 + A_y^2 + A_z^2)]+ \\
a^{v}_{A2c}[U_{31}"\sqrt{3}(A_x^2 - A_y^2) + V_{31}"(2A_z^2 - A_x^2 - A_y^2)]+ \\
+a^{v}_{A2d}[U_{44}"\sqrt{3}(A_x^2 - A_y^2) + V_{44}"(2A_z^2 - A_x^2 - A_y^2)]+ \\
+a^{v}_{A2e}[X_{51}"2A_yA_z + Y_{51}"2A_zA_x + Z_{51}"2A_xA_y]+ \\
+a^{v}_{A2f}[X_{44}"2A_yA_z + Y_{44}"2A_zA_x + Z_{44}"2A_xA_y].
\end{aligned}
\tag{3.2.32}
$$

Table 2.3 shows that terms quadratic in the perturbation components and transforming as (Γ_4, K^+) are all $\equiv 0$. A detailed numerical analysis is now necessary to decide, which of the remaining terms given in Eq. (3.2.32) are important and which can be omitted. Other perturbations, involving arbitrarily different symmetry-breaking properties can be constructed in the same way, considering their transformation properties under time reversal and combining them with the corresponding basis matrices of Eq. (3.2.3).

3.3 Pseudo-Spin Development of the Γ_7 Subspace of Valence-Band States

We have learned in the preceding chapters how to develop an effective Hamiltonian, which is adapted to the crystal symmetry. This development is more or less complicated depending on the number of electron states involved. It becomes complex if a large basis of interacting states has to be considered. The technique is interesting, however, if only a few, almost degenerate states are present, which are well separated in energy from other states. In the example discussed in the foregoing section one can simplify the situation if one takes advantage from the fact that the split-off valence-band states, which transform as Γ_7, are separated due to the spin-orbit coupling from the other valence-band states that transform as Γ_8. As stated above, the two blocks of valence-band states can be coupled by symmetry-breaking effects. If, however, the energy due to this coupling is small compared to the spin-orbit splitting, the two blocks are not considerably mixed by the perturbation. Then, it is sufficient to consider the influence of the symmetry-breaking effects only within the blocks and to neglect the coupling between different blocks. In this case, the Hamiltonian describing the valence-band states is block diagonal and different effective Hamiltonians can be developed in order to describe each of them separately.

Let us discuss first the simplest but rather atypical case in which the highest valence subband is only two-fold degenerate and has Γ_7 symmetry (see Fig. 3.9 and Ref. [3]). All other bands (including the valence subband with Γ_8 symmetry, which is separated by spin-orbit interaction from the considered states) and the conduction band are neglected. As discussed above, besides the spin contribution, the wave

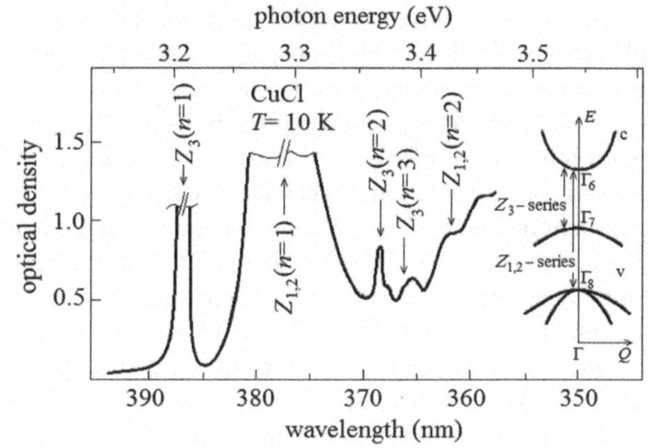

Fig. 3.9 Left: Part of optical absorption spectrum of a CuCl film at a temperature of 10 K, which comprises several absorption lines of both the Z_3- and $Z_{1,2}$-exciton series. Adapted from Ref. [4]. Right: Schematic band structure of CuCl. Notice the reverse order of the two uppermost valence bands Γ_7 and Γ_8 in comparison with most other zincblende semiconductors

functions of the valence-band electrons contain an angular-momentum part with $l_v = 1$, transforming as (x, y, z). The spin states are labeled α and β and indicate the spin-up and down states, respectively. The eigenvectors (v_1^v and v_2^v) of the valence band have been given in Eq. (3.2.14) by

$$\begin{aligned} v_1^v &= (-|0\rangle\beta + \sqrt{2}|-1\rangle\alpha)/\sqrt{3} = (-z\beta + (x - iy)\alpha)/\sqrt{3} \\ v_2^v &= (|0\rangle\alpha - \sqrt{2}|1\rangle\beta)/\sqrt{3} = (z\alpha + (x + iy)\beta)/\sqrt{3}. \end{aligned} \tag{3.3.1}$$

Considering the total angular momentum $\boldsymbol{j}_v = \boldsymbol{l}_v \oplus \boldsymbol{\sigma}_v$ and its projection component onto the z-axis j_v^z these functions are eigenfunctions of the total angular momentum $\boldsymbol{j}_v = 1/2$ of the valence-band states, with eigenvalues for $j_v^z = \mp 1/2$, respectively.

Conduction band and valence band are both two-fold degenerate. As discussed in Chap. 2, because of this two-fold degeneracy, we had chosen the Pauli-spin matrices σ_e^x, σ_{e^y}, and σ_{e^z} and the unit matrix 1_e as a basis to span the matrix describing the conduction-band states. We now construct an effective valence-band pseudo-spin Hamiltonian in the same way: This Hamiltonian describes the two-fold degenerate valence subband. The effective Hamiltonian shall remain invariant under all symmetry operations of the point group of the crystal and transform as a scalar (Γ_1). In addition, it is an even function under time reversal (K^+). Proceeding as in the case of the conduction band, in order to describe the Γ_7 valence-band states and their mutual interactions, we use the Pauli-spin matrices σ_v^x, σ_v^y, and σ_v^z and the unit matrix 1_v as a basis to span the matrix of the valence subband, i.e.

$$\sigma_v^x = \begin{pmatrix} 0 & 1 \\ 1 & 0 \end{pmatrix}; \sigma_v^y = \begin{pmatrix} 0 & -i \\ i & 0 \end{pmatrix}; \sigma_v^z = \begin{pmatrix} 1 & 0 \\ 0 & -1 \end{pmatrix}; 1_v = \begin{pmatrix} 1 & 0 \\ 0 & 1 \end{pmatrix}. \tag{3.3.2}$$

This set of matrices operates only on the valence-sub-band states v_1^v and v_2^v. We can now define in this subspace all symmetry-breaking interactions. Since the dimension of the conduction-band states and that of the valence sub-band are the same and since the symmetry of the basis matrices used to develop the pseudo-spin Hamiltonians are identical, the effective pseudo-spin Hamiltonian H_7^v describing the split-off valence-band takes exactly the same form as the one discussed for the conduction band in Chap. 2. Only the parametrized constants giving the strengths of interactions between the states due to internal or external perturbations are different for both Hamiltonians. This reflects the fact that the wave functions, used to calculate the interaction matrix elements, are different in both cases.

One can thus summarize from Chap. 2 that $\boldsymbol{Q}$-linear terms and a linear Stark effect do not exist here. This point shows also up in Eqs. (3.2.25) and (3.2.26) by the fact that the valence-sub-band states v_1^v and v_2^v are not coupled with each other. The valence-band dispersion also varies quadratically with the wave-vector of the valence-band electron, giving rise to an effective valence-band mass. This effective mass is isotropic and characterized by only one constant value. In the third order, the wave-vector components can be used to construct an effective intrinsic magnetic field that lifts the degeneracy of this valence sub-band, which remains degenerate at the Γ-point. Similar to the spin degenerate conduction band, effects varying quadratically with an electric or magnetic field or bilinear in the magnetic and electric field can also be constructed.

As shown already in connection with Eq. (2.3.2) for the magnetic field $\boldsymbol{B}$, a linear Zeeman Effect is present if the Γ_7 subspace is considered alone. It lifts the degeneracy of the Γ_7 sub-band states. The corresponding Hamiltonian H_{B17}^v is given by

$$H_{B17}^v = a_{B17}^v \boldsymbol{B} \cdot \boldsymbol{\sigma_v} = a_{B17}^v (B_x \sigma_v^x + B_y \sigma_v^y + B_z \sigma_v^z) \qquad (3.3.3)$$

where the coefficient a_{B17}^v characterizes the influence of the magnetic field on the Γ_7 valence band and leads to a splitting of the initially spin-degenerate states α and β. When comparing to Sect. 3.2, a_{B17}^v is resulting from the coefficients a_{B1a}^v to a_{B1d}^v in Eqs. (3.2.27)–(3.2.30), which all contribute to the linear splitting of the Γ_7 valence band. Considering now the split-off band separately, the different contributions lifting the spin-degeneracy are simulated by one single term.

It is interesting to mention explicitly that the full treatment of the valence band including spin-orbing coupling leads to anisotropic $\boldsymbol{Q}$-linear and effective-mass dispersion-terms not only for the states with Γ_8-symmetry at the Γ-point but also for the split-off band with Γ_7-symmetry. This anisotropy vanishes if the spin-orbit coupling is first diagonalized and the coupling between the valence bands with different symmetry neglected. This indicates that in the case of the split-off band a possible anisotropy of the dispersion is due to symmetry breaking, i.e. a coupling of states from different irreducible representations, which have still the full point-group symmetry. In addition, hybridization effects (as observed in Fig. 3.5 to Fig. 3.7) are eliminated if the spin-orbit coupling is only approximately considered as discussed above. Thus, the restriction to valence-band subspaces can introduce a modification of the energy dispersion and a loss of the fine-structure of the states under consideration.

3.4 Pseudo-Spin Development of the Γ_8 Subspace of Valence-Band States

Similarly to the split-off valence band discussed above we will now consider the remaining states of the valence band that give rise to the Γ_8 subspace. Due to spin-orbit splitting the split-off Γ_7 valence band has a significantly different energy than the Γ_8 valence band and is not further considered. Again, all other bands including the lowest lying conduction band are also neglected. Then, only the valence-band states v_3^v to v_6^v of Eq. (3.2.13) or Eq. (3.2.14) are considered in the following. They are eigenstates of the angular-momentum operator with $\boldsymbol{j}_v = 3/2$. This situation is realized in most simple binary III-V, II-VI, and I-VII semiconductors with zincblende structure (as e.g. GaAs, ZnSe, ZnTe or CuBr etc.) where the Γ_8 states represent the highest valence subband and the spin-orbit splitting is large compared to symmetry-breaking interactions, see also Fig. 3.2 and Ref. [5].

In order to construct the invariant representation of the Hamiltonian, which acts in the four-fold degenerate subspace of valence-band states with Γ_8 symmetry, we construct first the projection components onto the x-, y- , and z-axis of the total angular-momentum operator $\boldsymbol{j}_v = 3/2$. The valence-band states v_3^v to v_6^v are eigenfunctions of $\boldsymbol{j}_v$ and j_v^z. Following Ref. [4] we find for j_v^x, j_v^y, and j_v^z in this basis

$$j_v^x = \begin{pmatrix} 0 & \sqrt{3}/2 & 0 & 0 \\ \sqrt{3}/2 & 0 & 1 & 0 \\ 0 & 1 & 0 & \sqrt{3}/2 \\ 0 & 0 & \sqrt{3}/2 & 0 \end{pmatrix};$$

$$j_v^y = \begin{pmatrix} 0 & i\sqrt{3}/2 & 0 & 0 \\ -i\sqrt{3}/2 & 0 & i & 0 \\ 0 & -i & 0 & i\sqrt{3}/2 \\ 0 & 0 & -i\sqrt{3}/2 & 0 \end{pmatrix}; \tag{3.4.1}$$

$$j_v^z = \begin{pmatrix} -3/2 & 0 & 0 & 0 \\ 0 & -1/2 & 0 & 0 \\ 0 & 0 & 1/2 & 0 \\ 0 & 0 & 0 & 3/2 \end{pmatrix}.$$

The matrices given above can be used to construct the unit matrix 1_{v8} of the subspace of the valence-band states having Γ_8 symmetry. When calculating explicitly it follows from Eq. (3.4.1) that

$$((j_v^x)^2 + (j_v^y)^2 + (j_v^z)^2) = S_{v8} = (15/4)1_{v8} = (15/4)\begin{pmatrix} 1&0&0&0\\ 0&1&0&0\\ 0&0&1&0\\ 0&0&0&1 \end{pmatrix}$$

which can be seen as an analogy to Eq. (2.2.5).

From the matrices j_v^i with $i = (x, y, z)$ we construct the other basis matrices, which are necessary to span the four-dimensional subspace of the valence band and which are adapted to the crystal symmetry. We first form the product $(\Gamma_8 \otimes \Gamma_8)$. From the multiplication scheme given in Table 2.4 we obtain

$$\Gamma_8 \otimes \Gamma_8 = \Gamma_1 \oplus \Gamma_2 \oplus \Gamma_3 \oplus 2\Gamma_4 \oplus 2\Gamma_5\,. \tag{3.4.2}$$

From Table 2.3 we may calculate the sixteen matrices, which form a basis of our system. Accordingly we denote:

$$\begin{aligned} S_{v8} = (j_v^x)^2 + (j_v^y)^2 + (j_v^z)^2 \Leftrightarrow \text{transforming as} \Leftrightarrow (\Gamma_1, K^+)\\ T_{v8} = (j_v^x\{j_v^y, j_v^z\} + c.p.) \Leftrightarrow (\Gamma_2, K^-)\\ (U_{v8}, V_{v8}) = (\sqrt{3}((j_v^x)^2 - (j_v^y)^2), 2(j_v^z)^2 - (j_v^x)^2 - (j_v^y)^2) \Leftrightarrow (\Gamma_3, K^+)\\ (P_{v8}, Q_{v8}, R_{v8}) = (j_v^x, j_v^y, j_v^z) \Leftrightarrow (\Gamma_4, K^-)\\ (P'_{v8}, Q'_{v8}, R'_{v8}) = ((j_v^x)^3, (j_v^y)^3, (j_v^z)^3) \Leftrightarrow (\Gamma_4, K^-)\\ (X_{v8}, Y_{v8}, Z_{v8}) = (\{j_v^y, j_v^z\}, \{j_v^z, j_v^x\}, \{j_v^x, j_v^y\}) \Leftrightarrow (\Gamma_5, K^+)\\ (X'_{v8}, Y'_{v8}, Z'_{v8}) = (\{j_v^x, ((j_v^y)^2 - (j_v^z)^2)\}, \{j_v^y, ((j_v^z)^2 - (j_v^x)^2)\}, \{j_v^z, ((j_v^x)^2 - (j_v^y)^2)\}) \Leftrightarrow (\Gamma_5, K^-)\,, \end{aligned} \tag{3.4.3}$$

where

$$\{j_v^y, j_v^z\} = (1/2)(j_v^y j_v^z + j_v^z j_v^y) \tag{3.4.4}$$

and "c.p." denotes "cyclic permutation". We have added the subscript "8" to the matrix nomenclature in order to indicate that we are restricting the discussion to the Γ_8 subspace of the valence band.

Inspecting Eq. (3.4.3), we see that the basis matrices with Γ_1, Γ_3, and one with Γ_5 symmetry are even functions with respect to time reversal (indicated by K^+), the others transform as K^-. The matrices given in Eq. (3.4.3) are linearly independent of each other. They can therefore be used as symmetry-adapted basis-matrices to span our subspace of the Γ_8 valence-band states. We obtain explicitly for these matrices the following set:

$$S_{v8} = (15/4)\begin{pmatrix} 1&0&0&0\\ 0&1&0&0\\ 0&0&1&0\\ 0&0&0&1 \end{pmatrix}$$

$$T_{v8} = (3\sqrt{3}/4)\begin{pmatrix} 0&0&-i&0\\ 0&0&0&i\\ i&0&0&0\\ 0&-i&0&0 \end{pmatrix}$$

$$U_{v8} = 3\begin{pmatrix} 0&0&1&0\\ 0&0&0&1\\ 1&0&0&0\\ 0&1&0&0 \end{pmatrix}; V_{v8} = 3\begin{pmatrix} 1&0&0&0\\ 0&-1&0&0\\ 0&0&-1&0\\ 0&0&0&1 \end{pmatrix}$$

$$P_{v8} = \begin{pmatrix} 0&\sqrt{3}/2&0&0\\ \sqrt{3}/2&0&1&0\\ 0&1&0&\sqrt{3}/2\\ 0&0&\sqrt{3}/2&0 \end{pmatrix}; Q_{v8} = \begin{pmatrix} 0&i\sqrt{3}/2&0&0\\ -i\sqrt{3}/2&0&i&0\\ 0&-i&0&i\sqrt{3}/2\\ 0&0&-i\sqrt{3}/2&0 \end{pmatrix};$$

$$R_{v8} = (1/2)\begin{pmatrix} -3&0&0&0\\ 0&-1&0&0\\ 0&0&1&0\\ 0&0&0&3 \end{pmatrix}$$

$$P'_{v8} = \begin{pmatrix} 0&7\sqrt{3}/8&0&3/4\\ 7\sqrt{3}/8&0&5/2&0\\ 0&5/2&0&7\sqrt{3}/8\\ 3/4&0&7\sqrt{3}/8&0 \end{pmatrix}; Q'_{v8} = \begin{pmatrix} 0&7\sqrt{3}i/8&0&-3i/4\\ -7\sqrt{3}i/8&0&5i/2&0\\ 0&-5i/2&0&7\sqrt{3}i/8\\ 3i/4&0&-7\sqrt{3}i/8&0 \end{pmatrix};$$

$$R'_{v8} = (1/8)\begin{pmatrix} -27&0&0&0\\ 0&-1&0&0\\ 0&0&1&0\\ 0&0&0&27 \end{pmatrix}$$

$$X_{v8} = (\sqrt{3}/2)\begin{pmatrix} 0&-i&0&0\\ i&0&0&0\\ 0&0&0&i\\ 0&0&-i&0 \end{pmatrix}; Y_{v8} = (\sqrt{3}/2)\begin{pmatrix} 0&-1&0&0\\ -1&0&0&0\\ 0&0&0&1\\ 0&0&1&0 \end{pmatrix}; Z_{v8} = (\sqrt{3}/2)\begin{pmatrix} 0&0&i&0\\ 0&0&0&i\\ -i&0&0&0\\ 0&-i&0&0 \end{pmatrix}$$

$$X'_{v8} = (1/4)\begin{pmatrix} 0&-\sqrt{3}&0&-3\\ -\sqrt{3}&0&3&0\\ 0&3&0&-\sqrt{3}\\ -3&0&-\sqrt{3}&0 \end{pmatrix}; Y'_{v8} = (1/4)\begin{pmatrix} 0&\sqrt{3}i&0&-3i\\ -\sqrt{3}i&0&-3i&0\\ 0&3i&0&\sqrt{3}i\\ 3i&0&-\sqrt{3}i&0 \end{pmatrix};$$

$$Z'_{v8} = (\sqrt{3}/2)\begin{pmatrix} 0&0&-1&0\\ 0&0&0&1\\ -1&0&0&0\\ 0&1&0&0 \end{pmatrix}. \tag{3.4.5}$$

The symmetry-breaking interactions can now be constructed in the Γ_8 valence-band subspace. The procedure is the same as discussed above: the symmetry-breaking perturbations are developed in a power series, multiplied by the basis matrices given in Eq. (3.4.5), and interaction terms invariant under the symmetry operations of the crystal and under time reversal are considered for the Hamiltonian.

Let us consider in detail the effect of a finite wave-vector $\boldsymbol{Q} = (Q_x, Q_y, Q_z)$, which transforms as (Γ_5, K^-) in zincblende structure. When all valence-band states are considered, we have seen in Eqs. (3.2.23) and (3.2.24) that two $\boldsymbol{Q}$-linear terms exist. In the reduced subspace of valence-band states with Γ_8 symmetry, we see in Eq. (3.4.3) that only the basis matrices

$$(X'_{v8}, Y'_{v8}, Z'_{v8}) = (\{j_v^x, ((j_v^y)^2 - (j_v^z)^2)\}, \{j_v^y, ((j_v^z)^2 - (j_v^x)^2)\}, \{j_v^z, ((j_v^x)^2 - (j_v^y)^2)\})$$

fulfill the required condition and transform as (Γ_5, K^-). They can thus give rise to a $\boldsymbol{Q}$-linear term, which reads explicitly:

$$\begin{aligned} H^v_{Q1v8} &= a^v_{Q1v8}[X'_{v8}Q_x + Y'_{v8}Q_y + Z'_{v8}Q_z] \\ &= a^v_{Q1v8}Q_x \begin{pmatrix} 0 & -\sqrt{3}/4 & 0 & -3/4 \\ -\sqrt{3}/4 & 0 & 3/4 & 0 \\ 0 & 3/4 & 0 & -\sqrt{3}/4 \\ -3/4 & 0 & -\sqrt{3}/4 & 0 \end{pmatrix} + \\ &+ a^v_{Q1v8}Q_y \begin{pmatrix} 0 & i\sqrt{3}/4 & 0 & -i3/4 \\ -i\sqrt{3}/4 & 0 & -i3/4 & 0 \\ 0 & i3/4 & 0 & i\sqrt{3}/4 \\ i3/4 & 0 & -i\sqrt{3}/4 & 0 \end{pmatrix} + \\ &+ a^v_{Q1v8}Q_z \begin{pmatrix} 0 & 0 & -\sqrt{3}/2 & 0 \\ 0 & 0 & 0 & \sqrt{3}/2 \\ -\sqrt{3}/2 & 0 & 0 & 0 \\ 0 & \sqrt{3}/2 & 0 & 0 \end{pmatrix}. \end{aligned} \tag{3.4.6}$$

When comparing the block of Γ_8 valence-band states in Eq. (3.2.25) or Eq. (3.2.26) for the Q_x-component (and neglecting the split-off band with Γ_7 symmetry) with Eq. (3.4.6), we see that the $\boldsymbol{Q}$-linear term has exactly the same form in both cases. The numerical treatment of the interaction is, however, much easier in the subspace of states with Γ_8 symmetry than treating the entire valence-band.

The second $\boldsymbol{Q}$-linear term of Eqs. (3.2.25) and (3.2.26) is mainly important for the modification of the coupling between the blocks. The coupling between states with different symmetry is, however, neglected here. This is a good approximation if the spin-orbit splitting is large compared to the considered symmetry-breaking effect. The Schrödinger equation corresponding to Eq. (3.2.26) and the resulting dispersion relation for different crystal directions are shown in Figs. 3.5, 3.6 and 3.7 when restricting to the subspace of states with Γ_8 symmetry for small values of $Q1_{54-100}$ to $Q1_{54-111}$.

As stated above, $\boldsymbol{Q}$-squared terms are even terms under Kramers' conjugation. Inspecting Eq. (3.4.3) shows that three $\boldsymbol{Q}$-squared terms can be constructed in the block of Γ_8 valence-band states. (This has to be compared to the six terms existing in the six-fold degenerate valence band of Sect. 3.2.) The corresponding Hamiltonian defined in the Γ_8-subspace of valence-band states then reads

$$H^v_{Q2v8} = (h^2/2m_0)[\gamma_1 \boldsymbol{Q}^2 S_{v8} + \gamma_2 \left(\sqrt{3}(Q_x^2 - Q_y^2)U_{v8} + (2Q_z^2 - Q_x^2 - Q_y^2)V_{v8}\right) + \\ +2\gamma_3 \left(Q_y Q_z X_{v8} + Q_z Q_x Y_{v8} + Q_x Q_y Z_{v8}\right)] \ , \tag{3.4.7}$$

where γ_1, γ_2, and γ_3 denote the Luttinger parameters [6–8] and m_0 the free electron mass. In the absence of any $\boldsymbol{Q}$-linear terms (a situation that is considered usually in literature) the dispersion of the valence bands may be calculated analytically and one obtains

$$E^v_{1,2} = E^v_0 + (h^2/2m_0)\left(\gamma_1 \boldsymbol{Q}^2 \pm 2\sqrt{\gamma_2^2(Q_x^4 + Q_y^4 + Q_z^4) + 3(\gamma_3^2 - \gamma_2^2)(Q_x^2 Q_y^2 + Q_y^2 Q_z^2 + Q_z^2 Q_x^2)}\right). \tag{3.4.8}$$

At finite wave-vectors, if the parameters γ_2 or γ_3 are different from zero, the four-fold degenerate valence bands are split into two spin-degenerate bands of different curvature, which are usually called the "heavy"- and "light"-hole bands. (See as an example Fig. 3.2 or the left panels of Figs. 3.5–3.8.) They are characterized by the heavy and light hole effective masses m_{hh} and m_{lh}, respectively. If the Luttinger parameters γ_2 and γ_3 are different from each other, the dispersion is anisotropic, leading to a "warping". In the case that γ_2 and γ_3 are equal, one obtains [6]

$$m_{hh} = m_0/(\gamma_1 - 2\gamma_2) \text{ and } m_{lh} = m_0/(\gamma_1 + 2\gamma_2) \ . \tag{3.4.9}$$

The above discussion of symmetry breaking remains valid if a small externally applied electric field is considered. In this case the basis matrices $(X_{v8}, Y_{v8}, Z_{v8}) = (\{j_v^y, j_v^z\}, \{j_v^z, j_v^x\}, \{j_v^x, j_v^y\})$ transform as the electric field components $\boldsymbol{E} = (E_x, E_y, E_z)$ as (Γ_5, K^+) and can be used to describe the linear Stark effect acting in the Γ_8 subspace of the valence-band states. For $E_x = E_y = 0$ the electric field terms give now rise to the effective Hamiltonian

$$H^v_{E1v8} = a^v_{E1v8}[X_{v8}E_x + Y_{v8}E_y + Z_{v8}E_z] \\ = a^v_{E1v8}E_z \begin{pmatrix} 0 & 0 & i\sqrt{3}/2 & 0 \\ 0 & 0 & 0 & i\sqrt{3}/2 \\ -i\sqrt{3}/2 & 0 & 0 & 0 \\ 0 & -i\sqrt{3}/2 & 0 & 0 \end{pmatrix}. \tag{3.4.10}$$

The same matrix form is obtained when considering strain effects. When comparing Eq. (3.4.10) with Eq. (3.2.21) and Eq. (3.2.22), we see that this coupling scheme is the one obtained within the Γ_8-block when using the valence-band basis-functions in which the spin-orbit coupling is diagonalized. In this approximation, the effect of the applied electric field is isotropic, i.e. the splitting of the two doublets is independent of the orientation of the electric field with respect to the crystal axes. We conclude that a possible anisotropy is due to a modification of the mixing of the valence-band states with Γ_7 and Γ_8 symmetry by the electric field.

A linear magnetic field (causing the symmetry breaking terms of Eqs. (3.2.27)–(3.2.30)) gives now rise to two contributions:

$$H^v_{B1v8a} = a^v_{B1v8a}[P_{v8}B_x + Q_{v8}B_y + R_{v8}B_z] \tag{3.4.11}$$

and

$$H^v_{B1v8b} = a^v_{B1v8b}[P'_{v8}B_x + Q'_{v8}B_y + R'_{v8}B_z]\,, \tag{3.4.12}$$

respectively. Depending on the direction of the magnetic field, the degeneracy of the valence-band states can again be completely lifted. The pseudo-angular momentum matrices (P_{v8}, Q_{v8}, R_{v8}) and (P'_{v8}, Q'_{v8}, R'_{v8}) are given in Eq. (3.4.5). The terms in Eqs. (3.4.11) and (3.4.12) correspond to the parameters κ and q introduced by Luttinger [8]

$$\begin{aligned} a^v_{B1v8a} &= -2\mu_B\kappa \\ a^v_{B1v8b} &= -2\mu_B q\,, \end{aligned} \tag{3.4.13}$$

where μ_B represents the electron magneton of Bohr. These terms have the required symmetry properties and lead to a linear Zeeman splitting of the Γ_8 valence-band.

We have seen in this section that the pseudo-spin development reproduces exactly the coupling of states within the diagonal blocks of the complete system if the wave functions v^v_3 to v^v_6 of Eq. (3.2.13), in which the spin-orbit coupling is diagonalized, are used. As in Sect. 3.3, when applying this method, only the mixing of states from different irreducible representations is neglected while the mixing within the states belonging to same irreducible representation is correctly described. The procedure discussed above allows to determine the detailed fine-structure of states that are otherwise almost degenerate.

References

1. Cho, K.: Phys. Rev. B **14**, 4463 (1976)
2. Chiang, T.-C., Eastman, D.E.: Phys. Rev. B **22**, 2940 (1980)
3. Hönerlage, B., Lévy, R., Grun, J.B., Klingshirn, C., Bohnert, K.: Phys. Rep. **124**, 161 (1985)
4. Kato, Y., Goto, T., Fujii, T., Ueta, M.: J. Phys. Soc. Jpn. **36**, 169 (1974)
5. Yu, P.Y., Cardona, M.: Fundamentals of Semiconductors: Physics and Material Properties. Springer, Berlin, Heidelberg (1995)
6. Klingshirn, C.: Semiconductor Optics, 3rd edn. Springer, Berlin, Heidelberg (2005)
7. Dresselhaus, G.: Phys. Rev **100**, 580 (1955)
8. Luttinger, J.M.: Phys. Rev. **102**, 1030 (1956)

Chapter 4
Exciton Ground State in Zincblende-Type Semiconductors

4.1 Exciton Ground-State Energy

We are now interested in the following problem: Let us consider the ground state of an ideal direct gap semiconductor at low temperature, i.e. at $T = 0$ K, containing N electrons. As indicated in Fig. 2.2 the semiconductor is characterized by the presence of an energy gap between the lowest lying conduction band (noted "c") and the uppermost valence band (noted "v"). In addition the semiconductor is supposed to possess only one conduction and one valence band and the states are defined by spin-orbitals. The N electrons occupy all electronic valence-band states and all conduction-band states are unoccupied. This system shows no electric conductivity (or it behaves as an insulator) since, when applying a small electric field to it, no electric current can flow. The sample can be polarized by the field but no free electric charges can move in the material. This situation characterizes the crystal-ground state whose energy and charge distribution is supposed to be known.

When adding some energy (e.g. thermally, optically, by applying strong electric or magnetic fields, etc.) electrons can be excited from the valence band to the conduction band. Then, electronic states of the valence band are unoccupied (these states are in the following called "defect-electron states" and noted by the subscript "v") and some states in the conduction band are occupied by electrons (subscript "e"). Instead of using spin-orbitals, the occupied and unoccupied states can be fully characterized by their wave-vectors and spins, i.e. the variables $(\boldsymbol{k}_v, \boldsymbol{\sigma}_v)$ and $(\boldsymbol{k}_e, \boldsymbol{\sigma}_e)$, respectively.

We want to describe now the situation when only one single electron is excited from the valence band to the conduction band, leaving behind a defect electron in the valence band. We are interested in the electronic excitation, which has the lowest energy. We call this excited state of the crystal the "exciton ground state". Its energy is evidently superior to that of the crystal-ground state since for the excitation of the electron to the conduction band some energy has to be supplied to the crystal. In order to determine this energy, one can proceed in the following way:

© Springer International Publishing AG, part of Springer Nature 2018
B. Hönerlage and I. Pelant, *Symmetry and Symmetry-Breaking in Semiconductors*, Springer Tracts in Modern Physics 279,
https://doi.org/10.1007/978-3-319-94235-3_4

Let us first consider the calculation of the one-particle electron-band structure of the conduction band. When calculating this energy as function of wave-vector $\boldsymbol{k}_e$ one considers one additional conduction-band electron in the presence of the N electrons, which occupy all the valence-band states. This is an $(N+1)$-particle problem. The additional conduction-band electron has a spin $\boldsymbol{\sigma}_e$ and interacts via Coulomb interaction with the valence-band electrons. This interaction has two different contributions: First, there is the direct Coulomb interaction between the charged particles. Since the electrons have the same charge, the Coulomb interaction is repulsive and the energy one has to pay is of the order of the energy separation between the conduction and valence band. Second, electrons are indiscernible fermions, i.e. they are spin 1/2 particles. This leads to an exchange interaction between the additional electron and the electrons in the valence band if they have the same wave-vector and are in the same spin state.

The problem to describe an excited state of a semiconductor (corresponding to the N-particle problem), where one electron occupies a state in the conduction band instead of a valence-band state, can now be approximately solved using the one-particle band-structure calculation of the conduction band described above. The energy of this state, which is taken as reference, has to be reduced by the additional interactions, which show up in the $(N+1)$-particle problem when compared to N-particle problem for the excited excited state:

If, due to the excitation, only one defect electron is created in the valence band but if otherwise the charge distribution of the valence-band electrons remains unchanged, the additional interaction is the Coulomb interaction between the conduction-band electron that interacts with the defect electron, characterized by the variables $(\boldsymbol{k}_v, \boldsymbol{\sigma}_v)$. All other two-particle interactions are the same in both problems. Since the resulting energy is subtracted from the energy of the $(N+1)$-particle system (given by the energy of the conduction-band electron), this resembles to an attractive Coulomb interaction between conduction-band electron and defect electron. This leads to the fact that the energy of the exciton ground state is smaller than the reference energy, namely the band gap between the conduction band and valence band. This energy difference is called the "exciton-binding energy".

The dispersion of the valence band and the solution of the $(N+1)$-particle problem describing the conduction band being supposed to be known, the exciton problem is treated in the framework of the Hartree-Fock approximation. In this approach, after integration of the variables describing the $(N-1)$ valence-band electrons, which are not involved in the transition, the remaining matrix elements depend only on the wave-functions of the additional electron and of the defect-electron state. One thus obtains an effective two-particle Hamiltonian, which has to be solved to determine the exciton-binding energy. In the most simple case, when neglecting exchange interaction and details of the band structure, the problem becomes similar to that of hydrogen atom. Therefore, we expect to obtain bound states, called "excitons", whose energies follow more or less a Rydberg series and, apart from these, "continuum states" corresponding to free electrons in the conduction band and free defect-electrons in the valence bands.

The procedure outlined above shows that the model is quite crude and neglects a number of effects, which are due to the fact that the charge distribution of the valence-band electrons in the crystal-ground state and the exciton problem are not the same. But the model shows the physical origin of the observed exciton features. It also indicates that excitons are not only characterized by the conduction- and valence-band electron-wave functions, but, in addition, they posses an "envelope function". It describes the spatial correlation between conduction- and valence-band electrons and corresponds to the bound electron wave function in the "hydrogen atom model".

Since excitons are defined as quasi-particles within a crystal, the exciton states (and namely their envelope functions) have to be compatible with the crystal symmetry. Therefore, all functions are Bloch functions, i.e. periodic, delocalized functions respecting the full crystal symmetry. The envelope function is constructed from a wave packet of conduction- and valence-band states, from which the exciton is build. At the Γ-point (see Ref. [1, 2]) the total symmetry of exciton states Γ_{ex} is now given by the symmetry of the envelope function (labeled Γ_{env}) multiplied by the symmetry of the valence-band (Γ_v) and by the symmetry of the conduction-band states (Γ_e), i.e. $\Gamma_{ex} = \Gamma_{env} \otimes \Gamma_v \otimes \Gamma_e$ gives the total symmetry of exciton states.

We will consider in the following only the exciton ground state. Its envelope function has spherical symmetry (i.e. it transforms as Γ_1). As we have discussed in the foregoing chapters, the conduction-band states transform as Γ_6 and the valence-bands as $\Gamma_v = (\Gamma_8, \Gamma_7)$. Since the valence-band states with Γ_8 and Γ_7 symmetry are no longer degenerate, two exciton series split in energy and featuring different symmetries are expected. The symmetry properties of their ground-states are given by $\Gamma_{ex} = \Gamma_1 \otimes \Gamma_8 \otimes \Gamma_6 = \Gamma_5 \oplus \Gamma_4 \oplus \Gamma_3$ and $\Gamma_{ex} = \Gamma_1 \otimes \Gamma_7 \otimes \Gamma_6 = \Gamma_5 \oplus \Gamma_2$, respectively. This situation characterizes the exciton states, which we will discuss now.

If one is mainly interested in wave-vector regions $\boldsymbol{k}$ close to the maximum (minimum) of the valence (conduction) band, the dispersion of the conduction and valence band are mainly parabolic (we will see this and also some deviations from this rule more precisely in the following). This leads to the fact that the dispersions may be described by "effective masses". As known from hydrogen atom, relative and center-of-mass motion can be separated in two-particle problems. Since the center-of-mass motion is translational invariant this quantity is a constant of motion and the center-of-mass wave-vector is a good quantum number. The relative motion, i.e. the electron and defect-electron wave-vectors being not constants of motion in the two-particle problem, the exciton function has to be developed using a superposition of products of electron and defect-electron Bloch functions. Their wave-vectors define the exciton center-of-mass wave-vector $\boldsymbol{Q}$ according to

$$\boldsymbol{Q} = \boldsymbol{k}_e - \boldsymbol{k}_v \ . \tag{4.1.1}$$

For the discussion of exciton properties it is now possible and convenient to introduce instead of the term "defect-electron" (describing the fact that a place is vacant in an otherwise filled valence band) a new quasi-particle, a "hole". Its state is denoted by the index "h". Then, instead of solving the N-particle problem within the

Hartree-Fock approximation, one considers a two-particle problem that is equivalent to it. This is possible since coordinates of the other $(N-1)$ particles have been integrated in the Hartree-Fock approximation. Their influence persist only in the electron and hole Bloch wave-functions. One has to bear in mind, however, that the exciton treated in this approach is resulting from an N-particle problem. This point is e.g. important when discussing screening effects.

In this approach the hole state is derived from that of the defect-electron valence-band state (characterized by $(\boldsymbol{k}_v, \boldsymbol{\sigma}_v)$). It turns out that it is obtained from it by Kramers' conjugation, i.e. by time reversal of the defect-electron state. In our case this results in the fact that the one-particle Bloch wave-functions have to be replaced by their complex conjugated functions and the wave-vectors $\boldsymbol{k}_v \rightarrow$ by $-\boldsymbol{k}_h$. (This approach is slightly different from the term "Defektelektron", which has been used in Ref. [3] to denote the vacant electron state in the valence-band.)

In particular, for the valence-band states described by Eq. (3.1.2) the approach used throughout this book implies the replacements

$$|1\rangle = -(x+iy)/\sqrt{2} \rightarrow -(x-iy)/\sqrt{2} = -|-1\rangle$$

and

$$|-1\rangle = (x-iy)/\sqrt{2} \rightarrow (x+iy)/\sqrt{2} = -|1\rangle .$$

Introducing holes and considering the two-particle system, Eq. (4.1.1) reads

$$\boldsymbol{Q} = \boldsymbol{k}_e + \boldsymbol{k}_h \; . \tag{4.1.2}$$

Kramers' conjugation affects also spinors, which change their signs if the Kramers' conjugation operator is applied two-times to them. Then, concerning spins, the spin-up state of the defect-electron α_v is replaced by the spin-down state of the hole β_h, i.e. $\alpha_v \rightarrow \beta_h$, and the spin-down state $\beta_v \rightarrow -\alpha_h$.

Excitons and holes are "quasi-particles", i.e. these wave packets are during a certain time coherent excitations, which are characterized by an ensemble of quantum numbers and posses a certain energy, angular momentum, polarization, etc. During this time they behave like particles, which interact with their environment before they decay: their coherence can be destroyed through scattering before their energy is dissipated and the crystal returns to its ground state.

To summarize: the quasi particles named "holes" posses the following properties: A hole describes an unoccupied state in an otherwise filled valence band. The state is obtained by Kramers' conjugation of the valence-band state, i.e. by time reversal of the valence-band wave-function. A hole has a positive elementary charge $+e$ and an effective mass that is given by the curvature of the dispersion of the valence band. The effective mass is positive for negative band curvatures. In addition, matrix elements involving valence band states are replaced by (-1) times its complex conjugated term [1, 4] in the electron-hole representation. Since the hole state has its origin in the defect-electron state, hole states may be degenerate and electrons and holes may also interact due to exchange interaction. These points will differentiate the exciton problem from a simple hydrogen atom. Bound electron-hole pairs (the excitons) do not lead to electric conductivity since they are neutral with respect to their total charge. The pairs have to be broken to lead to free charges, which can then move from one occupied state to another, unoccupied one, and an electric current can be observed.

Applying the Kramers' conjugation operator to the set of basis electron-wave functions v_i^v of the valence band at the Γ-point given in Eq. (3.2.14) creates a similar set for the hole states. Using the replacements discussed above, the valence-band states now become:

$$\begin{aligned}
v_1^v &\rightarrow (z\alpha_h + (x+iy)\beta_h)/\sqrt{3} = |\phi_6^h\rangle \\
v_2^v &\rightarrow -(-z\beta_h + (x-iy)\alpha_h)/\sqrt{3} = |\phi_5^h\rangle \\
v_3^v &\rightarrow -(x+iy)\alpha_h/\sqrt{2} = |\phi_4^h\rangle \\
v_4^v &\rightarrow -(2z\alpha_h - (x+iy)\beta_h)/\sqrt{6} = |\phi_3^h\rangle \\
v_5^v &\rightarrow (2z\beta_h + (x-iy)\alpha_h)/\sqrt{6} = |\phi_2^h\rangle \\
v_6^v &\rightarrow -(x-iy)\beta_h/\sqrt{2} = |\phi_1^h\rangle .
\end{aligned} \tag{4.1.3}$$

Here the index "h" has been introduced to indicate that we are dealing with the "hole" representation for the valence-band states. Or, using the angular momentum functions for the hole states, one equally obtains, by applying Eq. (3.1.2), instead of Eq. (4.1.3):

$$\begin{aligned}
|\phi_1^h\rangle &= -|-1\rangle\beta_h \\
|\phi_2^h\rangle &= (\sqrt{2}|0\rangle\beta_h + |-1\rangle\alpha_h)/\sqrt{3} \\
|\phi_3^h\rangle &= -(\sqrt{2}|0\rangle\alpha_h + |1\rangle\beta_h)/\sqrt{3} \\
|\phi_4^h\rangle &= |1\rangle\alpha_h \\
|\phi_5^h\rangle &= -(-|0\rangle\beta_h + \sqrt{2}|-1\rangle\alpha_h)/\sqrt{3} \\
|\phi_6^h\rangle &= (|0\rangle\alpha_h - \sqrt{2}|1\rangle\beta_h)/\sqrt{3}.
\end{aligned} \tag{4.1.4}$$

Comparing Eqs. (4.1.3) and (4.1.4) with the definition given in Eq. (3.2.14) and dropping the index "h" in the spin assignment, we see that the $|\phi_i^h\rangle$ hole states with $i = (1, \dots, 6)$ are given essentially by the valence band states v_i^v. One obtains explicitly the identification:

$$\begin{aligned}
|\phi_1^h\rangle &= -v_3^v = -|3/2, -3/2\rangle \\
|\phi_2^h\rangle &= v_4^v = |3/2, -1/2\rangle \\
|\phi_3^h\rangle &= -v_5^v = -|3/2, 1/2\rangle \\
|\phi_4^h\rangle &= v_6^v = |3/2, 3/2\rangle \\
|\phi_5^h\rangle &= -v_1^v = -|1/2, -1/2\rangle \\
|\phi_6^h\rangle &= v_2^v = |1/2, 1/2\rangle .
\end{aligned} \tag{4.1.5}$$

This correspondence is useful to know if one wants to construct symmetry-adapted exciton-wave functions from electron-hole states following Ref. [2]. Equation (4.1.5) shows that hole states may differ from valence-band states by a phase factor, which appears due to Kramers' conjugation of the spinors. Using Eq. (4.1.5) one can formulate the exciton states in the electron-hole representation starting from symmetry adapted products of electron- and valence-band states.

As has already been mentioned, we are in the following interested in the exciton ground state. When compared to the simple hydrogen atom, a detailed discussion of excitons requires sometimes that care has to be taken concerning the following points:

- The states of this many-particle system are characterized by several quantum numbers. The main quantum number "n" is, to a large extent, analogous to that of the hydrogen atom. $n = 1$ denotes the exciton ground state; $n = 2$, $n = 3$, etc. indicate exciton excited states ; $n \rightarrow \infty$ denotes the electron-hole continuum. Other quantum numbers will be introduced in the following. The quantum numbers may develop in time due to relaxation processes.
- Exciton states are not eigenstates of the Hamiltonian describing free electrons or holes. Their wave functions are developed on the basis of electron and hole states and involve therefore many of these states. Then, besides internal or external perturbations, the full band structure and dispersion of conduction and valence bands (degeneracy, splitting, warping, etc.) has to be considered.
- There exists an exchange interaction between electron and defect-electron states, if the states are characterized by the same quantum numbers. This exchange interaction has no parallel in the simple model of a hydrogen atom and has sometimes to be considered in the exciton problem.
- Excitons are made of charges that polarize the material, in which they exist. Then, Coulomb interaction leads to a rearranging of the charge distribution, which can sometimes be approximated by screening effects. Screening is, however, different for direct and exchange interaction since the exchange interaction is a contact interaction while the Coulomb potential is of long range.
- Electrons, holes, and excitons may interact with each other and different other elementary excitations. For instance, two excitons may be bound together to form an excitonic molecule (in analogy to hydrogen molecule, see Fig. 4.1). - etc.

The points mentioned above can be incorporated into our invariant development of the effective Hamiltonian. It can thus describe excitons in semiconductors, including the electron spin and the crystal symmetry.

Let us first consider the symmetry of exciton states at the Γ-point [1, 2]. In order to describe electron states characterized by spin and orbital-angular momentum, one has to use the double-group representation. As mentioned above, conduction-band states are made up from s-type atomic orbitals and are only spin degenerate in zincblende-type material. They transform as Γ_6.

Concerning the matrices spanning the conduction-band electron subspace in the pseudo-spin formalism we will use similar to Eq. (3.2.2) the following notation:

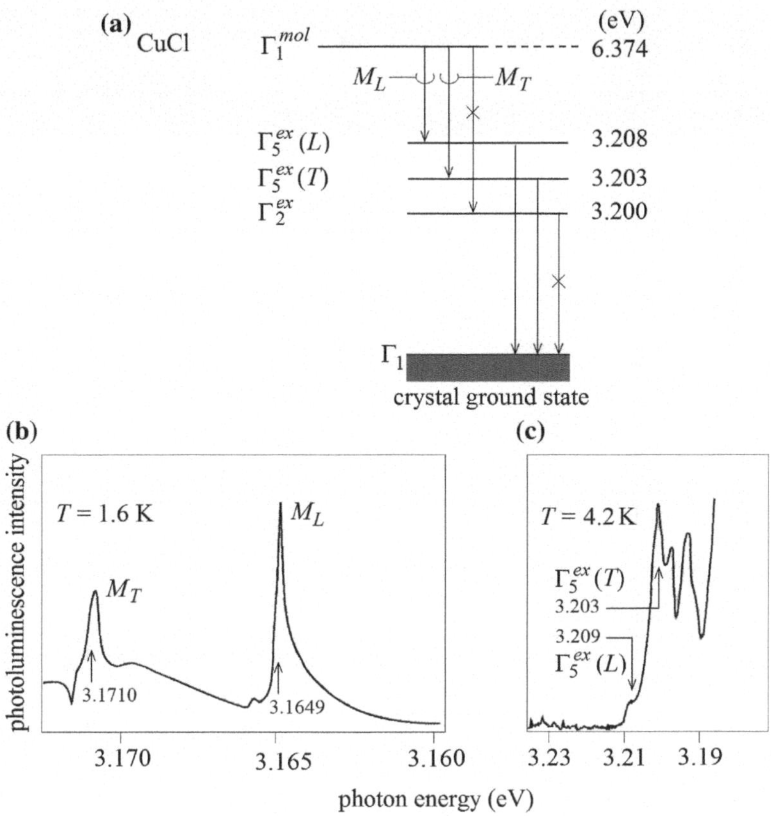

Fig. 4.1 Experimental illustration of how excitons (Z_3-series) manifest themselves in low-temperature luminescence spectra of CuCl crystals. Panel (**a**) displays energy levels of a CuCl excitonic molecule, i.e. a quasi-particle composed of two electrons and two holes (two excitons bound together). The molecule state labeled Γ_1^{mol} decays radiatively, one electron-hole pair recombining with emission of a luminescence photon while leaving the second exciton to survive. These radiative transitions terminate on the exciton levels Γ_5^{ex} (longitudinal (L) and transverse (T) exciton states). These excitons, recoiled in the recombination act, are the ones that posses a dipole moment. Corresponding experimental emission lines M_L and M_T are displayed in panel (**b**). Notice that there is no emission line that could be ascribed to the transition $\Gamma_1^{mol} \rightarrow \Gamma_2^{ex}$ at ~ 3.174 eV. Similarly, panel (**c**) shows luminescence lines originating in (single) exciton resonance radiative decays $\Gamma_5^{ex} \rightarrow \Gamma_1$ to the crystal ground state. Here, also a weak line due to partially allowed (owing to crystal imperfections) Γ_2^{ex}-exciton radiative decay at ~ 3.20 eV seems to be present. Concerning the emission at ~ 3.209 eV attributed to a not dipole-active $\Gamma_5^{ex}(L)$ recombination, a possible explanation may be as follows: The crystal is not perfect (the lattice can be deformed or one is close to a surface or a dislocation, etc.) and therefore the exciton wave-vector $\boldsymbol{Q}$ is not well defined. Then longitudinal excitons, which have definitely been created by the biexciton recombination (see the strong—here even stimulated—M_L emission line in panel (**c**)) perceive this perturbation (which is localized and not delocalized, contrary to free excitons) and recombine radiatively, even if in an ideal infinite bulk crystal they are not dipole-active. In these cases, however, unlike the situation shown in panel (**b**), unambiguous identification of the lines is not straightforward because of the presence of reabsorption/polariton effects (see Chap. 8). Adapted after Ref. [5–7]

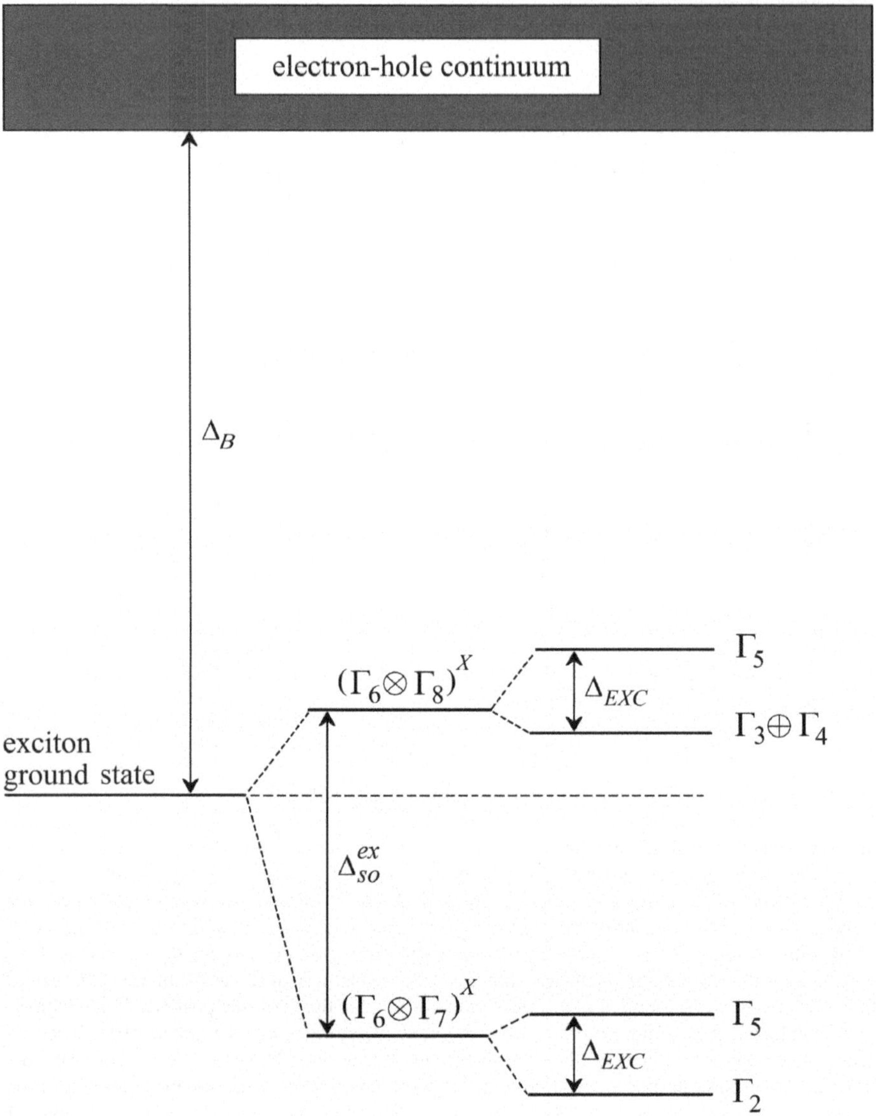

Fig. 4.2 Energy level scheme of excitons arising from holes in the valence bands with Γ_7 and Γ_8 symmetry and conduction-band electrons with Γ_6-symmetry in zincblende-type semiconductors. Δ_B and Δ_{so}^{ex} denote the exciton binding energy (determined with respect to the minimum of the continuum of electron-hole states) and the spin-orbit coupling energy, respectively. The spin-orbit coupling separates the $(\Gamma_6 \otimes \Gamma_8)^X$-exciton block from the so-called "split-off" exciton block with $(\Gamma_6 \otimes \Gamma_7)^X$-symmetry. Splitting within the blocks (indicated by Δ_{EXC}) is induced by the different exchange interactions, including the analytic, the non-analytic, and the cubic exchange. See text

$$\begin{aligned} 3 \cdot 1_e &\rightarrow S_e \\ \sigma_e^x &\rightarrow P_e \\ \sigma_e^y &\rightarrow Q_e \\ \sigma_e^z &\rightarrow R_e \,. \end{aligned} \tag{4.1.6}$$

As mentioned before, S_e transforms according to (Γ_1, K^+) and (P_e, Q_e, R_e) represent the pseudo-spin matrices that transform as (Γ_4, K^-).

As we will see, it is convenient in our invariant expansion to construct exciton-wave functions first from conduction-band states and the valence-band electron states. Excitons are in this approach the eigenstates of an effective Hamiltonian, given in a matrix representation. The eigenvectors of this exciton matrix are obtained by the Kronecker product of conduction-band electron states and valence-band states $|j, m_j\rangle$. This approach allows to identify the exciton states easily. Then, hole-wave functions defined above are introduced only at the end of the discussion of symmetry-breaking effects using Eq. (4.1.5). They can then be used when determining the values of the interaction-matrix elements.

Similarly to the procedure, which we have introduced when considering the spin-orbit interaction of the valence-band states, let us first discuss the exciton-energy structure in the case of the full point-group symmetry, i.e. in the absence of any symmetry-breaking effect. The uppermost filled valence-band states are six-fold degenerate at the Γ-point. They are made up from atomic p-orbitals with some admixtures of d-orbitals. Their degeneracy is partly lifted by spin-orbit interaction, giving rise to a set of four-fold degenerate states, transforming as Γ_8, and the two-fold degenerate Γ_7 states, [1, 2] as discussed in Sect. 3.2 and displayed in Figs. 2.2 and 3.2. In order to construct exciton states, we calculate again the Kronecker product of the valence-state matrices (which are given in Eq. (3.2.3)) with the pseudo-spin matrices $\boldsymbol{\sigma_e} = (\sigma_e^x, \sigma_e^y, \sigma_e^z)$ and 1_e. If one limits oneself to excitons with envelope functions of spherical symmetry, one obtains with Table 2.4 two exciton series (labeled Z_{12} and Z_3 in copper halides), namely

$$\begin{aligned} Z_{12} &\rightarrow \Gamma_6 \otimes \Gamma_8 = \Gamma_3 \oplus \Gamma_4 \oplus \Gamma_5 \\ &\qquad\qquad \text{and} \\ Z_3 &\rightarrow \Gamma_6 \otimes \Gamma_7 = \Gamma_2 \oplus \Gamma_5 \,. \end{aligned} \tag{4.1.7}$$

The electron-hole product-space, in which the exciton ground state is defined, is of dimension twelve, and all direct, exchange and symmetry-breaking perturbations can now be formulated in this subspace and calculated within a good approximation.

According to the transformation properties of the valence-state matrices given in Eqs. (3.2.3) and (3.2.4) and the electron-state matrices (Eq. (4.1.6)), we see that the two-particle Hamiltonian H^{ex} describing the exciton ground state at the Γ-point may have three different contributions, transforming as (Γ_1, K^+):

$$H^{ex} = H_d^{ex} + H_{so}^{ex} + H_{ech}^{ex} \,. \tag{4.1.8}$$

The first

$$H_d^{ex} = a_d^{ex} S_{11}\text{''} \otimes S_e \tag{4.1.9}$$

and the second term

$$H_{so}^{ex} = a_{so}^{ex} S_{44}\text{''} \otimes S_e \tag{4.1.10}$$

are the direct term H_d^{ex} (describing the Coulomb interaction between electron and hole) and the spin-orbit interaction H_{so}^{ex} of the valence band, which shows also up in the exciton problem. Since electron and hole have different charges, H_d^{ex} is an attractive interaction. Since the conduction-band states are made from s-type atomic orbitals, H_{so}^{ex} is originating only from the hole in the six-fold degenerate valence band. The electron spin, and therefore the indiscernibility of electrons does not play any role in these two terms.

The last contribution H_{ech}^{ex} in Eq. (4.1.8) is due to the exchange interaction between the electron in the conduction band and the hole in the valence band. It affects the exciton fine structure and modifies not only the direct electron-hole interaction but also the spin-orbit coupling. The resulting level scheme is sketched in Fig. 4.2, where Δ_B and Δ_{so}^{ex} denote the exciton-binding energy and the spin-orbit coupling energy, respectively. The binding energy is measured with respect to the energy minimum of the electron-hole continuum. Δ_{EXC} stands for the different electron-hole exchange energies, which further split the energies of the exciton states with Γ_5-symmetry and may separate in energy the exciton states with Γ_3 and Γ_4-symmetry.

It might be useful to stress here that admittedly both the spin-orbit coupling and the electron-hole exchange interaction give rise to the energy level scheme of Figs. 4.3 and 4.4, but while the spin-orbit coupling is already present in an unexcited semiconductor, the exchange interaction becomes relevant only in an electronically excited material.

As will be discussed below, four different exchange interaction terms of the exciton Hamiltonian can now be constructed from the matrices in Eqs. (3.2.4) and (4.1.6). These terms determine completely the energy fine structure of the twelve-fold degenerate exciton ground state in zincblende-type semiconductors. Let us now discuss the different terms.

4.2 Direct Electron-Hole Interaction, Spin-Orbit Coupling, and Symmetry-Adapted Exciton-Wave Functions

The direct term H_d^{ex} and the spin-orbit splitting H_{so}^{ex} of the exciton states give the contributions of the kinetic energy and the Coulomb interaction to the exciton binding energy. After coupling, the term H_d^{ex} gives a constant contribution to all exciton states. Due to H_{so}^{ex} the exciton ground state preserves the same energy fine structure as the valence band with the energies given in Eq. (3.2.11). When compared to the valence-band states, the exciton states are now doubly degenerate, however, reflecting the spin degeneracy of the conduction-band states. Using the basis matrices for conduction and valence-band states given in Eqs. (2.2.5) and (3.2.10), respectively, one calculates their Kronecker product Eq. (4.1.10). One then explicitly obtains for the matrix describing the exciton spin-orbit coupling :

$$H_{so}^{ex} = a_{so}^{ex} \begin{pmatrix} 3 & 0 & 0 & 0 & 0 & 0 & 0 & 0 & 0 & 0 & 0 & 0 \\ 0 & 3 & 0 & 0 & 0 & 0 & 0 & 0 & 0 & 0 & 0 & 0 \\ 0 & 0 & -3 & 0 & 3\sqrt{2} & 0 & 0 & 0 & 0 & 0 & 0 & 0 \\ 0 & 0 & 0 & -3 & 0 & 3\sqrt{2} & 0 & 0 & 0 & 0 & 0 & 0 \\ 0 & 0 & 3\sqrt{2} & 0 & 0 & 0 & 0 & 0 & 0 & 0 & 0 & 0 \\ 0 & 0 & 0 & 3\sqrt{2} & 0 & 0 & 0 & 0 & 0 & 0 & 0 & 0 \\ 0 & 0 & 0 & 0 & 0 & 0 & 0 & 0 & 3\sqrt{2} & 0 & 0 & 0 \\ 0 & 0 & 0 & 0 & 0 & 0 & 0 & 0 & 0 & 3\sqrt{2} & 0 & 0 \\ 0 & 0 & 0 & 0 & 0 & 0 & 3\sqrt{2} & 0 & -3 & 0 & 0 & 0 \\ 0 & 0 & 0 & 0 & 0 & 0 & 0 & 3\sqrt{2} & 0 & -3 & 0 & 0 \\ 0 & 0 & 0 & 0 & 0 & 0 & 0 & 0 & 0 & 0 & 3 & 0 \\ 0 & 0 & 0 & 0 & 0 & 0 & 0 & 0 & 0 & 0 & 0 & 3 \end{pmatrix} . \tag{4.2.1}$$

As discussed in the previous chapter in connection with Eq. (3.2.1) we use again the convention for the matrix products, and exciton states are expanded as products of the angular-momentum valence-band electron wave-functions $|m_j\rangle$, valence-band electron spin (α or β) and the conduction-band electron spin (α_e or β_e). The exciton states v_i^{ex} are given in this basis in the form

$$\begin{aligned} v_i^{ex} = a_{i,1}|1\rangle\alpha\alpha_e + a_{i,2}|1\rangle\alpha\beta_e + a_{i,3}|1\rangle\beta\alpha_e + a_{i,4}|1\rangle\beta\beta_e + a_{i,5}|0\rangle\alpha\alpha_e + \\ + a_{i,6}|0\rangle\alpha\beta_e + \cdots + a_{i,12}|-1\rangle\beta\beta_e \qquad (i = 1, \ldots, 12). \end{aligned} \tag{4.2.2}$$

In order to be more explicit, we first remember that the rows and columns of the basis matrices (given in Eq. (3.2.4)) describing the interactions in-between the valence-band states were ordered according to ($|1\rangle\alpha$, $|1\rangle\beta$, $|0\rangle\alpha$, $|0\rangle\beta$, $|-1\rangle\alpha$, $|-1\rangle\beta$). In the exciton problem, their direct product with the conduction-band electron-states (α_e and β_e) is formed, and the rows and columns of the corresponding exciton matrices of dimension 12 are ordered as indicated in Eq. (4.2.2). Remembering this convention, Eq. (4.2.1) directly shows, which exciton-basis states are coupled through the spin-orbit coupling. This coupling is now diagonalized by a well-defined linear combination of the exciton-basis states. In other words: One can diagonalize the Hamiltonian Eq. (4.2.1) by making use of the recipe explained in Appendix A (which, of course, can hardly be done now "by hand") and finding simultaneously the corresponding eigenfunctions (i.e. Eq. (4.2.3)) as being linear combinations of the basis functions $|1\rangle\alpha\alpha_e$, $|1\rangle\alpha\beta_e$, $|1\rangle\beta\alpha_e, \ldots, |-1\rangle\beta\beta_e$.

One then obtains for the exciton ground state the following set of normalized exciton-wave functions v_i^{ex} (with $i = 1$ to 12):

$$
v_i^{ex} = \begin{pmatrix}
0 & 0 & 0 & 0 & 0 & 0 & 0 & 0 & 0 & 0 & 0 & 1 \\
0 & 0 & 0 & 0 & 0 & 0 & 0 & 0 & 0 & 0 & 1 & 0 \\
0 & 0 & 0 & -\sqrt{6}/3 & 0 & 0 & 0 & 0 & 0 & \sqrt{3}/3 & 0 & 0 \\
0 & 0 & -\sqrt{6}/3 & 0 & 0 & 0 & 0 & 0 & \sqrt{3}/3 & 0 & 0 & 0 \\
0 & 0 & 0 & \sqrt{3}/3 & 0 & 0 & 0 & 0 & 0 & \sqrt{6}/3 & 0 & 0 \\
0 & 0 & \sqrt{3}/3 & 0 & 0 & 0 & 0 & 0 & \sqrt{6}/3 & 0 & 0 & 0 \\
0 & -\sqrt{3}/3 & 0 & 0 & 0 & 0 & 0 & \sqrt{6}/3 & 0 & 0 & 0 & 0 \\
-\sqrt{3}/3 & 0 & 0 & 0 & 0 & 0 & \sqrt{6}/3 & 0 & 0 & 0 & 0 & 0 \\
0 & \sqrt{6}/3 & 0 & 0 & 0 & 0 & 0 & \sqrt{3}/3 & 0 & 0 & 0 & 0 \\
\sqrt{6}/3 & 0 & 0 & 0 & 0 & 0 & \sqrt{3}/3 & 0 & 0 & 0 & 0 & 0 \\
0 & 0 & 0 & 0 & 0 & 1 & 0 & 0 & 0 & 0 & 0 & 0 \\
0 & 0 & 0 & 0 & 1 & 0 & 0 & 0 & 0 & 0 & 0 & 0
\end{pmatrix} . \tag{4.2.3}
$$

This set of exciton-wave functions is explicitly given by:

$$
\begin{aligned}
v_1^{ex} &= (-|0\rangle\beta + \sqrt{2}|-1\rangle\alpha)\beta_e/\sqrt{3} = v_1^v\beta_e \\
v_2^{ex} &= (-|0\rangle\beta + \sqrt{2}|-1\rangle\alpha)\alpha_e/\sqrt{3} = v_1^v\alpha_e \\
v_3^{ex} &= (|0\rangle\alpha - \sqrt{2}|1\rangle\beta)\beta_e/\sqrt{3} = v_2^v\beta_e \\
v_4^{ex} &= (|0\rangle\alpha - \sqrt{2}|1\rangle\beta)\alpha_e/\sqrt{3} = v_2^v\alpha_e \\
v_5^{ex} &= |-1\rangle\beta\beta_e = v_3^v\beta_e \\
v_6^{ex} &= |-1\rangle\beta\alpha_e = v_3^v\alpha_e \\
v_7^{ex} &= (\sqrt{2}|0\rangle\beta + |-1\rangle\alpha)\beta_e/\sqrt{3} = v_4^v\beta_e \\
v_8^{ex} &= (\sqrt{2}|0\rangle\beta + |-1\rangle\alpha)\alpha_e/\sqrt{3} = v_4^v\alpha_e \\
v_9^{ex} &= (\sqrt{2}|0\rangle\alpha + |1\rangle\beta)\beta_e/\sqrt{3} = v_5^v\beta_e \\
v_{10}^{ex} &= (\sqrt{2}|0\rangle\alpha + |1\rangle\beta)\alpha_e/\sqrt{3} = v_5^v\alpha_e \\
v_{11}^{ex} &= |1\rangle\alpha\beta_e = v_6^v\beta_e \\
v_{12}^{ex} &= |1\rangle\alpha\alpha_e = v_6^v\alpha_e .
\end{aligned} \tag{4.2.4}
$$

The result on the right hand side of Eq. (4.2.4) is easy to understand: The valence-band wave-functions v_i^v with i =(1 to 6) given in Eq. (3.2.13) diagonalize the valence-band spin-orbit interaction. Equally, α_e and β_e are eigenstates of the conduction-band electron-states, which are not influenced by the spin-orbit coupling. Therefore, their direct products of the forms (v_i^v α_e) and (v_i^v β_e) lead to a basis for exciton states, in which the direct electron-hole as well as the spin-orbit interaction are diagonalized. As we will see this situation changes as soon as the electron-hole exchange-interaction is included in the exciton problem.

Because of the spin-orbit splitting of the valence band one obtains two exciton series. The eigenvalues of the four states obtained from the direct product of the conduction-band states with that of the valence band with Γ_7-symmetry (v_i^{ex} states with i = 1 to 4) are given by $-6a_{so}^{ex}$. This is the ground-state energy of an exciton series originating from the "split-off" valence band. For the eight states obtained from the the conduction-band states and the valence-band states having Γ_8-symmetry (v_i^{ex} states with $i = 5$ to 12) one obtains an exciton ground state having the eigenvalue $3a_{so}^{ex}$.

At this point it is useful to give explicitly the exciton-wave functions defined for the conduction- and valence-band basis together with those given in the electron-hole basis. Substituting the hole states as defined in Eq. (4.1.5) into Eq. (4.2.4) one obtains:

$$
\begin{aligned}
v_1^{ex} &= (-|0\rangle\beta + \sqrt{2}|-1\rangle\alpha)\beta_e/\sqrt{3} = v_1^v\beta_e = -|\phi_5^h\rangle\beta_e = -|\Phi_3\rangle \\
v_2^{ex} &= (-|0\rangle\beta + \sqrt{2}|-1\rangle\alpha)\alpha_e/\sqrt{3} = v_1^v\alpha_e = -|\phi_5^h\rangle\alpha_e = -|\Phi_1\rangle \\
v_3^{ex} &= (|0\rangle\alpha - \sqrt{2}|1\rangle\beta)\beta_e/\sqrt{3} = v_2^v\beta_e = |\phi_6^h\rangle\beta_e = |\Phi_4\rangle \\
v_4^{ex} &= (|0\rangle\alpha - \sqrt{2}|1\rangle\beta)\alpha_e/\sqrt{3} = v_2^v\alpha_e = |\phi_6^h\rangle\alpha_e = |\Phi_2\rangle \\
v_5^{ex} &= |-1\rangle\beta\beta_e = v_3^v\beta_e = -|\phi_1^h\rangle\beta_e = -|\Psi_5\rangle \\
v_6^{ex} &= |-1\rangle\beta\alpha_e = v_3^v\alpha_e = -|\phi_1^h\rangle\alpha_e = -|\Psi_1\rangle \\
v_7^{ex} &= (\sqrt{2}|0\rangle\beta + |-1\rangle\alpha)\beta_e/\sqrt{3} = v_4^v\beta_e = |\phi_2^h\rangle\beta_e = |\Psi_6\rangle \\
v_8^{ex} &= (\sqrt{2}|0\rangle\beta + |-1\rangle\alpha)\alpha_e/\sqrt{3} = v_4^v\alpha_e = |\phi_2^h\rangle\alpha_e = |\Psi_2\rangle \\
v_9^{ex} &= (\sqrt{2}|0\rangle\alpha + |1\rangle\beta)\beta_e/\sqrt{3} = v_5^v\beta_e = -|\phi_3^h\rangle\beta_e = -|\Psi_7\rangle \\
v_{10}^{ex} &= (\sqrt{2}|0\rangle\alpha + |1\rangle\beta)\alpha_e/\sqrt{3} = v_5^v\alpha_e = -|\phi_3^h\rangle\alpha_e = -|\Psi_3\rangle \\
v_{11}^{ex} &= |1\rangle\alpha\beta_e = v_6^v\beta_e = |\phi_4^h\rangle\beta_e = |\Psi_8\rangle \\
v_{12}^{ex} &= |1\rangle\alpha\alpha_e = v_6^v\alpha_e = |\phi_4^h\rangle\alpha_e = |\Psi_4\rangle.
\end{aligned}
\tag{4.2.5}
$$

In addition, we have introduced the following notations labeled $|\Psi_i\rangle$ and $|\Phi_i\rangle$:

$$
\begin{aligned}
|\Psi_1\rangle &= -v_3^v\alpha_e = |\phi_1^h\rangle\alpha_e = -|3/2, -3/2\rangle\alpha_e \\
|\Psi_2\rangle &= v_4^v\alpha_e = |\phi_2^h\rangle\alpha_e = |3/2, -1/2\rangle\alpha_e \\
|\Psi_3\rangle &= -v_5^v\alpha_e = |\phi_3^h\rangle\alpha_e = -|3/2, 1/2\rangle\alpha_e \\
|\Psi_4\rangle &= v_6^v\alpha_e = |\phi_4^h\rangle\alpha_e = |3/2, 3/2\rangle\alpha_e \\
|\Psi_5\rangle &= -v_3^v\beta_e = |\phi_1^h\rangle\beta_e = -|3/2, -3/2\rangle\beta_e \\
|\Psi_6\rangle &= v_4^v\beta_e = |\phi_2^h\rangle\beta_e = |3/2, -1/2\rangle\beta_e \\
|\Psi_7\rangle &= -v_5^v\beta_e = |\phi_3^h\rangle\beta_e = -|3/2, 1/2\rangle\beta_e \\
|\Psi_8\rangle &= v_6^v\beta_e = |\phi_4^h\rangle\beta_e = |3/2, 3/2\rangle\beta_e.
\end{aligned}
\tag{4.2.6}
$$

where the exciton states ($|\Psi_1\rangle$ to $|\Psi_8\rangle$) result from the direct product of the electrons in the conduction band and the holes in the valence band, which transforms along Γ_8. On the other hand one has:

$$\begin{aligned}
|\Phi_1\rangle &= -v_1^v\alpha_e = |\phi_5^h\rangle\alpha_e = -|1/2, -1/2\rangle\alpha_e \\
|\Phi_2\rangle &= v_2^v\alpha_e = |\phi_6^h\rangle\alpha_e = |1/2, 1/2\rangle\alpha_e \\
|\Phi_3\rangle &= -v_1^v\beta_e = |\phi_5^h\rangle\beta_e = -|1/2, -1/2\rangle\beta_e \\
|\Phi_4\rangle &= v_2^v\beta_e = |\phi_6^h\rangle\beta_e = |1/2, 1/2\rangle\beta_e.
\end{aligned} \tag{4.2.7}$$

These exciton states ($|\Phi_1\rangle$ to $|\Phi_4\rangle$) are obtained from the valence-band states transforming as Γ_7. The notations ($|\Psi_1\rangle$ to $|\Psi_8\rangle$) and ($|\Phi_1\rangle$ to $|\Phi_4\rangle$) were introduced in Ref. [1] and are commonly used for the exciton states.

Using the exciton basis functions of Eq. (4.2.5) the direct electron-hole interaction and the spin-orbit coupling are diagonalized. ($|\Psi_1\rangle$ to $|\Psi_8\rangle$) and ($|\Phi_1\rangle$ to $|\Phi_4\rangle$) give two exciton blocks, the latter being called the split-off exciton series. These two exciton blocks are separated in energy by the value

$$\Delta_{so}^{ex} = 9a_{so}^{ex} \tag{4.2.8}$$

as shown in Fig. 4.2. One should mention here that the valence-band spin-orbit splitting Δ_{so}^{v} given in Eq. (3.2.15) is at this stage equal to the exciton spin-orbit splitting Δ_{so}^{ex}, since including the electron spin and forming excitons does not modify the spin-orbit coupling and the energetic separation of the valence bands. (Δ_{so}^{ex} results from Eq. (4.1.10) where, according to Eq. (4.1.6), the matrix $S_e = 3 \cdot 1_e$. Since S_e is not the unit matrix one finds the parameter a_{so}^{ex} = (1/3) a_{so}^{v} when comparing Eq. (4.2.8) to Eq. (3.2.15).) In general, Δ_{so}^{ex} can, however, be slightly different from Δ_{so}^{v} since the splitting can be modified by the electron-hole exchange interaction, which will be discussed in the following. This possibility is accounted for by the fact that a_{so}^{v} and a_{so}^{ex} are parameters in the formalism of invariant expansion.

4.3 Electron-Hole Exchange Interaction

The last contribution in Eq. (4.1.8) H_{ech}^{ex} concerns the exchange interaction between the electron in the conduction band and the hole in the valence band. According to their transformation properties (Eqs. (3.2.4) and (4.1.6)) four different terms can be constructed in zincblende type semiconductors. They are:

$$\begin{aligned}
H_{ech1}^{ex} &= a_{ech}^1(P_{41}\text{''} \otimes P_e + Q_{41}\text{''} \otimes Q_e + R_{41}\text{''} \otimes R_e) \\
H_{ech2}^{ex} &= a_{ech}^2(P_{54}\text{''} \otimes P_e + Q_{54}\text{''} \otimes Q_e + R_{54}\text{''} \otimes R_e) \\
H_{ech3}^{ex} &= a_{ech}^3(P_{34}\text{''} \otimes P_e + Q_{34}\text{''} \otimes Q_e + R_{34}\text{''} \otimes R_e) \\
H_{ech4}^{ex} &= a_{ech}^4(P_{14}\text{''} \otimes P_e + Q_{14}\text{''} \otimes Q_e + R_{14}\text{''} \otimes R_e).
\end{aligned} \tag{4.3.1}$$

As discussed in connection with Eqs. (3.2.3) and (3.2.4) the interaction matrices (P_{41}", Q_{41}", R_{41}") and (P_{14}", Q_{14}", R_{14}") are directly originating from the valence-band spin- or angular-momentum matrices, while the matrices (P_{54}", Q_{54}", R_{54}") and

(P_{34}", Q_{34}", R_{34}") are generated through the spin-orbit coupling. The importance of the terms H^{ex}_{ech2} and H^{ex}_{ech3} should therefore diminish with decreasing spin-orbit coupling.

Let us discuss the physical origin of the different exchange-interaction terms. As can be seen from Eq. (4.3.1) together with Eq. (3.2.4), H^{ex}_{ech4} describes the direct exchange interaction between valence-band and conduction-band electrons, which are in the same spin state. It is important to notice that the angular momentum of these states is not involved. On the other hand, H^{ex}_{ech1} describes the spin-orbit coupling that comprises the angular momentum of the valence band, which interacts directly with the spin of the conduction-band electron. This term is the signature of the indiscernibility of electrons: if a spin-orbit coupling exists for valence-band electrons, the spin-orbit coupling also influences the electrons in the conduction band.

One could point out here that H^{ex}_{ech1} and H^{ex}_{ech4} are normally the most important contributions to the electron-hole exchange interaction H^{ex}_{ech}. The other terms are of higher order and involve the exchange interaction and the spin-orbit coupling simultaneously. This is obvious in the terms H^{ex}_{ech2} and H^{ex}_{ech3}: In their case the total angular momentum of the valence-band states $\boldsymbol{j_v}$ has first to be formed through spin-orbit coupling. Then, the spin part of the total angular momentum couples through exchange interaction to the spin of the electron $\boldsymbol{\sigma_e}$ in the conduction band. Since this is a higher order process, it is less important for the determination of the exciton energies. Both terms have cubic symmetry since they involve the orientation of the total angular momentum with respect to the cubic axes.

Let us now consider the exchange interaction in detail. According to Eq. (4.3.1) we calculate the Kronecker product of the matrices described by Eqs. (3.2.4) and (2.2.3). After transformation to the basis given in Eq. (4.2.3), i.e. the exciton states, in which the spin-orbit interaction is diagonalized, we obtain for H^{ex}_{ech4} the matrix form

$$H^{ex}_{ech4} = a^4_{ech}$$

$$\begin{pmatrix}
-2/3 & 0 & 0 & 0 & 0 & 4\sqrt{6}/3 & -4\sqrt{2}/3 & 0 & 0 & 0 & 0 & 0 \\
0 & 2/3 & -4/3 & 0 & 0 & 0 & 0 & 4\sqrt{2}/3 & -4\sqrt{2}/3 & 0 & 0 & 0 \\
0 & -4/3 & 2/3 & 0 & 0 & 0 & 0 & 4\sqrt{2}/3 & -4\sqrt{2}/3 & 0 & 0 & 0 \\
0 & 0 & 0 & -2/3 & 0 & 0 & 0 & 0 & 0 & 4\sqrt{2}/3 & -4\sqrt{6}/3 & 0 \\
0 & 0 & 0 & 0 & 2 & 0 & 0 & 0 & 0 & 0 & 0 & 0 \\
4\sqrt{6}/3 & 0 & 0 & 0 & 0 & -2 & 4\sqrt{3}/3 & 0 & 0 & 0 & 0 & 0 \\
-4\sqrt{2}/3 & 0 & 0 & 0 & 0 & 4\sqrt{3}/3 & 2/3 & 0 & 0 & 0 & 0 & 0 \\
0 & 4\sqrt{2}/3 & 4\sqrt{2}/3 & 0 & 0 & 0 & 0 & -2/3 & 8/3 & 0 & 0 & 0 \\
0 & -4\sqrt{2}/3 & -4\sqrt{2}/3 & 0 & 0 & 0 & 0 & 8/3 & -2/3 & 0 & 0 & 0 \\
0 & 0 & 0 & 4\sqrt{2}/3 & 0 & 0 & 0 & 0 & 0 & 2/3 & 4\sqrt{3}/3 & 0 \\
0 & 0 & 0 & -4\sqrt{6}/3 & 0 & 0 & 0 & 0 & 0 & 4\sqrt{3}/3 & -2 & 0 \\
0 & 0 & 0 & 0 & 0 & 0 & 0 & 0 & 0 & 0 & 0 & 2
\end{pmatrix} \tag{4.3.2}$$

and for H^{ex}_{ech1}:

$$H^{ex}_{ech1} = a^1_{ech} \begin{pmatrix} 2 & 0 & 0 & 0 & 0 & -\sqrt{6} & \sqrt{2} & 0 & 0 & 0 & 0 & 0 \\ 0 & -2 & 4 & 0 & 0 & 0 & 0 & -\sqrt{2} & \sqrt{2} & 0 & 0 & 0 \\ 0 & 4 & -2 & 0 & 0 & 0 & 0 & -\sqrt{2} & \sqrt{2} & 0 & 0 & 0 \\ 0 & 0 & 0 & 2 & 0 & 0 & 0 & 0 & 0 & -\sqrt{2} & \sqrt{6} & 0 \\ 0 & 0 & 0 & 0 & 3 & 0 & 0 & 0 & 0 & 0 & 0 & 0 \\ -\sqrt{6} & 0 & 0 & 0 & 0 & -3 & 2\sqrt{3} & 0 & 0 & 0 & 0 & 0 \\ \sqrt{2} & 0 & 0 & 0 & 0 & 2\sqrt{3} & 1 & 0 & 0 & 0 & 0 & 0 \\ 0 & -\sqrt{2} & -\sqrt{2} & 0 & 0 & 0 & 0 & -1 & 4 & 0 & 0 & 0 \\ 0 & \sqrt{2} & \sqrt{2} & 0 & 0 & 0 & 0 & 4 & -1 & 0 & 0 & 0 \\ 0 & 0 & 0 & -\sqrt{2} & 0 & 0 & 0 & 0 & 0 & 1 & 2\sqrt{3} & 0 \\ 0 & 0 & 0 & \sqrt{6} & 0 & 0 & 0 & 0 & 0 & 2\sqrt{3} & -3 & 0 \\ 0 & 0 & 0 & 0 & 0 & 0 & 0 & 0 & 0 & 0 & 0 & 3 \end{pmatrix}. \tag{4.3.3}$$

In the matrices given in Eqs. (4.3.2) and (4.3.3) the blocks of four rows and columns in the upper left corner determine the coupling within the split-off exciton states having

$$\Gamma_6 \otimes \Gamma_7 = \Gamma_2 \oplus \Gamma_5$$

symmetry. The block of rows and columns 5 to 12 in the lower right corner determine that within the exciton block of

$$\Gamma_6 \otimes \Gamma_8 = \Gamma_3 \oplus \Gamma_4 \oplus \Gamma_5$$

symmetry. The remaining non-diagonal blocks give the interaction between states of the different exciton blocks obeying to different irreducible representations.

It is an important point to remark that the product of electron- and hole-wave functions given in Eqs. (4.2.6) and (4.2.7) are no longer eigenfunctions of the exciton Hamiltonian (i.e. they diagonalize no longer the Hamiltonian describing the exciton ground state) if the electron-hole exchange interaction is included. The modification is twofold: First, within the different exciton blocks (formed due to the spin-orbit coupling) the exchange interaction may lead to energy splittings between the different irreducible representations of the blocks as evoked in Eq. (4.1.7). Then the eigenfunctions of the exciton Hamiltonian are linear combinations of the electron-hole wave functions Eqs. (4.2.6) and (4.2.7). Second, the exchange interaction leads to a coupling in-between the different exciton blocks that are separated by the spin-orbit coupling, giving rise to a mixing of their states. In zincblende type material, the exchange interaction is, however, much smaller than the spin-orbit coupling. Therefore, one works in general with the basis given by the wave functions of Eq. (4.2.6) or Eq. (4.2.7) and treats—if necessary—the non-diagonalized part of the exchange interaction as a perturbation.

When comparing Eqs. (4.3.2) and (4.3.3) we see that the terms H^{ex}_{ech1} and H^{ex}_{ech4} have in their matrix representation the same structure, giving rise to the same coupling scheme of the exciton states. The coefficients a^1_{ech} and a^4_{ech} are, however, different and both matrices are linearly independent: a^1_{ech} and a^4_{ech} enter with different signs and weights into the two diagonal blocks, which are derived from the valence-band states with Γ_7 and Γ_8-symmetry, and into the non-diagonal blocks, which describe a coupling of exciton states belonging to these different blocks.

Let us now consider H^{ex}_{ech3} and H^{ex}_{ech2}. H^{ex}_{ech2} has the form

$$H^{ex}_{ech2} = a^2_{ech} \begin{pmatrix} -2/3 & 0 & 0 & 0 & 0 & -\sqrt{6}/6 & \sqrt{2}/6 & 0 & 0 & 0 & 0 & 0 \\ 0 & 2/3 & -4/3 & 0 & 0 & 0 & 0 & -\sqrt{2}/6 & \sqrt{2}/6 & 0 & 0 & 0 \\ 0 & -4/3 & 2/3 & 0 & 0 & 0 & 0 & -\sqrt{2}/6 & \sqrt{2}/6 & 0 & 0 & 0 \\ 0 & 0 & 0 & -2/3 & 0 & 0 & 0 & 0 & 0 & -\sqrt{2}/6 & \sqrt{6}/6 & 0 \\ 0 & 0 & 0 & 0 & 0 & 0 & 0 & 0 & 0 & 0 & 0 & -1 \\ -\sqrt{6}/6 & 0 & 0 & 0 & 0 & 0 & \sqrt{3}/3 & 0 & 0 & 0 & 0 & 0 \\ \sqrt{2}/6 & 0 & 0 & 0 & 0 & \sqrt{3}/3 & 2/3 & 0 & 0 & 0 & 0 & 0 \\ 0 & -\sqrt{2}/6 & -\sqrt{2}/6 & 0 & 0 & 0 & 0 & -2/3 & -1/3 & 0 & 0 & 0 \\ 0 & \sqrt{2}/6 & \sqrt{2}/6 & 0 & 0 & 0 & 0 & -1/3 & -2/3 & 0 & 0 & 0 \\ 0 & 0 & 0 & -\sqrt{2}/6 & 0 & 0 & 0 & 0 & 0 & 2/3 & \sqrt{3}/3 & 0 \\ 0 & 0 & 0 & \sqrt{6}/6 & 0 & 0 & 0 & 0 & 0 & \sqrt{3}/3 & 0 & 0 \\ 0 & 0 & 0 & 0 & -1 & 0 & 0 & 0 & 0 & 0 & 0 & 0 \end{pmatrix} \tag{4.3.4}$$

and H^{ex}_{ech3} reads explicitly

$$H^{ex}_{ech3} = a^3_{ech} \begin{pmatrix} -8/3 & 0 & 0 & 0 & 0 & -2\sqrt{6}/3 & 2\sqrt{2}/3 & 0 & 0 & 0 & 0 & 0 \\ 0 & 8/3 & -16/3 & 0 & 0 & 0 & 0 & -2\sqrt{2}/3 & 2\sqrt{2}/3 & 0 & 0 & 0 \\ 0 & -16/3 & 8/3 & 0 & 0 & 0 & 0 & -2\sqrt{2}/3 & 2\sqrt{2}/3 & 0 & 0 & 0 \\ 0 & 0 & 0 & -8/3 & 0 & 0 & 0 & 0 & 0 & -2\sqrt{2}/3 & 2\sqrt{6}/3 & 0 \\ 0 & 0 & 0 & 0 & 2 & 0 & 0 & 0 & 0 & 0 & 0 & 6 \\ -2\sqrt{6}/3 & 0 & 0 & 0 & 0 & -2 & -2\sqrt{3}/3 & 0 & 0 & 0 & 0 & 0 \\ 2\sqrt{2}/3 & 0 & 0 & 0 & 0 & -2\sqrt{3}/3 & -10/3 & 0 & 0 & 0 & 0 & 0 \\ 0 & -2\sqrt{2}/3 & -2\sqrt{2}/3 & 0 & 0 & 0 & 0 & 10/3 & 14/3 & 0 & 0 & 0 \\ 0 & 2\sqrt{2}/3 & 2\sqrt{2}/3 & 0 & 0 & 0 & 0 & 14/3 & 10/3 & 0 & 0 & 0 \\ 0 & 0 & 0 & -2\sqrt{2}/3 & 0 & 0 & 0 & 0 & 0 & -10/3 & -2\sqrt{3}/3 & 0 \\ 0 & 0 & 0 & 2\sqrt{6}/3 & 0 & 0 & 0 & 0 & 0 & -2\sqrt{3}/3 & -2 & 0 \\ 0 & 0 & 0 & 0 & 6 & 0 & 0 & 0 & 0 & 0 & 0 & 2 \end{pmatrix}. \tag{4.3.5}$$

The exchange interaction terms H^{ex}_{ech3} and H^{ex}_{ech2} have a similar but slightly different structure than H^{ex}_{ech1} and H^{ex}_{ech4}:

Concerning the non-diagonal exciton blocks, all four interaction terms lead exactly to the same coupling scheme in-between the states of the two exciton blocks. The states become thus mixed in the same way by the four electron-hole exchange interaction terms H^{ex}_{ech1} to H^{ex}_{ech4}.

Concerning the diagonal blocks, it is interesting to notice that the terms H^{ex}_{ech2} and H^{ex}_{ech3} show in their matrix representation not exactly the same coupling structure as H^{ex}_{ech1} and H^{ex}_{ech4}. An important exception is that the diagonal elements of H^{ex}_{ech2} and

H^{ex}_{ech3} for the states v^{ex}_5 and v^{ex}_{12} are modified and that these states are now coupled. As we will see, this coupling leads to a modification of the energy fine structure of the exciton states of the block, derived from the valence-band states with Γ_8-symmetry. It gives rise to an energy splitting between the exciton states with Γ_3 and Γ_4 symmetry.

Let us first discuss the coupling in-between the two exciton blocks (i.e. the non-diagonal exciton blocks) in more detail. As mentioned above, the non-diagonal exciton blocks given in Eq. (4.3.2) to Eq. (4.3.5) have identical structure, leading to the same mixing of states. In all cases the split-off exciton states (states v^{ex}_1 to v^{ex}_4) are coupled to the same states (v^{ex}_5 to v^{ex}_{12}) from the $\Gamma_6 \otimes \Gamma_8$-exciton block. This may be easily understood when remembering that the exchange interaction has its origin in the indiscernibility of electrons. This means that matrix elements involving two different exciton wave functions and the Coulomb interaction between electron and hole may be different from zero. This is possible if the two wave functions have the electron spin indexes [(α_e and α) and/or (β_e and β)] exchanged and if they involve the same eigenfunctions of the angular-momentum operator.

Let us consider as an example the exciton wave-function

$$v^{ex}_1 = (-|0\rangle\beta + \sqrt{2}|-1\rangle\alpha)\beta_e/\sqrt{3}$$

We remember (Eq. (4.2.4)) that

$$\begin{aligned}
v^{ex}_5 &= |-1\rangle\beta\beta_e \\
v^{ex}_6 &= |-1\rangle\beta\alpha_e \\
v^{ex}_7 &= (\sqrt{2}|0\rangle\beta + |-1\rangle\alpha)\beta_e/\sqrt{3} \\
v^{ex}_8 &= (\sqrt{2}|0\rangle\beta + |-1\rangle\alpha)\alpha_e/\sqrt{3} \\
v^{ex}_9 &= (\sqrt{2}|0\rangle\alpha + |1\rangle\beta)\beta_e/\sqrt{3} \\
v^{ex}_{10} &= (\sqrt{2}|0\rangle\alpha + |1\rangle\beta)\alpha_e/\sqrt{3} \\
v^{ex}_{11} &= |1\rangle\alpha\beta_e \\
v^{ex}_{12} &= |1\rangle\alpha\alpha_e.
\end{aligned}$$

In order to take into account the indiscernibility of electrons we drop the subscript "e". We now see that the combination $|0\rangle\beta\beta$ of v^{ex}_1 is also present in v^{ex}_7 but not in the other exciton wave functions. Similarly, $|-1\rangle\alpha\beta$ becomes (after the electron-spin of the conduction-band and of the valence-band states are exchanged) $|-1\rangle\beta\alpha$. This function appears only in v^{ex}_6 and nowhere else. Therefore, due to the exchange interaction, the exciton wave-function v^{ex}_1 is coupled to the states v^{ex}_6 and v^{ex}_7 but not to the other exciton wave-functions. In addition, it results from the wave functions that the coupling to the state v^{ex}_6 is by a factor $\sqrt{3}$ stronger than to v^{ex}_7. All other couplings between the different states given in the matrices Eq. (4.3.2) to Eq. (4.3.5) may be interpreted in the same way.

Let us now consider the two diagonal exciton blocks, resulting from the spin-orbit splitting of the valence-band states. We first consider the subspace of exciton states, originating from the electron and hole states transforming as Γ_6 and Γ_7, respectively. (This is the so-called "split-off exciton block".) These exciton states belong to the irreducible representations $\Gamma_6 \otimes \Gamma_7 = \Gamma_2 \oplus \Gamma_5$. Since the exchange interaction couples the conduction and valence-band states, the wave functions given in Eq. (4.2.5) are no longer eigenfunctions of the exciton Hamiltonian, but the exciton states are given by a linear combination of these functions.

As shown in Eq. (4.3.2) to Eq. (4.3.5), all exchange interactions have the same coupling structure within the split-off exciton subspace and lead to an energy splitting of the split-off exciton states. It turns out that the different exchange interaction terms and the spin-orbit coupling can be diagonalized by the following exciton-wave functions:

$$|0,0\rangle_c = (v_3^{ex} - v_2^{ex})/\sqrt{2} \tag{4.3.6}$$

and

$$\begin{aligned} |1,1\rangle_c &= v_4^{ex} \\ |1,0\rangle_c &= (v_3^{ex} + v_2^{ex})/\sqrt{2} \\ |1,-1\rangle_c &= v_1^{ex}. \end{aligned} \tag{4.3.7}$$

In these equations we have used the convention that usually exciton states are classified according to the system's total angular momentum $\boldsymbol{J} = \boldsymbol{j_v} \oplus \boldsymbol{\sigma_e}$ and its projection component onto the z-axis, J_z. The exciton states are then characterized by the exciton quantum numbers J and M_J and denoted by $|J, M_J\rangle$.

Defining for simplicity

$$a_{ech} = 2a_{ech}^1 - (2/3)a_{ech}^2 - (8/3)a_{ech}^3 - (2/3)a_{ech}^4 \tag{4.3.8}$$

the energies of the four split-off exciton states (and thus the fine structure of this exciton block) are given by

$$E_{25}^i = a_{ech}(-3, 1, 1, 1) \tag{4.3.9}$$

with $i = (1 \text{ to } 4)$, respectively.

The irreducible representation with Γ_2 symmetry is a one-dimensional representation (i.e. it contains only one exciton state), Γ_5 is a 3-dimensional representation. As given in Eq. (4.3.9) the degeneracy of the split-off exciton states is lifted by the exchange interaction. Since the Hamiltonian does not contain any symmetry-breaking interactions, its eigenfunctions (i.e. the exciton-wave functions) must have the full crystal symmetry and belong to irreducible representations. We can therefore identify the exciton state $(v_3^{ex} - v_2^{ex})/\sqrt{2}$ having the energy $E_{25}^1 = -3a_{ech}$ with the exciton state of Γ_2 symmetry. It is given in Eq. (4.3.6). The states (v_4^{ex}, $(v_3^{ex} + v_2^{ex})/\sqrt{2}$, and v_1^{ex}) given in Eq. (4.3.7) are then the Γ_5 states that are triply degenerated in energy, namely $E_{25}^2 = E_{25}^3 = E_{25}^4 = a_{ech}$.

The symmetry adapted exciton-wave functions can also be obtained by calculating the product states $|1/2, \pm 1/2\rangle \alpha_e$ and $|1/2, \pm 1/2\rangle \beta_e$ of Eq. (4.2.7), using the multiplication Tables of Ref. [2]. One has only to remember that the valence-band states $|1/2, \pm 1/2\rangle$ have Γ_7 symmetry and the conduction-band states ($\alpha_e = |1/2, +1/2\rangle$ and $\beta_e = |1/2, -1/2\rangle$) have Γ_6 symmetry. Up to an insignificant phase factor appearing in the product states (i.e. the exciton states) with Γ_5 symmetry the functions given in Ref. [2] are identical to those given in Eqs. (4.3.6) and (4.3.7).

Using the notation $|J, M_J\rangle$ the state $|0, 0\rangle_c$ with Γ_2 symmetry is a state with total angular momentum $J = 0$. The states ($|1, 1\rangle_c$; $|1, 0\rangle_c$; $|1, -1\rangle_c$) are $J = 1$ exciton states. It is now convenient to define exciton states in the electron-hole representation by using the correspondence between valence-band and hole states given in Eq. (4.2.5). The wave functions Eqs. (4.3.6) and (4.3.7) then transform into

$$|0, 0\rangle_c = (|\Phi_4\rangle + |\Phi_1\rangle)/\sqrt{2} \tag{4.3.10}$$

and

$$\begin{aligned} |1, 1\rangle_c &= |\Phi_2\rangle \\ |1, 0\rangle_c &= (|\Phi_4\rangle - |\Phi_1\rangle)/\sqrt{2} \\ |1, -1\rangle_c &= -|\Phi_3\rangle. \end{aligned} \tag{4.3.11}$$

Following Ref. [1, 4] or Eq. (3.1.2) ($\boldsymbol{x}, \boldsymbol{y}, \boldsymbol{z}$) polarized pair states can be easily constructed from Eq. (4.3.11) and one obtains:

$$\begin{aligned} |x\rangle_c &= (-|1, 1\rangle_c + |1, -1\rangle_c)/\sqrt{2} \\ |y\rangle_c &= i(|1, 1\rangle_c + |1, -1\rangle_c)/\sqrt{2} \\ |z\rangle_c &= |1, 0\rangle_c. \end{aligned} \tag{4.3.12}$$

Equation (4.3.12) shows that such $J = 1$ states have a dipole moment. They are of Γ_5 symmetry in zincblende-type crystals. (It is interesting to notice that according to Ref. [2] the vectors ($\boldsymbol{x}, \boldsymbol{y}, \boldsymbol{z}$) are the basis functions of the irreducible representation of Γ_5 symmetry. Therefore, it is important to stress at this point that excitons with Γ_5 symmetry are the only excitons, which have a dipole moment. All other excitons with a different symmetry do not transform as the vectors ($\boldsymbol{x}, \boldsymbol{y}, \boldsymbol{z}$).) In zincblende-type crystals the electric-dipole transition from the crystal ground state is non-vanishing only to $J = 1$ exciton states and the exciton states of Eq. (4.3.12) have non-vanishing transition-dipole matrix elements with linearly ($\boldsymbol{x}, \boldsymbol{y}, \boldsymbol{z}$)-polarized light. $J = 1$ states are said to be "dipole active" since they couple within the dipole approximation to the electromagnetic radiation field, which is also characterized by a dipole moment [1, 4]. As we will see in detail in Chap. 8, these exciton states are said to carry a finite "oscillator strength".

Let us now consider the block of exciton states originating from conduction-band electrons of Γ_6 symmetry and the hole states transforming as Γ_8. As discussed in the foregoing Sect. 4.2 this $\Gamma_6 \otimes \Gamma_8$ exciton-block is separated in energy from the split-off exciton ground state by $9a_{so}^{ex}$ due to the spin-orbit interaction. One can now identify the symmetry of the exciton states and their energy by applying the same procedure as detailed above for the split-off exciton states.

Concerning the $\Gamma_6 \otimes \Gamma_8$ excitons one considers again only the diagonal block of the exciton-exchange interaction matrices defined through the states υ_5^{ex} to υ_{12}^{ex}. The exciton states transform as $\Gamma_6 \otimes \Gamma_8 = \Gamma_3 \oplus \Gamma_4 \oplus \Gamma_5$. The symmetry adapted exciton-wave functions can be obtained using the multiplication tables of Ref. [2] and calculating the product states $|3/2, \pm 3/2\rangle \alpha_e$, $|3/2, \pm 1/2\rangle \alpha_e$, $|3/2, \pm 3/2\rangle \beta_e$ and $|3/2, \pm 1/2\rangle \beta_e$ of Eq. (4.2.6). In the electron-hole representation these exciton states are characterized by a total angular momentum $J = 2$ and $J = 1$, respectively.

As discussed above, the $J = 1$ exciton states can be identified to have Γ_5 symmetry. These states are triply degenerate in energy. The $J = 2$ states posses a quadrupole moment and may in principle be five-fold degenerate. The $J = 1$ states being identified also within the $\Gamma_6 \otimes \Gamma_8$ exciton-block as states with Γ_5 symmetry, the remaining $J = 2$ states must be formed from states having Γ_3 and Γ_4 symmetry.

We first consider the exchange interactions matrices resulting from H_{ech1}^{ex} and H_{ech4}^{ex}. Treating the $\Gamma_6 \otimes \Gamma_8$-exciton block separately from the split-off exciton block (see Eq. (4.3.2) and Eq. (4.3.3)), we see that these exchange interaction matrices are linearly dependent on each other, i.e. the reduced matrix of H_{ech4}^{ex} is obtained from H_{ech1}^{ex} by multiplying it by a factor of (2/3). Therefore, both terms describe the same exchange interaction within this approximation.

In the basis of conduction-band and valence-band electron states υ_i^{ex} of Eq. (4.2.3), the reduced exchange-interaction term H_{ech1}^{ex} of Eq. (4.3.3) can be diagonalized by the following normalized basis of exciton states υ_{iel}^{ex} (with $i = 5$ to 12):

$$\upsilon_{iel}^{ex} = \begin{pmatrix} 0 & 0 & 0 & 0 & 0 & 0 & 0 & 1 \\ 0 & 0 & -\sqrt{3}/2 & 0 & 0 & 0 & 1/2 & 0 \\ 0 & 0 & 1/2 & 0 & 0 & 0 & \sqrt{3}/2 & 0 \\ 0 & -\sqrt{2}/2 & 0 & 0 & 0 & \sqrt{2}/2 & 0 & 0 \\ 0 & \sqrt{2}/2 & 0 & 0 & 0 & \sqrt{2}/2 & 0 & 0 \\ -1/2 & 0 & 0 & 0 & \sqrt{3}/2 & 0 & 0 & 0 \\ \sqrt{3}/2 & 0 & 0 & 0 & 1/2 & 0 & 0 & 0 \\ 0 & 0 & 0 & 1 & 0 & 0 & 0 & 0 \end{pmatrix}. \tag{4.3.13}$$

One obtains from Eq. (4.3.3) the eigenvalues

$$E_{exiel} = a_{ech}^1(-5, -5, -5, 3, 3, 3, 3, 3). \tag{4.3.14}$$

As mentioned before, the exchange interaction matrix H_{ech4}^{ex} is diagonalized by the same set of wave functions and leads to the same energy-level scheme having the energies

$$E_{exie4} = (2/3)a_{ech}^4(-5, -5, -5, 3, 3, 3, 3, 3). \tag{4.3.15}$$

Equations (4.3.14) and (4.3.15) show that the exchange interactions H_{ech1}^{ex} and H_{ech4}^{ex} lift partly the degeneracy of the $\Gamma_6 \otimes \Gamma_8$-exciton block. The states υ_{5el}^{ex} to υ_{7el}^{ex}

are triply degenerate and thus correspond to the $J = 1$ states. The states v^{ex}_{8e1} to v^{ex}_{12e1} are fivefold degenerate and are separated in energy from the $J = 1$ exciton states by $8(a^1_{ech} + (2/3)\, a^4_{ech})$. These $J = 2$ states have partially mixed Γ_3 and Γ_4 symmetry, i.e. these states are not yet symmetry adapted. After having diagonalized the exchange interaction terms H^{ex}_{ech1} and H^{ex}_{ech4} we obtain for the $J = 1$ states when using the electron-hole notation of Eq. (4.2.5):

$$\begin{aligned} |1,1\rangle &= (|\Psi_3\rangle + \sqrt{3}|\Psi_8\rangle)/2 \\ |1,0\rangle &= -(|\Psi_2\rangle + |\Psi_7\rangle)/\sqrt{2} \\ |1,-1\rangle &= (|\Psi_6\rangle + \sqrt{3}|\Psi_1\rangle)/2 \end{aligned} \tag{4.3.16}$$

and for the $J = 2$ states

$$\begin{aligned} |2,2\rangle &= |\Psi_4\rangle \\ |2,1\rangle &= (|\Psi_8\rangle - \sqrt{3}|\Psi_3\rangle)/2 \\ |2,0\rangle &= (|\Psi_2\rangle - |\Psi_7\rangle)/\sqrt{2} \\ |2,-1\rangle &= (\sqrt{3}|\Psi_6\rangle - |\Psi_1\rangle)/2 \\ |2,-2\rangle &= -|\Psi_5\rangle. \end{aligned} \tag{4.3.17}$$

These states are the same as those introduced in Ref. [1, 4]. The $J = 1$ states $|1,1\rangle, |1,0\rangle$, and $|1,-1\rangle$ have the same characteristics as the states ($|1,1\rangle_c$; $|1,0\rangle_c$; $|1,-1\rangle_c$) of the split-off band discussed above, i.e. they are dipole active. The $J = 2$ states remain degenerate when considering only H^{ex}_{ech1} and H^{ex}_{ech4}. As we will show now, inclusion of the exchange interaction terms H^{ex}_{ech2} and H^{ex}_{ech3} results in an energy separation between states with Γ_3 and Γ_4-symmetry. The exciton wave functions given in Eq. (4.3.17) are then used to construct the missing symmetry-adapted exciton states.

Let us consider for example the reduced exchange-interaction matrix H^{ex}_{ech2} of Eq. (4.3.4). It can be diagonalized using the following normalized basis of the v^{ex}_{ie2} exciton states (with $i = 5$ to 12) of Eq. (4.2.2):

$$v^{ex}_{ie2} = \begin{pmatrix} \sqrt{2}/2 & 0 & 0 & 0 & 0 & -\sqrt{2}/2 & 0 & 0 \\ 0 & 0 & 0 & 0 & -\sqrt{3}/2 & 0 & 0 & 1/2 \\ 0 & 0 & 0 & 0 & 1/2 & 0 & 0 & \sqrt{3}/2 \\ 0 & \sqrt{2}/2 & 0 & -\sqrt{2}/2 & 0 & 0 & 0 & 0 \\ 0 & \sqrt{2}/2 & 0 & \sqrt{2}/2 & 0 & 0 & 0 & 0 \\ 0 & 0 & -1/2 & 0 & 0 & 0 & \sqrt{3}/2 & 0 \\ 0 & 0 & \sqrt{3}/2 & 0 & 0 & 0 & 1/2 & 0 \\ \sqrt{2}/2 & 0 & 0 & 0 & 0 & \sqrt{2}/2 & 0 & 0 \end{pmatrix} \tag{4.3.18}$$

and one obtains from Eq. (4.3.4) the eigenvalues

$$E_{exie2} = a_{ech}^2(-1, -1, -1/3, -1/3, -1/3, 1, 1, 1). \tag{4.3.19}$$

(Attention has to be payed to the fact that the order of the states given in Eq. (4.3.18) is different from that of Eq. (4.3.13) and that they are labeled υ_{ie2}^{ex} with $i = 5$ to 12 in the following.) As it is immediately seen in Eq. (4.3.19), the states υ_{5e2}^{ex} and υ_{6e2}^{ex} are doubly degenerate. Since the exchange-interaction matrix H_{ech2}^{ex} has the full point-group symmetry and since the states υ_{5e2}^{ex} and υ_{6e2}^{ex} belong to the $\Gamma_6 \otimes \Gamma_8$-exciton block, these $J = 2$ states correspond to states with Γ_3 symmetry.

The states υ_{7e2}^{ex} to υ_{9e2}^{ex} are the triply degenerate $J = 1$ states with Γ_5 symmetry, which we have already discussed above in connection with H_{ech1}^{ex} and which are given in Eqs. (4.3.13) and (4.3.16). The exchange interaction term H_{ech2}^{ex} under consideration has no further influence on the wave functions and the degeneracy of these states but may only shift them together in energy.

States υ_{10e2}^{ex} to υ_{12e2}^{ex} are triply degenerate and are separated by $\Delta_{34} = 2a_{ech}^2$ in energy from the states with Γ_3 symmetry. Following the same argumentation as above, these remaining $J = 2$ states have to be the states with Γ_4 symmetry. The υ_{ie2}^{ex} states are now fully adapted to the crystal symmetry and we see that some of the states given in Eq. (4.3.18) are different from those of Eq. (4.3.13). The exchange interactions H_{ech2}^{ex} and H_{ech3}^{ex} both act differently on exciton states with Γ_3 and Γ_4 symmetry and may partially lift their degeneracy. They give rise to the so-called "cubic exchange interaction".

Using the notation introduced in Ref. [1, 4] one finds for the symmetry-adapted Γ_3 states:

$$\begin{aligned} |2,+\rangle = (|2,2\rangle + |2,-2\rangle)/\sqrt{2} &= (|\Psi_4\rangle - |\Psi_5\rangle)/\sqrt{2} \\ |2,0\rangle &= (|\Psi_2\rangle - |\Psi_7\rangle)/\sqrt{2} \end{aligned} \tag{4.3.20}$$

and for those with Γ_4 symmetry

$$\begin{aligned} |1,+\rangle = -(|2,1\rangle + |2,-1\rangle)/\sqrt{2} &= -((|\Psi_8\rangle - \sqrt{(3)}|\Psi_3\rangle)/2 + (\sqrt{3}|\Psi_6\rangle - |\Psi_1\rangle)/2)/\sqrt{2} \\ |1,-\rangle = i(|2,-1\rangle - |2,1\rangle)/\sqrt{2} &= i((\sqrt{3}|\Psi_6\rangle - |\Psi_1\rangle)/2 - (|\Psi_8\rangle - \sqrt{3}|\Psi_3\rangle)/2)/\sqrt{2} \\ |2,-\rangle = (|2,2\rangle - |2,-2\rangle)/\sqrt{2} &= (|\Psi_4\rangle + |\Psi_5\rangle)/\sqrt{2}. \end{aligned} \tag{4.3.21}$$

As stated above, the reduced exchange interaction H_{ech3}^{ex} is diagonalized by the same wave functions as those of H_{ech2}^{ex} given in Eq. (4.3.18). They lead to the same energy level scheme and exciton fine structure as given above, only the coupling coefficients are different from those obtained from H_{ech2}^{ex}. The eigenvalues are given in this case by

$$E_{exie3} = a_{ech}^3(8, 8, -4/3, -4/3, -4/3, -4, -4, -4) \tag{4.3.22}$$

for the multiplets having Γ_3, Γ_5, and Γ_4 symmetry, respectively.

We have seen that the electron-hole exchange interaction couples states, which belong to different exciton blocks and mixes their wave functions. In the example discussed above in connection with the Hamiltonian described by Eq. (4.3.4) the exciton state $v_1^{ex} = (-|0\rangle\beta + \sqrt{2}|-1\rangle\alpha)\beta_e/\sqrt{3}$ couples to the exciton states v_6^{ex} and v_7^{ex}. This means that the states $|1,-1\rangle_c = -|\Phi_3\rangle$ and $|1,-1\rangle = (|\Psi_6\rangle + \sqrt{3}|\Psi_1\rangle)/2$ are coupled. The same result is found for the exciton state $|1,0\rangle_c$ which is coupled to the state $|1,0\rangle$ while the state $|1,1\rangle_c$ is coupled to $|1,1\rangle$. These are the non-diagonal exciton blocks which are responsible for this coupling. An important consequence of the coupling is that oscillator strength of dipole-active excitons is transferred by the electron-hole exchange interaction in-between excitons, originating from the valence band with Γ_7 and that with Γ_8 symmetry if they are characterized by $J = 1$ and the same J_z quantum number. As mentioned above, the exchange interaction is generally much smaller than the spin-orbit coupling. Therefore, one can work with the basis given by the wave functions of Eq. (4.2.6) or Eq. (4.2.7) and treat the non-diagonalized part of the exchange interaction as a perturbation.

Let us neglect in the rest of this chapter the non-diagonal blocks, which we have just discussed. Then the matrices H_{ech1}^{ex} to H_{ech4}^{ex} reduce to two block matrices, which can be diagonalized using the states indicated in (Eq. (4.3.10) and Eq. (4.3.11)), and in (Eq. (4.3.16), Eq. (4.3.20), and Eq. (4.3.21)).

We will first discuss the nature of the exchange interaction for the split-off exciton ground state, i.e. the exciton block arising from the electrons and holes that transform as Γ_6 and Γ_7, respectively. The calculation of the exchange interaction energy is discussed in detail in Ref. [1]. It turns out that its value depends on the dipole moment of the state. The exciton states with Γ_2 symmetry have no dipole moment and are therefore not at all affected by exchange interaction. On the other hand, as given in Eq. (4.3.12) for zincblende-type semiconductors, exciton states with Γ_5 symmetry have a dipole moment and experience exchange interaction. If we take "z" as the direction of propagation characterizing the wave-vector $\boldsymbol{Q}$, the Γ_5-exciton states ($|1,1\rangle_c$ and $|1,-1\rangle_c$) have (as follows from Eq. (4.3.12)) their dipole moments orientated perpendicular to "z" and they are called "transverse excitons". The dipole moment of the $|1,0\rangle_c$ exciton state is orientated parallel to "z", i.e. it is a "longitudinal" exciton. The exchange interaction can now be divided into two parts: first, the analytic part, which is independent of the orientation of the dipole. It influences in the same way transverse and longitudinal Γ_5-exciton states and lifts the degeneracy between excitons with Γ_2 and Γ_5 symmetry. Second, the exchange interaction has a non-analytic contribution. This part of the exchange interaction depends on the orientation of the dipole moment with respect to the direction of propagation of the exciton. The non-analytic part of the exchange interaction acts only on the longitudinal-exciton state and thus lifts the degeneracy of the longitudinal and transverse states. We clearly see in this fact that the finite center-of-mass wave-vector of excitons (which defines the terms "longitudinal" and "transverse") is a symmetry-breaking perturbation.

Up to now we have identified exciton states by their total angular momentum and discussed, whether excitons were dipole active or not, i.e. whether excitons couple to the electromagnetic radiation field. This approach is interesting since excitons often show up in the optical properties of semiconductors. Historically, excitons were first discussed as being analog to atoms or small molecules and the symmetry of their spin-orbitals was considered separately from their total angular momentum. In this context expressions like"singlet-" and "triplet-" or "ortho-" and "para-" excitons are then used.

In order to clarify these expressions let us consider for the moment a simple atom or molecule, e.g. a He-atom or H_2-molecule. It is immersed into an electromagnetic radiation field (a photon field) and a transition from an occupied electronic state to an unoccupied state is studied. The electromagnetic radiation field is a transverse field, characterized by its dipole moment, and has a total angular momentum $J = 1$. If we take again "z" as direction of propagation, a photon corresponds to a quasi-particle, characterized by $|J = 1, M_J = \pm 1\rangle$. A photon can induce an optical transition between two electron states characterized by $\Delta L = \pm 1$ and $\Delta M_L = \pm 1$ (L being the quantum number of the orbital-momentum operator and M_L that of its magnetic moment, respectively). On the other hand, a photon does not change the spin state of the considered electron. Then, in an optical transition, the spin of the electron in its initial state and in its excited state are the same: i.e. the spins of the excited electron and of the missing electron are parallel.

In the language used in the preceding chapters for semiconductors this signifies that the spin-states of the electron in the conduction band and of the generated defect-electron in the valence band are the same. Excitons are defined, however, as electron-hole pair states, where the hole state is obtained from the defect-electron state by Kramer's conjugation. (Using this description in terms of "electron" and "hole" one easily understands the resulting electron-hole attraction, leading to bound states of excitons situated within the energy gap of the semiconductor.) Since Kramer's conjugation reverses the spin-state of the defect-electron in the valence band when forming the hole, optically excited excitons are characterized by the fact that electron and hole spins are in opposite states. The electron-hole pair states have anti-parallel spins (with a total spin $S = 0$) and these excitons are called in analogy with atomic physics "spin-singlet excitons" or in short "singlet excitons". This nomenclature was later also applied to characterize dipole-active $J = 1$ exciton states or excitons with Γ_5 symmetry in zincblende-type semiconductors since they can be optically excited.

If the electron in the conduction band or the hole in the valence band undergoes after the excitation by the light field a spin-flip, electron and hole spins are parallel and have a total spin $S = 1$. Such excitons are called "spin-triplet excitons" or in short "triplet excitons". Similar to the atomic or molecular model the singlet-exciton states have a higher energy than the triplet state due to the exchange interaction. In general, the expression "triplet excitons" characterizes exciton states that have no dipole moment as for example the state $|0, 0\rangle_c$ given in Eq. (4.3.6), which transforms as Γ_2.

But attention has to be payed to the fact that this exciton state is non-degenerate and it is not a multiplet. In addition, neither the total spin S nor the total angular momentum J are good quantum numbers in semiconductors. They thus do not indicate the multiplicity of a subgroup of states.

In some materials also the nomenclature "para" and "ortho" exciton is used instead of "singlet" or "triplet excitons" (e.g. in Cu_2O or other semiconductor oxides). In these materials transitions to exciton states with S-type envelope functions are dipole forbidden and an optical transition is accompanied with a spin-flip process of an electron. (This situation is somewhat similar to the excitation of electrons in He atoms.) If there an electron and a hole are created to build an exciton by resonant one-photon absorption, they have parallel spins. Thus, such excitons are in these substances (unlike to zincblende material) the ortho (= triplet) excitons rather than the para (= singlet) excitons. To crown the confusion the ortho-excitons have in Cu_2O and related substances *higher energy* than the optically forbidden para-excitons. One must therefore be very cautious when using the terms "singlet" and "triplet" excitons because their behavior and properties might be very different in different semiconductors.

Since excitons with Γ_2 symmetry and longitudinal Γ_5 excitons (i.e. the $|1, 0\rangle_c$ exciton states) do not couple to the electromagnetic radiation field they are also called "dark" exciton states . This nomenclature is used in opposition to "bright" excitons that couple to the light field and are observed e.g. in photoluminescence. This distinction, however, cannot be taken literally. Perturbations, which destroy the point-group symmetry, can transfer oscillator strength from dipole-active exciton-states to dipole-forbidden states, which may thus also show up (weakly) in luminescence measurements (see e.g. panel (c) in Fig. 4.1). In any case, attention has to be payed when using these expressions since they use atomic models to describe exciton states, which have a much more complex structure.

Let us come back to the exchange interaction of the split-off exciton-ground state and its fine structure. As discussed with Eqs. (4.3.8) and (4.3.9) the four exchange interaction terms H^{ex}_{ech1} to H^{ex}_{ech4} are diagonalized by the exciton states in Eqs. (4.3.6) and (4.3.7). These interaction terms influence the split-off exciton states in exactly the same way. When considering only their analytical parts by introducing, in accordance with Eq. (4.3.8),

$$a^A_{ech} = 2a^{1A}_{ech} - (2/3)a^{2A}_{ech} - (8/3)a^{3A}_{ech} - (2/3)a^{4A}_{ech} \qquad (4.3.23)$$

one obtains for the eigenvalues of the split-off states

$$E_{sti} = a^A_{ech}(-3, 1, 1, 1) \qquad (4.3.24)$$

with $i = (1 \text{ to } 4)$, respectively.

The triplet exciton state $|0, 0\rangle_c$ with Γ_2 symmetry has the lowest energy of $-3a^A_{ech}$ and the singlet states, i.e. ($|1, 1\rangle_c$; $|1, 0\rangle_c$; $|1, -1\rangle_c$), which transform as Γ_5, are shifted to higher energies by a^A_{ech}. Thus, the exchange-interaction terms H^{ex}_{ech1} to H^{ex}_{ech4} lift the degeneracy between singlet and triplet states. One therefore introduces the singlet-triplet splitting Δ_{st} between exciton states with Γ_5 and with Γ_2 symmetry as

$$\Delta_{st} = 4a^A_{ech}. \tag{4.3.25}$$

The degeneracy of dipole active exciton states is further lifted by the non-analytical electron-hole exchange interaction (Ref. [1, 4]) that shifts longitudinal excitons to higher energies by a^{NA}_{ech} and does not influence the other states. This energy separation is called the "longitudinal-transverse splitting energy" $\Delta_{LT} = a^{NA}_{ech}$ of dipole active excitons. The resulting energy scheme is shown in Fig. 4.3 where the quantization axes is taken parallel to the center-of-mass wave-vector $\boldsymbol{Q}$. Various means of $\Gamma_6 \otimes \Gamma_7$-exciton luminescence manifestation in CuCl crystals, illustrating experimentally the above discussion are shown in Fig. 4.1.

Figure 4.4 shows the scheme of exciton levels, arising from the holes in the valence band with Γ_8-symmetry and conduction-band electrons in zincblende-type semiconductors. The z-axis of quantization, which is used to identify the $|J, M_J\rangle$ states, is parallel to the wave-vector $\boldsymbol{Q}$. Similar to the split-off band, splitting and a fine structure of the exciton states is induced by the different exchange interactions given in Eq. (4.3.1). Schematically, the exchange interaction terms a^1_{ech} and a^4_{ech} separate the $J = 1$ and $J = 2$ states by an energy difference $\Delta_{34,5}$. Then, the cubic exchange interaction terms a^2_{ech} and a^3_{ech} shift the Γ_5-states and lifts the degeneracy of exciton states with Γ_3 from those with Γ_4-symmetry by the amount of Δ_{34}. As discussed above, this "cubic-exchange" term is small compared to the terms a^1_{ech} and a^4_{ech} since it has its origin in the modification of the spin-orbit coupling through the electron-hole exchange-interaction. It is therefore usually neglected. In addition, the longitudinal-transverse splitting Δ_{LT} is induced by the nonanalytic parts of the exchange interactions on the dipole active $J = 1$ exciton states. All these features are documented by the experimental reflectance curve of a CuBr crystal in Fig. 4.5.

The energies $E(\Gamma_j)$ of the states with symmetry Γ_j (with $j = 3, 4, 5T, 5L$) are given in terms of the exchange interactions a^i_{ech} (with $i = 1$ to 4) by

$$\begin{aligned}
E(\Gamma_3) &= 3a^1_{ech} - a^2_{ech} + 8a^3_{ech} + 2a^4_{ech} \\
E(\Gamma_4) &= 3a^1_{ech} + a^2_{ech} - 4a^3_{ech} + 2a^4_{ech} \\
E(\Gamma_{5T}) &= -5a^1_{ech} - (1/3)a^2_{ech} - (4/3)a^3_{ech} - (10/3)a^4_{ech} \\
E(\Gamma_{5L}) &= -5a^1_{ech} - (1/3)a^2_{ech} - (4/3)a^3_{ech} - (10/3)a^4_{ech} + \Delta_{LT}.
\end{aligned} \tag{4.3.26}$$

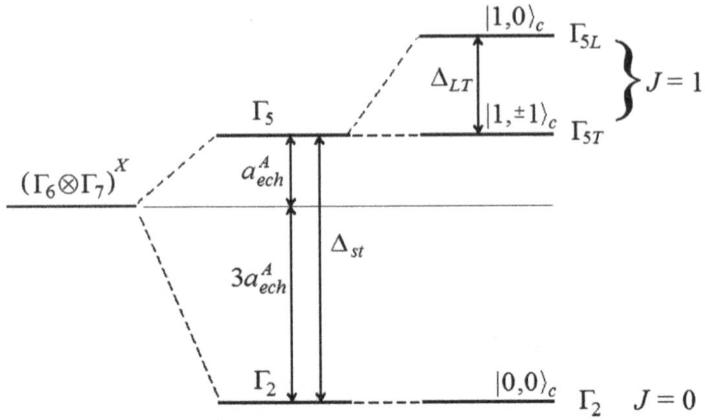

Fig. 4.3 Energy level scheme of excitons arising from holes in the valence band with Γ_7 symmetry and conduction-band electrons with Γ_6 symmetry in zincblende-type semiconductors. Two modes of manifestation of the exchange interaction are visualized. The singlet-triplet energy splitting Δ_{st} within this "split-off" exciton block $(\Gamma_6 \otimes \Gamma_7)^X$ is induced by the analytic (index "A") exchange interaction a^A_{ech}. The energies of the states and their dependence on the different contributions to the exchange interactions are given in Eq. (4.3.24) and Eq. (4.3.23), respectively. The non-analytic exchange interaction gives rise to the longitudinal-transverse splitting Δ_{LT}. $|1,+1\rangle_c$ and $|1,-1\rangle_c$ denote the two transverse exciton states, which are dipole-active and thus couple to the electromagnetic radiation field. $|1,0\rangle_c$ labels the longitudinal exciton state. The z-axis of quantization, which is used to identify the $|J, M_J\rangle_c$ states, is parallel to the wave-vector $\boldsymbol{Q}$. See text

In order to summarize the properties of the $\Gamma_6 \otimes \Gamma_8$ exciton states: When using the wave functions v^{ex}_{ie2} given in Eq. (4.3.18) or Eqs. (4.3.20) and (4.3.21), an effective exciton Hamiltonian formed from electrons in the conduction band transforming as Γ_6 and holes in the valence band with Γ_8 symmetry is described by a diagonal matrix. It includes the spin-orbit coupling and all different exchange interactions. The exciton ground state has still the full crystal symmetry, to which the wave functions are adapted. The same statement is valid for the split-off $\Gamma_6 \otimes \Gamma_7$ exciton states when using the states in Eq. (4.3.6) and Eq. (4.3.7). One has to remember, however, that some non-diagonal matrix elements due to the exchange interaction couple in principle the different exciton blocks. This coupling was neglected above and can be considered using perturbation theories. For dipole-active excitons such a coupling can lead to a transfer of oscillator strength between both exciton series.

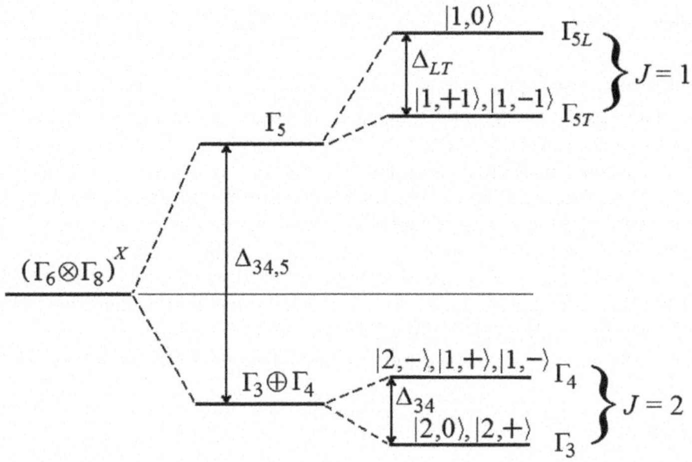

Fig. 4.4 Energy level scheme of excitons arising from holes in the valence band with Γ_8 symmetry and conduction-band electrons with Γ_6 symmetry in zincblende-type semiconductors. The energy splitting is induced by the different exchange interactions, giving rise to a separation $\Delta_{34,5}$ of states with (Γ_3 or Γ_4) and Γ_5 symmetry, the non-analytic (giving rise to the longitudinal-transverse splitting Δ_{LT}), and the cubic exchange (Δ_{34} splitting). The energies of the states and their dependence on the different contributions to the exchange interactions are given in Eq. (4.3.26). Concerning the states with Γ_5 symmetry, $|1, +1\rangle$ and $|1, -1\rangle$ denote the two transverse exciton states, which are dipole-active and thus couple to the electromagnetic radiation field. $|1, 0\rangle$ labels the longitudinal exciton state. The z-axis of quantization, which is used to identify the $|J, M_J\rangle$ states, is parallel to the wave-vector $\boldsymbol{Q}$. See text. In particular, for the notation of the Γ_3 or Γ_4 states see Eqs. (4.3.20) and (4.3.21)

Fig. 4.5 Normal reflectance spectrum of CuBr in the $\Gamma_6 \otimes \Gamma_8$ exciton region. Here Δ_{st} = 1.1 meV means the singlet-triplet $[(\Gamma_5 - (\Gamma_3, \Gamma_4)]$ splitting due to exchange interaction. Besides, it can be seen that the "cubic-exchange" terms and the corresponding splitting Δ_{34} of exciton states with Γ_3 and Γ_4 symmetry are negligible. On the other hand, the longitudinal-transverse splitting Δ_{LT} = 12.3 meV of the Γ_5 (i.e. J = 1) state is considerably larger here than in most of the common semiconductors. After Ref. [8]. See text

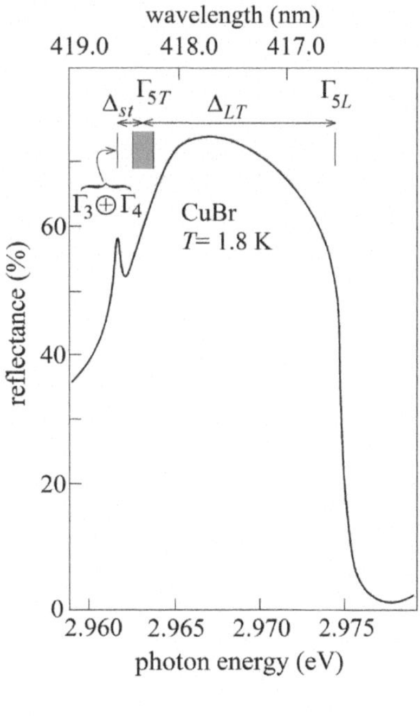

References

1. Cho, K.: Phys. Rev. B **14**, 4463 (1976)
2. Koster, J.F., Dimmock, J.O., Wheeler, R.G., Statz, H.: Properties of the Thirty-two Point Groups. MIT Press, Cambridge, Mass. (1963)
3. Spenke, E.: Elektronische Halbleiter. Springer-Verlag, Berlin (1956)
4. Cho, K.: Excitons, Topics in Current Physics 14, Springer Verlag, Berlin, Heidelberg (1979)
5. Hanamura, E., Inoue M.: Physics of highly excited states in solids. In: Ueta, M., Nishina, Y. (eds.) Lecture Notes in Physics vol. 57, p. 25. Springer-Verlag, Berlin, Heidelberg (1976)
6. Ueta, M., Nagasawa, N.: In: Physics of highly excited states in solids. Ueta, M., Nishina, Y. (eds.) Lecture Notes in Physics vol. 57, p. 1. Springer-Verlag, Berlin, Heidelberg (1976)
7. Goto, T., Takahashi, T., Ueta, M.: J. Phys. Soc. Jpn **24**, 314 (1968)
8. Wecker, C., Daunois, A., Deiss, J.L., Florini, P., Merle, J.C.: Solid State Commun. **31**, 649 (1979)

Chapter 5
Pseudo-Spin Development of the Exciton Ground State in Zincblende-Type Semiconductors

We have seen in the foregoing chapters that the development of an effective Hamiltonian, which is adapted to the crystal symmetry, is an efficient tool to describe electronic excitations in semiconductors. The development procedure becomes, however, quite tedious and inefficient if a large basis of electron states is considered when describing a system. On the contrary, this technique is interesting if only a couple of almost degenerate states, which are well separated in energy from other states, has to be analyzed. This is the case for example in the exciton problem that has been discussed in the previous chapter. Here, the split-off exciton states (resulting from electrons and holes that transform as Γ_6 and Γ_7, respectively) are separated by the spin-orbit coupling from the exciton block obtained from electrons and holes that transform as Γ_6 and Γ_8, respectively. As we have already discussed, the two exciton blocks are coupled by the electron-hole exchange interaction, which is, however, small compared to the spin-orbit coupling. In this case, the exciton blocks are approximately decoupled and different effective Hamiltonians that are independent from each other can be developed for the two exciton blocks. This will be done in the present chapter. We understand this as a "Pseudo-Spin Development" of excitons in a subspace of almost degenerated states.

5.1 The $\Gamma_6 \otimes \Gamma_7$ Subspace of the Exciton Ground State

Let us now discuss the simplest case of a two-band bulk semiconductor, where the lowest conduction band of Γ_6 symmetry is only spin degenerate. As we have seen in Sect. 3.3, the sixfold degenerate highest valence band is split by spin-orbit interaction into two bands having Γ_7 and Γ_8 symmetry. Let the uppermost valence band be the band with Γ_7 symmetry, which is also only twofold degenerate. The semiconductor has zincblende structure (T_d point-group symmetry) and a direct band gap at the center of the Brillouin zone. This situation is realized in CuCl to which our discussion fully applies.

© Springer International Publishing AG, part of Springer Nature 2018
B. Hönerlage and I. Pelant, *Symmetry and Symmetry-Breaking in Semiconductors*, Springer Tracts in Modern Physics 279,
https://doi.org/10.1007/978-3-319-94235-3_5

In copper halides the considered exciton series is labeled the "Z_3-exciton series". The spin degenerate lowest conduction band and the highest valence band with Γ_7 symmetry are considered in the following, all other bands (including the valence band with Γ_8 symmetry together with the appertaining ($\Gamma_6 \otimes \Gamma_8$) exciton series called "$Z_{1,2}$-exciton series") are neglected. The overall situation relevant for optical transitions is sketched in the right half panel of Fig. 3.9, where the left half presents part of the corresponding excitonic absorption spectrum of a high-quality CuCl thin film.

In the case considered here the electron states in the conduction band are made up from s-type spin orbitals, i.e. they have an angular momentum $l_e = 0$. The total-angular momentum of the electron states is therefore fully determined by the electron spin. Because of this twofold degeneracy, the total angular momentum operator $\boldsymbol{j}_e = 1/2$ is now used for the pseudo-spin development of the conduction-band electron-subspace. The eigenfunctions of its z-component j_e^z are the electron spin-up and down states α_e and β_e, respectively.

As discussed above, the wave functions of the valence-band electrons contain, besides the spin contribution, an angular-momentum part transforming as (x, y, z). This originates from the fact that the valence band is due to atomic-spin orbitals with a spin $\sigma_v = 1/2$ and an angular momentum $l_v = 1$. The spin states are labeled α and β and indicate again the spin-up and down states of the valence-band defect-electrons, respectively. The eigenvectors (v_1^v and v_2^v) of the valence band transforming as Γ_7 are given in Eq. (3.3.1) by

$$\begin{aligned} v_1^v &= (-|0\rangle\beta + \sqrt{2}|-1\rangle\alpha)/\sqrt{3} = (-z\beta + (x - iy)\alpha)/\sqrt{3} \\ v_2^v &= (|0\rangle\alpha - \sqrt{2}|1\rangle\beta)/\sqrt{3} = (z\alpha + (x + iy)\beta)/\sqrt{3}. \end{aligned} \tag{5.1.1}$$

Considering the total-angular momentum $\boldsymbol{j}_v = \boldsymbol{l}_v \oplus \boldsymbol{\sigma}_v$ and its projection component onto the z-axis j^z, the functions Eq. (5.1.1) are also eigenfunctions of the total-angular momentum operator $\boldsymbol{j}_v$ of the valence-band states with $j_v^z = \mp 1/2$, respectively. By analogy to the conduction band discussed above, because of its twofold degeneracy, the total angular-momentum operator $\boldsymbol{j}_v = 1/2$ is now used for the pseudo-spin description of the valence-band subspace. Using Eq. (4.1.3) or Eq. (4.1.4) we can then introduce hole states and establish their correspondence to the valence-band states of the split-off band using Eq. (4.1.5).

As discussed in Chap. 4, excitons are formed in the direct product space of electron and hole states, i.e. we build the Kronecker product $\boldsymbol{j}_v \otimes \boldsymbol{j}_e$. The exciton ground state is thus only fourfold degenerate. We now construct in this subspace an effective exciton Hamiltonian, which has the same symmetry properties as the full Hamilton operator, i.e. which remains invariant under all symmetry operations of the point group of the crystal. It transforms as a scalar (which has Γ_1 symmetry) and is an even function under time reversal (K^+). Since the conduction band is only twofold degenerate, we choose, as discussed in Chap. 2, the Pauli-spin matrices, which are the $j = 1/2$ matrices σ_e^x, σ_e^y, and σ_e^z together with the unit matrix 1_e as a basis to span the matrix describing the interacting conduction-band states. These matrices are given in Eqs. (2.2.3) and (2.2.5) by

$$\sigma_e^x = \begin{pmatrix} 0 & 1 \\ 1 & 0 \end{pmatrix}; \sigma_e^y = \begin{pmatrix} 0 & -i \\ i & 0 \end{pmatrix}; \sigma_e^z = \begin{pmatrix} 1 & 0 \\ 0 & -1 \end{pmatrix}; 1_e = \begin{pmatrix} 1 & 0 \\ 0 & 1 \end{pmatrix}. \tag{5.1.2}$$

We proceed likewise in order to describe the Γ_7 valence-band states and use for these states the Pauli-spin matrices σ_{v7}^x, σ_{v7}^y, and σ_{v7}^z and the unit matrix 1_{v7} as basis matrices, i.e.

$$\sigma_{v7}^x = \begin{pmatrix} 0 & 1 \\ 1 & 0 \end{pmatrix}; \sigma_{v7}^y = \begin{pmatrix} 0 & -i \\ i & 0 \end{pmatrix}; \sigma_{v7}^z = \begin{pmatrix} 1 & 0 \\ 0 & -1 \end{pmatrix}; 1_{v7} = \begin{pmatrix} 1 & 0 \\ 0 & 1 \end{pmatrix}. \tag{5.1.3}$$

These two sets of Pauli matrices operate only on the conduction (subscript "e") or valence (subscript "$v7$") split-off band states, respectively. The transformation properties of these matrices are (Γ_1, K^+) in the case of the unit matrix and (Γ_4, K^-) for the Pauli matrices, respectively. The Kronecker product of conduction- and valence-band electron-states spans a basis of the fourfold degenerate exciton ground state.

Excluding all symmetry breaking interactions, the effective exciton Hamiltonian H_7^{ex} now reads in this pseudo-spin space:

$$H_7^{ex} = \Delta_{07} 1_{v7} \otimes 1_e + \Delta_{17}(\sigma_{v7}^x \otimes \sigma_e^x + \sigma_{v7}^y \otimes \sigma_e^y + \sigma_{v7}^z \otimes \sigma_e^z). \tag{5.1.4}$$

Equation (5.1.4) gives the only terms that can contribute to the Hamiltonian in this subspace, which has a lower dimensionality compared to that treated in Sect. 4.1. No other terms that have the required transformation properties (Γ_1, K^+) for a Hamiltonian under the symmetry operations of the point group and under time reversal can be constructed from the basis matrices given in Eqs. (5.1.2) and (5.1.3). In analogy with Eqs. (4.1.8) and (5.1.4) gives the fine structure of the split-off exciton states in a parametrized form. It depends only on two parameters: The first one, Δ_{07}, accounts for the exciton binding energy. It is determined by H_d^{ex} of Eq. (4.1.9) and a part of H_{so}^{ex} of Eq. (4.1.10), since now the spin-orbit interaction is already considered. Similarly, the second parameter Δ_{17} stands for the different electron-hole exchange interaction terms discussed in Sect. 4.3. One remarks that only one parameter describes the electron-hole exchange interaction now, which acts within the states of the split-off band.

Explicitly, the Hamiltonian describing the conduction- and valence-band electron-exchange interaction

$$H_{ech7}^{ex} = \Delta_{17}(\sigma_{v7}^x \otimes \sigma_e^x + \sigma_{v7}^y \otimes \sigma_e^y + \sigma_{v7}^z \otimes \sigma_e^z)$$

takes (as can be easily calculated by using Eqs. (5.1.2) and (5.1.3)) the matrix form

$$H^{ex}_{ech7} = \Delta_{17} \begin{pmatrix} 1 & 0 & 0 & 0 \\ 0 & -1 & 2 & 0 \\ 0 & 2 & -1 & 0 \\ 0 & 0 & 0 & 1 \end{pmatrix}. \tag{5.1.5}$$

It is important to notice that the matrix form of H^{ex}_{ech7} given in Eq. (5.1.5) reproduces exactly (up to constant factors) the matrix forms of the diagonal Γ_7 exciton blocks of the full Hamiltonians H^{ex}_{ech1} to H^{ex}_{ech4} that are given in Eqs. (4.3.2)–(4.3.5). This means that to include all valence-band states when treating the electron-hole exchange interaction introduces only a coupling between the different diagonal blocks. The additional terms of the exchange interaction do not, however, change the internal structure of the exciton block, which is separated from the other block by the spin-orbit coupling.

As we have seen, the excitons transform like Γ_2 and Γ_5, respectively. As discussed in connection with Eq. (4.3.9) for the split-off exciton block, the exchange interaction given in Eq. (5.1.5) can be diagonalized. As in Eq. (4.3.9) one then obtains the eigenvalues

$$E_{sti7} = \Delta_{17}(-3, 1, 1, 1) \tag{5.1.6}$$

with $i = (1 \text{ to } 4)$, respectively. The factor Δ_{17} of E_{sti7} in Eq. (5.1.6) has a different meaning than the coefficient a_{ech} in Eq. (4.3.8): In Sect. 4.3 a_{ech} is given in terms of the different electron-hole exchange interaction terms of the full exciton states. Here, Δ_{17} is a parameter defined only in the electron-hole subspace considered: the split-off exciton ground state.

The exciton states in Eq. (5.1.6) can be classified according to their total angular momentum (or pseudo-spin) $J = 0$ and $J = 1$. The spin-triplet-exciton state, which does not carry a dipole moment, corresponds to the eigenstate of pseudo-spin $J = 0$ and is not degenerate. The spin-singlet exciton states correspond to eigenstates with a pseudo-spin $J = 1$ and are threefold degenerate.

According to Eqs. (4.3.10) and (4.3.11) and using the notation $|J, M_J\rangle$, one obtains again for these exciton states in the electron-hole representation:

$$|0, 0\rangle_{c7} = (|\Phi_4\rangle + |\Phi_1\rangle)/\sqrt{2} \tag{5.1.7}$$

and

$$\begin{aligned} |1, 1\rangle_{c7} &= |\Phi_2\rangle \\ |1, 0\rangle_{c7} &= (|\Phi_4\rangle - |\Phi_1\rangle)/\sqrt{2} \\ |1, -1\rangle_{c7} &= -|\Phi_3\rangle. \end{aligned} \tag{5.1.8}$$

The calculation of the energies E_{sti7} was discussed in detail in Sect. 4.3. We have mentioned that their values depend on the dipole moment of the state and its orientation relative to the direction of the exciton propagation, given by their center-of-mass wave-vector $\boldsymbol{Q}$. Let us take this direction as quantization axis "z". The exciton states with Γ_2 symmetry do not carry a dipole moment and are not affected by exchange interaction. Concerning the $J = 1$ exciton states with Γ_5 symmetry, which posses a dipole moment, the analytical part of the exchange interaction is independent of the orientation of the dipole and lifts only the degeneracy between excitons with Γ_2 and Γ_5 symmetry, i.e. induces the singlet-triplet splitting. The non-analytic contribution of the exchange interaction, however, affects only longitudinal excitons and lifts the degeneracy of the longitudinal and transverse Γ_5 states [1, 2].

Let us separate the exchange interaction term Δ_{17} into its analytic part Δ_{17}^a and the longitudinal-transverse splitting Δ_{17}^{LT} in the form

$$\Delta_{17} = \Delta_{17}^a + \Delta_{17}^{LT} . \tag{5.1.9}$$

When using the symmetry-adapted exciton-basis functions ($|0, 0\rangle_{c7}$, $|1, 1\rangle_{c7}$, $|1, 0\rangle_{c7}$, $|1, -1\rangle_{c7}$) given in Eqs. (5.1.7) and (5.1.8), the exciton state $|1, 0\rangle_{c7}$ corresponds (as it has been explained in Sect. 4.3) to the longitudinal exciton. The Hamiltonian given in Eq. (5.1.5) is diagonal in this basis and takes the matrix form:

$$H_7^{ex} = \begin{pmatrix} \Delta_{07} & 0 & 0 & 0 \\ 0 & \Delta_{07} + 4\Delta_{17}^a & 0 & 0 \\ 0 & 0 & \Delta_{07} + 4\Delta_{17}^a + \Delta_{17}^{LT} & 0 \\ 0 & 0 & 0 & \Delta_{07} + 4\Delta_{17}^a \end{pmatrix} . \tag{5.1.10}$$

Using the present approach one obtains for the singlet-triplet splitting Δ_{st7} between exciton states

$$\Delta_{st7} = 4\Delta_{17}^a \tag{5.1.11}$$

from the analytical part of the exchange interaction. The splitting between longitudinal and transverse states is resulting from its non-analytical part. As in simple atoms, the electron-hole exchange interaction is a repulsive interaction in this subspace and therefore Δ_{17}^a and Δ_{17}^{LT} are ≥ 0.

Comparing with Sect. 4.3 one sees from this discussion that the reduction of the number of valence band states has simplified a lot the determination of the effective Hamiltonian, while the energy level scheme remains the same as given in Fig. 4.3. The Hamiltonian describes, however, only a reduced exciton subspace.

5.2 Symmetry-Breaking Effects in the $\Gamma_6 \otimes \Gamma_7$ Exciton Ground State

Let us discuss now symmetry-breaking effects on the exciton ground state, which is defined in the $\Gamma_6 \otimes \Gamma_7$ electron-hole subspace as introduced in the preceding section. We have mentioned that direct, exchange, and spin-orbit interactions between electrons and holes possess the full crystal symmetry. These interactions can be accounted for by using symmetry-adapted exciton-wave functions as basis functions in which the exciton Hamiltonian is diagonal.

In general an exciton Hamiltonian depends, however, on an additional set of physical quantities, which break the full point-group symmetry of the crystal. Changing the value of these quantities modifies not only the exciton eigenvalues and the exciton-fine structure, but also the exciton-wave functions.

The procedure to develop the exciton Hamiltonian may now be continued, considering symmetry-breaking perturbations. Such perturbations can be treated in detail in the frame of the pseudo-spin formalism. One thus obtains an insight in the effect of the perturbations on the different exciton states: Symmetry-breaking perturbations can mix the exciton states and lead to a shift of their energies.

Let us consider in the following some of these physical quantities and discuss the influence of their symmetry properties on the exciton states.

5.2.1 Magnetic-Field Dependence

According to the transformation properties (Γ_4, K^-) of the magnetic field $\boldsymbol{B}$ in T_d point-group symmetry, the linear magnetic-field dependence of the Hamiltonian, labeled H_{B1}^{ex7} in the $\Gamma_6 \otimes \Gamma_7$ electron-hole subspace, takes the form

$$\begin{aligned} H_{B1}^{ex7} &= g^c 1_{v7} \otimes (\boldsymbol{B} \cdot \boldsymbol{\sigma_e}) + g^{v7} (\boldsymbol{B} \cdot \boldsymbol{\sigma_{v7}}) \otimes 1_e = \\ &= g^c 1_{v7} \otimes (B_x \sigma_e^x + B_y \sigma_e^y + B_z \sigma_e^z) + g^{v7} (B_x \sigma_{v7}^x + B_y \sigma_{v7}^y + B_z \sigma_{v7}^z) \otimes 1_e, \end{aligned} \tag{5.2.1}$$

where g^c and g^{v7} are the Landé (or g-)factors of electrons and holes, respectively. The energy shifts that they induce correspond to the linear Zeeman effect. The interaction terms have the required symmetry properties of a Hamiltonian and lead to a splitting of the degenerate conduction and valence bands.

Since the operators $\boldsymbol{B}$, $\boldsymbol{\sigma_v}$, and $\boldsymbol{\sigma_e}$ have all odd symmetry under time reversal, their product $(\boldsymbol{B} \cdot \boldsymbol{\sigma_v} \otimes \boldsymbol{\sigma_e})$ has also odd time-reversal symmetry and, consequently, does not appear in Eq. (5.2.1). Thus, the electron-hole exchange interaction cannot depend linearly on the magnetic field and it does not influence the linear Zeeman effect when restricting to this subspace. This indicates that measurements of the exciton Zeeman effect lead directly to a combination of the electron and hole Landé factors, which are not modified by the exchange interaction. Terms linear in $\boldsymbol{B}$ couple to the electron or hole-spin states and can lead to a spin flip of one of the carriers, while the other particle remains in its former state.

All second order terms in $\boldsymbol{B}$ are symmetric under time reversal (K^+ symmetry). They contain direct and exchange interactions according to

$$\begin{aligned} H_{B2}^{ex7} = \beta_0 B^2 1_{v7} \otimes 1_e + \beta_1 B^2(\sigma_{v7}^x \otimes \sigma_e^x + \sigma_{v7}^y \otimes \sigma_e^y + \sigma_{v7}^z \otimes \sigma_e^z) + \\ + \beta_2[3(\sigma_{v7}^x \otimes \sigma_e^x - \sigma_{v7}^y \otimes \sigma_e^y)(B_x^2 - B_y^2) + \\ (2\sigma_{v7}^z \otimes \sigma_e^z - \sigma_{v7}^x \otimes \sigma_e^x - \sigma_{v7}^y \otimes \sigma_e^y)(2B_z^2 - B_x^2 - B_y^2)] + \\ + \beta_3[(\sigma_{v7}^y \otimes \sigma_e^z + \sigma_{v7}^z \otimes \sigma_e^y)B_y B_z + c.p.], \end{aligned} \tag{5.2.2}$$

where c. p. stands for cyclic permutation. The coefficients β_i are constants.

H_{B1}^{ex} + H_{B2}^{ex} describe all interactions varying up to the second order in the magnetic field $\boldsymbol{B}$. The term proportional to β_0 gives rise to the quadratic Zeeman effect, the terms proportional to β_1 to β_3 are due to the magnetic field dependence of the exchange interaction. While the term proportional to β_1 is independent of the field orientation with respect to the crystal axis, the symmetry is broken by the terms proportional to β_2 and β_3. These terms may mix different exciton states at finite magnetic fields and depend on the orientation of the magnetic field with respect to the crystal axis. They then describe a simultaneous spin-flip of electrons and holes, induced by the terms quadratic in $\boldsymbol{B}$. Similarly, all odd orders of $\boldsymbol{B}$ lead to one-particle spin flips, all even orders to simultaneous electron-hole spin flips. In addition, all higher order terms lead to the same coupling schemes between the states as those given in Eqs. (5.2.1) and (5.2.2).

5.2.2 *Wave-Vector Dependent Interactions*

Concerning the wave-vector $\boldsymbol{Q}$ dependent terms (transforming as (Γ_5, K^-)) or its power of n-th order noted $\boldsymbol{Q}^n$, their structure and importance are discussed in detail in Ref. [3]. We give it here for completeness. A term linear in $\boldsymbol{Q}$ is absent in the Hamiltonian since it is forbidden by symmetry. The term quadratic in $\boldsymbol{Q}^2$ reads:

$$\begin{aligned} H_{Q2}^{ex7} = G_0 Q^2 1_{v7} \otimes 1_e + \delta_1 Q^2(\sigma_{v7}^x \otimes \sigma_e^x + \sigma_{v7}^y \otimes \sigma_e^y + \sigma_{v7}^z \otimes \sigma_e^z) + \\ + \delta_2[3(\sigma_{v7}^x \otimes \sigma_e^x - \sigma_{v7}^y \otimes \sigma_e^y)(Q_x^2 - Q_y^2) + \\ (2\sigma_{v7}^z \otimes \sigma_e^z - \sigma_{v7}^x \otimes \sigma_e^x - \sigma_{v7}^y \otimes \sigma_e^y)(2Q_z^2 - Q_x^2 - Q_y^2)] + \\ + \delta_3[(\sigma_{v7}^y \otimes \sigma_e^z + \sigma_{v7}^z \otimes \sigma_e^y)Q_y Q_z + c.p.]. \end{aligned} \tag{5.2.3}$$

The direct exciton effective mass is isotropic, but an anisotropy may arise due to the wave-vector dependence of the exchange interaction. The exciton center of mass motion is governed by the effective exciton mass M_{ex}, which is simply given by the sum of the electron effective mass in the conduction band (m_e) and the effective hole mass (m_h). Thus G_0 is given by

$$G_0 = \hbar^2/2M_{ex} = \hbar^2/2(m_e + m_h). \tag{5.2.4}$$

It is interesting to notice that, although $\boldsymbol{B}$ and $\boldsymbol{Q}$ have different spatial transformation properties, the interaction terms given in Eqs. (5.2.2) and (5.2.3) have the same structure, leading therefore to the same coupling scheme of the exciton states.

Similar to β_2 and β_3, the point-group symmetry is broken by the exchange-interaction terms proportional to δ_2 and δ_3. These terms describe simultaneous spin-flips of electrons and holes, which are induced by the finite wave-vector and depend on the propagation direction of the exciton.

Similar to the electron or hole g-factor, the term discussed in Eq. (2.5.6) is a term cubic in $\boldsymbol{Q}$, which reads

$$H_{Q3}^{ex7} = K_{v7}((Q_x^2 - Q_y^2)Q_z\sigma_{v7}^z + c.p.) \otimes 1_e + K_e 1_{v7} \otimes ((Q_x^2 - Q_y^2)Q_z\sigma_e^z + c.p.), \tag{5.2.5}$$

K_e and K_{v7} being again arbitrary constants. As pointed out by Dresselhaus in Ref. [4] this term is due to the wave-vector dependence of the spin-orbit coupling in crystals with zincblende structure. For a fixed wave-vector $\boldsymbol{Q}$ Eq. (5.2.5) has the same structure as Eq. (5.2.1). Therefore, the $\boldsymbol{Q}^2$ and $\boldsymbol{Q}^3$ terms in Eqs. (5.2.3) and (5.2.5) may be looked upon as an effective magnetic field. In contrast to H_{B1}^{ex7}, however, the H_{Q3}^{ex7} terms lead to an intrinsic coupling of the different hole- (and therefore) exciton-spin states.

5.2.3 Electric-Field and Strain-Dependent Interactions

Let us denote by ϵ_{ij} the components of the strain tensor. They transform, similar to electric fields components, as (Γ_5, K^+) in crystals with T_d point-group symmetry. Since the Pauli-spin matrices $\boldsymbol{\sigma}_{v7}$ and $\boldsymbol{\sigma}_e$ are odd functions under time reversal (K^-), interaction terms proportional to only one of the spin matrices $\boldsymbol{\sigma}_{v7}$ or $\boldsymbol{\sigma}_e$ and linear in E or ϵ_{ij} (with $i \neq j$) cannot occur. Thus, single spin-flip processes cannot show up but the exciton spin has to be returned in a block by linear electric-field or strain perturbations. Again, the electron-hole exchange-interaction leads to terms, which have the same structure as Eq. (5.2.3). That is, in the case of strains

$$\begin{aligned} H_{eij}^{ex7} = \gamma_0(\epsilon_{xx} + \epsilon_{yy} + \epsilon_{zz})1_{v7} \otimes 1_e + \gamma_1(\epsilon_{xx} + \epsilon_{yy} + \epsilon_{zz})(\sigma_{v7}^x \otimes \sigma_e^x + \\ \sigma_{v7}^y \otimes \sigma_e^y + \sigma_{v7}^z \otimes \sigma_e^z) + \\ + \gamma_2[3(\sigma_{v7}^x \otimes \sigma_e^x - \sigma_{v7}^y \otimes \sigma_e^y)(\epsilon_{xx} - \epsilon_{yy}) + \\ (2\sigma_{v7}^z \otimes \sigma_e^z - \sigma_{v7}^x \otimes \sigma_e^x - \sigma_{v7}^y \otimes \sigma_e^y)(2\epsilon_{zz} - \epsilon_{xx} - \epsilon_{yy})] + \\ + \gamma_3[(\sigma_{v7}^y \otimes \sigma_e^z + \sigma_{v7}^z \otimes \sigma_e^y)\epsilon_{yz} + c.p.]. \end{aligned} \tag{5.2.6}$$

Fig. 5.1 demonstrates the effect of applied stress on the $\Gamma_6 \otimes \Gamma_7$ excitons in CuCl.

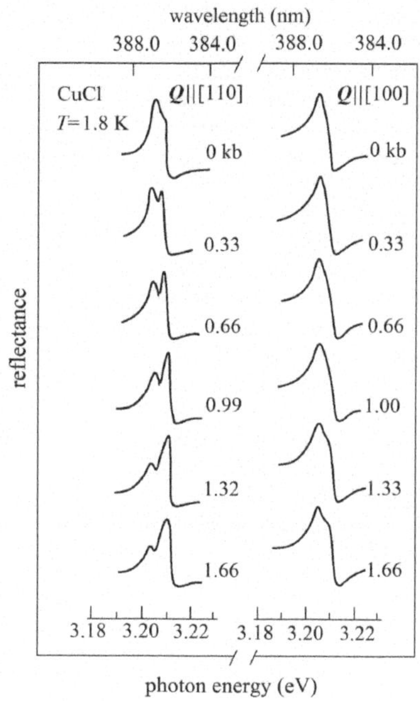

Fig. 5.1 Splitting of the $\Gamma_6 \otimes \Gamma_7$-exciton states ($Z_3$-exciton ground state) in uniaxially deformed zincblende CuCl single crystals, as revealed in stress-induces changes of the optical reflection spectra. In the left part the stress $\boldsymbol{P}$ was applied in the [001]-direction and the wave-vector of the incident light was $\boldsymbol{Q} \parallel [110]$, while for the right part $\boldsymbol{P} \parallel [001]$ and $\boldsymbol{Q} \parallel [100]$ hold. The polarization vector of the incident light $\boldsymbol{e} \parallel \boldsymbol{P}$ in both cases. The values of applied stress (in kb) are indicated at each curve. Adapted after Ref. [8]

Since $E_i E_j$ and ϵ_{ij} have the same transformation properties, the interaction terms quadratic in $\boldsymbol{E}^2$ have the same structure as those given in Eq. (5.2.6) for the components of the strain tensor. One has only to replace the strain-tensor elements ϵ_{ij} by $E_i E_j$. The corresponding constants are named α_i. In addition, there exists a term linear in $\boldsymbol{E}$, which takes the form

$$H_{E1}^{ex7} = \alpha_3[(\sigma_{v7}^y \otimes \sigma_e^z + \sigma_{v7}^z \otimes \sigma_e^y)E_x + c.p.]. \tag{5.2.7}$$

Notice that, as discussed above, a similar term linear in the wave-vector $\boldsymbol{Q}$ does not show up. The difference is due to the different transformation properties of both interactions under time reversal. Thus, the exchange interaction may vary linearly with an external or internal electric field. The experimentally observed effect of an external electric field on optical absorption spectra of CuCl is shown in Fig. 5.2.

Equations (5.2.6) and (5.2.7) give the complete set of perturbations linear in E and ϵ_{ij} and quadratic in $E_i E_j$. Obviously, there exist also exchange terms bilinear in wave-vector, magnetic field, electric field, and strain, etc., in which we are not further interested here.

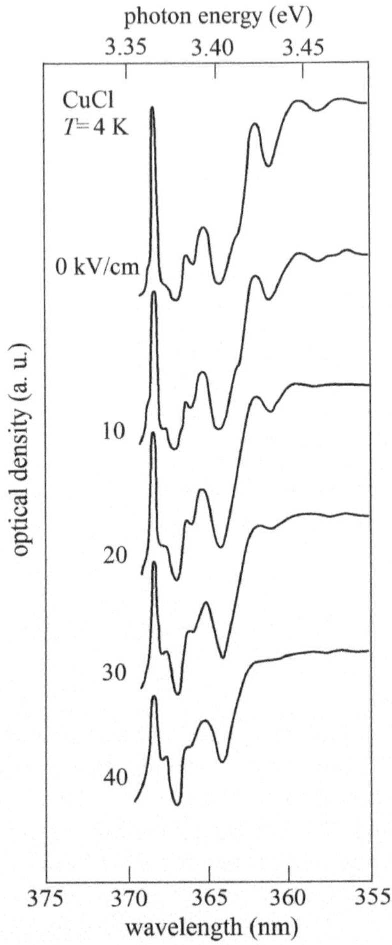

Fig. 5.2 Electric-field symmetry-breaking effects demonstrated by $\Gamma_6 \otimes \Gamma_7$ exciton absorption spectra of CuCl. The most distinct exciton line at ~368.5 nm belongs to the Z_3-series, $n = 2$ state (as can be recognized e.g. via comparison with Fig. 3.9). Both, an overall red shift of the absorption continuum (due to diminution of the exciton binding energy) and a splitting of the $n = 2$ line (Stark effect) can be clearly seen. Approximate values of applied electric-field are indicated at each curve. Adapted after Ref. [9]

5.3 The $\Gamma_6 \otimes \Gamma_8$ Subspace of the Exciton Ground State

Analogously to the split-off excitons discussed above we will now consider the remaining states of the exciton ground state that are defined in the $\Gamma_6 \otimes \Gamma_8$ subspace. We recall that the spin degenerate lowest conduction band has Γ_6 symmetry and the uppermost valence band Γ_8 symmetry. The split-off valence band is supposed to have an energy very different from that of the Γ_8 valence band and is not further considered. Again, all other bands are also neglected. α_e and β_e denote the electron spin-up and down states, respectively, and only the valence band states v_3^v to v_6^v of those in Eq. (3.2.13) or Eq. (3.2.14) are considered in the following. They are eigenstates of the angular momentum operator with $\boldsymbol{j}_v = 3/2$. The exciton ground state is formed from theses states. This situation is realized in most simple binary III-V, II-VI, and

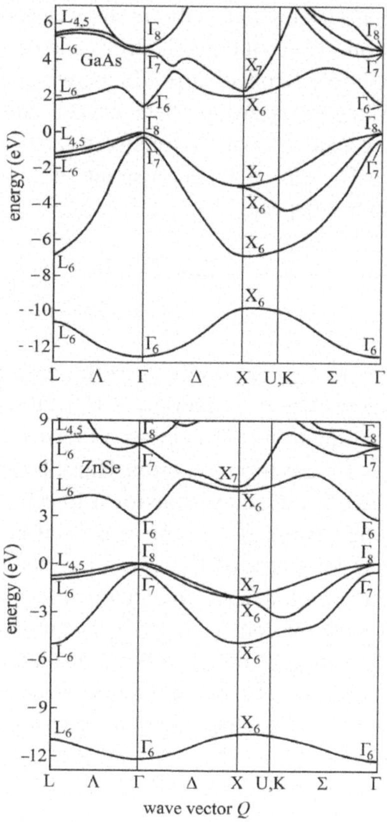

Fig. 5.3 Calculated energy band structure of GaAs and ZnSe crystals along the principal symmetry directions in the first Brillouin zone. According to Ref. [10]

I-VII semiconductors with zincblende structure (as e.g. GaAs, ZnSe, ZnTe or CuBr etc.). Two examples (for GaAs and ZnSe) are shown in Fig. 5.3. In copper halides this exciton series is called the "Z_{12}-exciton series" to which this discussion nicely applies since the spin-orbit splitting is large compared to the exchange interaction.

The valence-band wave-functions (v_3^v to v_6^v) are eigenfunctions of the total angular momentum operator $\boldsymbol{j}_v = 3/2$, with eigenvalues for $j_v^z = \mp 1/2$ and $j_v^z = \mp 3/2$, respectively. As already repeatedly stressed, using Eq. (4.1.3) or Eq. (4.1.4) one can introduce hole states and establish their correspondence to the valence-band states by Eq. (4.1.5).

We first construct the invariant representation of the effective exciton Hamiltonian, which acts in the eightfold degenerate subspace that is spanned by the valence-band states with Γ_8 symmetry and the conduction-band states with Γ_6 symmetry. We start from the symmetry adapted matrices given in Eq. (3.4.5) for the valence-band states and the set of Pauli matrices given in Eq. (5.1.2) for the conduction-band states, respectively. The transformation properties of the basis matrices for the valence-band states are given in Eq. (3.4.3), those for the conduction band are (Γ_1, K^+) in the case of the unit matrix 1_e and (Γ_4, K^-) for the Pauli matrices. The two sets of matrices operate only on the valence- and conduction-band states, respectively, and span a basis of both electron- and defect-electron subspaces. As above, we build their possible Kronecker products, which remain invariant under all symmetry operations of the point group of the crystal and which are even functions with respect to time reversal, i.e. which transforms as (Γ_1, K^+).

The effective Hamiltonian of the exciton ground states H_8^{ex} reads in this pseudo-spin subspace:

$$\begin{aligned} H_8^{ex} = \Delta_{08} 1_{v8} \otimes 1_e + \Delta_{18}(j_v^x \otimes \sigma_e^x + j_v^y \otimes \sigma_e^y + j_v^z \otimes \sigma_e^z) + \\ + \Delta_{28}\left[(j_v^x)^3 \otimes \sigma_e^x + (j_v^y)^3 \otimes \sigma_e^y + (j_v^z)^3 \otimes \sigma_e^z\right], \end{aligned} \tag{5.3.1}$$

where 1_{v8} denotes the four-dimensional unit matrix of the valence sub-band with Γ_8 symmetry. It is given by 4/15 of the matrix S_{v8} in Eq. (3.4.3).

Equation (5.3.1) gives all terms that have the full point-group symmetry and can contribute to the exciton Hamiltonian in the $\Gamma_6 \otimes \Gamma_8$ subspace. As we have seen, the exciton states transform like Γ_3, Γ_4 and Γ_5. H_8^{ex} has a reduced dimensionality when compared to the full exciton Hamiltonian given in Eq. (4.1.8). It gives the fine structure of these exciton states in a parametrized form. It depends only on three parameters: Δ_{08} that accounts for the exciton binding energy (including now a contribution from the spin-orbit interaction) and Δ_{18} and Δ_{28}, which stand for the different electron-hole exchange interaction terms discussed in Sect. 4.3. One remarks that two parameters describe the electron-hole exchange interaction here while four are necessary to describe it using the complete set of valence-band states. Explicitly, the Hamiltonian describing the conduction- and valence-band electron exchange interaction $H_{ech8}^{ex} = H_{ech18}^{ex} + H_{ech28}^{ex}$ is given by

$$H_{ech18}^{ex} = \Delta_{18}(j_v^x \otimes \sigma_e^x + j_v^y \otimes \sigma_e^y + j_v^z \otimes \sigma_e^z) \tag{5.3.2}$$

and

$$H_{ech28}^{ex} = \Delta_{28}\left[(j_v^x)^3 \otimes \sigma_e^x + (j_v^y)^3 \otimes \sigma_e^y + (j_v^z)^3 \otimes \sigma_e^z\right]. \tag{5.3.3}$$

Let us first investigate the Hamiltonian H_{ech18}^{ex}. When performing the direct products indicated in Eq. (5.3.2) by using Eqs. (3.4.1) and (2.2.3), H_{ech18}^{ex} takes the matrix form

$$H_{ech18}^{ex} = \Delta_{18} \begin{pmatrix} -3/2 & 0 & 0 & \sqrt{3} & 0 & 0 & 0 & 0 \\ 0 & 3/2 & 0 & 0 & 0 & 0 & 0 & 0 \\ 0 & 0 & -1/2 & 0 & 0 & 2 & 0 & 0 \\ \sqrt{3} & 0 & 0 & 1/2 & 0 & 0 & 0 & 0 \\ 0 & 0 & 0 & 0 & 1/2 & 0 & 0 & \sqrt{3} \\ 0 & 0 & 2 & 0 & 0 & -1/2 & 0 & 0 \\ 0 & 0 & 0 & 0 & 0 & 0 & 3/2 & 0 \\ 0 & 0 & 0 & 0 & \sqrt{3} & 0 & 0 & -3/2 \end{pmatrix}. \tag{5.3.4}$$

The exchange interaction given in Eq. (5.3.4) can be diagonalized and one then obtains the eigenvalues E_{sti18}

$$E_{sti18} = \Delta_{18}(-5/2, -5/2, -5/2, 3/2, 3/2, 3/2, 3/2, 3/2) \tag{5.3.5}$$

with $i = (1 \text{ to } 8)$, respectively.

The states 1 to 3 are triply degenerate, corresponding to the $J = 1$ states. As discussed above, they have Γ_5 symmetry. States 4 to 8 are 5 times degenerate and correspond to the $J = 2$ states with partially mixed Γ_3 and Γ_4 symmetry. We thus learn that H^{ex}_{ech18} is directly determined by the Hamiltonians H^{ex}_{ech1} and H^{ex}_{ech4} displayed in Eqs. (4.3.3) and (4.3.2), respectively. H^{ex}_{ech18} gives normally the main contribution to the exchange interaction between electron and hole.

The eigenstates v^{exd}_{i18} leading to the diagonalization of the Hamiltonian are given by

$$v^{exd}_{i18} = \begin{pmatrix} 0 & 0 & -\sqrt{3}/2 & 0 & 0 & 0 & 1/2 & 0 \\ 0 & 0 & 0 & 0 & 0 & 0 & 0 & 1 \\ 0 & -\sqrt{2}/2 & 0 & 0 & 0 & \sqrt{2}/2 & 0 & 0 \\ 0 & 0 & 1/2 & 0 & 0 & 0 & \sqrt{3}/2 & 0 \\ -1/2 & 0 & 0 & \sqrt{3}/2 & 0 & 0 & 0 & 0 \\ 0 & \sqrt{2}/2 & 0 & 0 & 0 & \sqrt{2}/2 & 0 & 0 \\ 0 & 0 & 0 & 0 & 1 & 0 & 0 & 0 \\ \sqrt{3}/2 & 0 & 0 & 1/2 & 0 & 0 & 0 & 0 \end{pmatrix}. \tag{5.3.6}$$

Let us now identify the eigenstates v^{exd}_{i18} of Eq. (5.3.6) in terms of the exciton eigenstates v^{ex}_{i18} defined for the $\Gamma_6 \otimes \Gamma_8$ subspace. According to the construction of the Kronecker product of the total angular momentum matrices for $j_v = 3/2$ with the Pauli-spin matrices, the exciton eigenstates v^{ex}_{i18} are given in the conduction- and valence-band electron basis by (see Eq. (4.2.4)):

$$\begin{aligned} v^{ex}_{118} &= |-1\rangle\beta\alpha_e = v^v_3\alpha_e \\ v^{ex}_{218} &= |-1\rangle\beta\beta_e = v^v_3\beta_e \\ v^{ex}_{318} &= (\sqrt{2}|0\rangle\beta + |-1\rangle\alpha)\alpha_e/\sqrt{3} = v^v_4\alpha_e \\ v^{ex}_{418} &= (\sqrt{2}|0\rangle\beta + |-1\rangle\alpha)\beta_e/\sqrt{3} = v^v_4\beta_e \\ v^{ex}_{518} &= (\sqrt{2}|0\rangle\alpha + |1\rangle\beta)\alpha_e/\sqrt{3} = v^v_5\alpha_e \\ v^{ex}_{618} &= (\sqrt{2}|0\rangle\alpha + |1\rangle\beta)\beta_e/\sqrt{3} = v^v_5\beta_e \\ v^{ex}_{718} &= |1\rangle\alpha\alpha_e = v^v_6\alpha_e \\ v^{ex}_{818} &= |1\rangle\alpha\beta_e = v^v_6\beta_e. \end{aligned} \tag{5.3.7}$$

In order to transform these states to the exciton electron-hole basis we have to use the correspondence between valence-band and hole states, which is given in Eq. (4.1.5), namely:

$$
\begin{aligned}
|\phi_1^h\rangle &= -v_3^v \\
|\phi_2^h\rangle &= v_4^v \\
|\phi_3^h\rangle &= -v_5^v \\
|\phi_4^h\rangle &= v_6^v.
\end{aligned}
\tag{5.3.8}
$$

Using Eq. (4.2.6) we then obtain the following states:

$$
\begin{aligned}
v_{118}^{ex} &= |-1\rangle\beta\alpha_e = v_3^v\alpha_e = -|\phi_1^h\rangle\alpha_e = -|\Psi_1\rangle \\
v_{218}^{ex} &= |-1\rangle\beta\beta_e = v_3^v\beta_e = -|\phi_1^h\rangle\beta_e = -|\Psi_5\rangle \\
v_{318}^{ex} &= (\sqrt{2}|0\rangle\beta + |-1\rangle\alpha)\alpha_e/\sqrt{3} = v_4^v\alpha_e = |\phi_2^h\rangle\alpha_e = |\Psi_2\rangle \\
v_{418}^{ex} &= (\sqrt{2}|0\rangle\beta + |-1\rangle\alpha)\beta_e/\sqrt{3} = v_4^v\beta_e = |\phi_2^h\rangle\beta_e = |\Psi_6\rangle \\
v_{518}^{ex} &= (\sqrt{2}|0\rangle\alpha + |1\rangle\beta)\alpha_e/\sqrt{3} = v_5^v\alpha_e = -|\phi_3^h\rangle\alpha_e = -|\Psi_3\rangle \\
v_{618}^{ex} &= (\sqrt{2}|0\rangle\alpha + |1\rangle\beta)\beta_e/\sqrt{3} = v_5^v\beta_e = -|\phi_3^h\rangle\beta_e = -|\Psi_7\rangle \\
v_{718}^{ex} &= |1\rangle\alpha\alpha_e = v_6^v\alpha_e = |\phi_4^h\rangle\alpha_e = |\Psi_4\rangle \\
v_{818}^{ex} &= |1\rangle\alpha\beta_e = v_6^v\beta_e = |\phi_4^h\rangle\beta_e = |\Psi_8\rangle.
\end{aligned}
\tag{5.3.9}
$$

When comparing with the exciton wave functions given in Eq. (4.2.4) or Eq. (4.2.5) we see that we have found again the states $|\Psi_1\rangle$ to $|\Psi_8\rangle$. As discussed above, they result from the direct product of the electron conduction-band states with the valence-band states that transform according to Γ_8.

Considering the electron-hole basis functions, which diagonalize the exchange interaction H_{ech18}^{ex}, one obtains (by combining Eq. (5.3.6) with the basis functions Eq. (5.3.7) and then applying Eq. (5.3.9) together with Eq. (4.3.16)) again for the $J = 1$ states (having Γ_5 symmetry)

$$
\begin{aligned}
v_{118}^{exd} &= (|\Psi_3\rangle + \sqrt{3}|\Psi_8\rangle)/2 = |1, 1\rangle \\
v_{218}^{exd} &= -(|\Psi_2\rangle + |\Psi_7\rangle)/\sqrt{2} = |1, 0\rangle \\
v_{318}^{exd} &= (|\Psi_6\rangle + \sqrt{3}|\Psi_1\rangle)/2 = |1, -1\rangle
\end{aligned}
\tag{5.3.10}
$$

and for the $J = 2$ states, which are used to construct in a similar way exciton states with Γ_3 and Γ_4 symmetry,

$$
\begin{aligned}
v_{418}^{exd} &= (|\Psi_8\rangle - \sqrt{3}|\Psi_3\rangle)/2 = |2, 1\rangle \\
v_{518}^{exd} &= |\Psi_4\rangle = |2, 2\rangle \\
v_{618}^{exd} &= (|\Psi_2\rangle - |\Psi_7\rangle)/\sqrt{2} = |2, 0\rangle \\
v_{718}^{exd} &= (\sqrt{3}|\Psi_6\rangle - |\Psi_1\rangle)/2 = |2, -1\rangle \\
v_{818}^{exd} &= -|\Psi_5\rangle = |2, -2\rangle.
\end{aligned}
\tag{5.3.11}
$$

Let us now examine the second exchange interaction Hamiltonian given by Eq. (5.3.3). After performing the matrix products indicated in Eq. (5.3.3) H^{ex}_{ech28} takes the matrix form

$$H^{ex}_{ech28} = \Delta_{28} \begin{pmatrix} -27/8 & 0 & 0 & 7\sqrt{3}/4 & 0 & 0 & 0 & 0 \\ 0 & 27/8 & 0 & 0 & 0 & 0 & 3/2 & 0 \\ 0 & 0 & -1/8 & 0 & 0 & 5 & 0 & 0 \\ 7\sqrt{3}/4 & 0 & 0 & 1/8 & 0 & 0 & 0 & 0 \\ 0 & 0 & 0 & 0 & 1/8 & 0 & 0 & 7\sqrt{3}/4 \\ 0 & 0 & 5 & 0 & 0 & -1/8 & 0 & 0 \\ 0 & 3/2 & 0 & 0 & 0 & 0 & 27/8 & 0 \\ 0 & 0 & 0 & 0 & 7\sqrt{3}/4 & 0 & 0 & -27/8 \end{pmatrix}. \tag{5.3.12}$$

Again, as discussed in connection with Eq. (4.3.9) for this exciton block, the exchange interaction given in Eq. (5.3.12) can be diagonalized. As in Eq. (4.3.19) one then obtains the eigenvalues E_{sti28}

$$E_{sti28} = \Delta_{28}(39/8, 39/8, -41/8, -41/8, -41/8, 15/8, 15/8, 15/8) \tag{5.3.13}$$

with $i = (1 \text{ to } 8)$, respectively. This is exactly the energy level scheme given in Fig. 4.4. The eigenstates, which diagonalize the Hamiltonian matrix, are given by

$$v^{exd}_{i28} = \begin{pmatrix} 0 & 0 & 0 & 0 & -\sqrt{3}/2 & 0 & 0 & 1/2 \\ \sqrt{2}/2 & 0 & 0 & 0 & 0 & 0 & -\sqrt{2}/2 & 0 \\ 0 & \sqrt{2}/2 & 0 & -\sqrt{2}/2 & 0 & 0 & 0 & 0 \\ 0 & 0 & 0 & 0 & 1/2 & 0 & 0 & \sqrt{3}/2 \\ 0 & 0 & -1/2 & 0 & 0 & \sqrt{3}/2 & 0 & 0 \\ 0 & \sqrt{2}/2 & 0 & \sqrt{2}/2 & 0 & 0 & 0 & 0 \\ \sqrt{2}/2 & 0 & 0 & 0 & 0 & 0 & \sqrt{2}/2 & 0 \\ 0 & 0 & \sqrt{3}/2 & 0 & 0 & 1/2 & 0 & 0 \end{pmatrix}. \tag{5.3.14}$$

To go further into details, let us recall the electron-hole interactions H^{ex}_{ech2} and H^{ex}_{ech3} defined in Eqs. (4.3.4) and (4.3.5). As we have seen in connection with Eq. (4.3.19) the exchange interactions H^{ex}_{ech2} and H^{ex}_{ech3} have no influence on the degeneracy of the $J = 1$ exciton states but may only shift them. Since H^{ex}_{ech2} and H^{ex}_{ech3} depend on the direction of the cubic axis, they act differently on exciton states with Γ_3 and Γ_4 symmetry and may partially lift the degeneracy of the $J = 2$ states. These terms are called "cubic-exchange interactions". The same result is now obtained from H^{ex}_{ech28}, which splits the Γ_3 from the Γ_4 states.

The exciton states v^{exd}_{128} and v^{exd}_{228} are twofold degenerate. According to Eq. (4.1.7) there is only one twofold degenerate irreducible representation (here Γ_3) that appears in the product subspace of $(\Gamma_6 \otimes \Gamma_8)$ exciton states. Therefore the exciton states v^{exd}_{128}

and v_{228}^{exd} are identified to have Γ_3 symmetry, v_{328}^{exd} to v_{528}^{exd} are triply degenerate. They correspond to the $J = 1$ states with Γ_5 symmetry, which are given in Eq. (5.3.10). The remaining states v_{628}^{exd}, v_{728}^{exd}, and v_{828}^{exd} must therefore be the states having Γ_4 symmetry.

The $J = 2$ states, which are adapted to the crystal symmetry, have been given in Eqs. (4.3.20) and (4.3.21). We obtain here for the reduced subspace the same states, namely, for the states with Γ_3 symmetry:

$$\begin{aligned} |2,+\rangle = (|2,2\rangle + |2,-2\rangle)/\sqrt{2} &= (|\Psi_4\rangle - |\Psi_5\rangle)/\sqrt{2} \\ |2,0\rangle &= (|\Psi_2\rangle - |\Psi_7\rangle)/\sqrt{2} \end{aligned} \tag{5.3.15}$$

and for those with Γ_4 symmetry:

$$\begin{aligned} |1,+\rangle &= -(|2,1\rangle + |2,-1\rangle)/\sqrt{2} = -((|\Psi_8\rangle - \sqrt{3}|\Psi_3\rangle)/2 + (\sqrt{3}|\Psi_6\rangle - |\Psi_1\rangle)/2)/\sqrt{2} \\ |1,-\rangle &= i(|2,-1\rangle - |2,1\rangle)/\sqrt{2} = i((\sqrt{3}|\Psi_6\rangle - |\Psi_1\rangle)/2 - (|\Psi_8\rangle - \sqrt{3}|\Psi_3\rangle)/2)/\sqrt{2} \\ |2,-\rangle &= (|2,2\rangle - |2,-2\rangle)/\sqrt{2} = (|\Psi_4\rangle + |\Psi_5\rangle)/\sqrt{2}. \end{aligned} \tag{5.3.16}$$

Normally, the influence of the electron-hole exchange-interaction term H_{ech28}^{ex} is very small and no splitting of exciton states with Γ_3 and Γ_4 symmetry has been observed in zincblende-type semiconductors. (See Fig. 4.5.) This is in accordance with our earlier remark that this term describes the influence of the exchange interaction on the spin-orbit splitting.

In general, a Hamiltonian of a system may depend on an additional set of physical quantities, which can be due to extrinsic perturbations (as magnetic fields B, electric fields E, strain e, and so on) or to intrinsic perturbations (as the exciton center-of-mass wave-vector $\boldsymbol{Q}$), which break the full point-group symmetry. Such perturbations will be discussed in the next section.

To discuss only the $\Gamma_6 \otimes \Gamma_8$ subspace of the exciton ground state is usually sufficient since in most simple binary semiconductors the spin-orbit coupling is the dominant interaction. Then the valence bands with Γ_7 and Γ_8 symmetry are well separated in energy and the different exciton series, which are formed from these bands together with the lowest conduction-band states, do not mix.

5.4 Symmetry-Breaking Effects in the $\Gamma_6 \otimes \Gamma_8$ Exciton Ground State

5.4.1 Magnetic-Field Dependence

Let us now consider the invariant expansion of the exciton Hamiltonian in the $\Gamma_6 \otimes \Gamma_8$ subspace. We will follow the same lines as for the split-off band, i.e. the exciton

Hamiltonian that was constructed in the $\Gamma_6 \otimes \Gamma_7$ subspace discussed above. Then the linear magnetic field $\boldsymbol{B}$ dependence takes the form

$$\begin{aligned} H_{B1}^{ex8} &= a_B^e 1_{v8} \otimes (\boldsymbol{B} \cdot \boldsymbol{\sigma_e}) + \left(a_{B1v8a}^v (\boldsymbol{B} \cdot \boldsymbol{j_v}) + a_{B1v8b}^v (B_x (j_v^x)^3 + B_y (j_v^y)^3 + B_z (j_v^z)^3)\right) \otimes 1_e = \\ &= (1/2) g^c \mu_B 1_{v8} \otimes (B_x \sigma_e^x + B_y \sigma_e^y + B_z \sigma_e^z) + \\ &+ \left(a_{B1v8a}^v (B_x j_v^x + B_y j_v^y + B_z j_v^z) + a_{B1v8b}^v (B_x (j_v^x)^3 + B_y (j_v^y)^3 + B_z (j_v^z)^3)\right) \otimes 1_e, \end{aligned} \tag{5.4.1}$$

where a_{B1v8a}^v and a_{B1v8b}^v are the g-factors of the holes in the valence band with Γ_8 symmetry. The pseudo-angular momentum matrices j_v^i and $(j_v^i)^3$ describe the coupling of states in the valence band with Γ_8 symmetry. The pseudo-angular momentum matrices j_v^i are given in Eq. (3.4.1). As mentioned above, the coefficients a_{B1v8a}^v and a_{B1v8b}^v correspond to the parameters κ and q introduced by Luttinger [5]

$$\begin{aligned} a_{B1v8a}^v &= -2\mu_B \kappa \\ a_{B1v8b}^v &= -2\mu_B q, \end{aligned} \tag{5.4.2}$$

where μ_B represents the electron magneton of Bohr.

In the case of the split-off band, the terms given in Eq. (5.4.1) originate directly from the conduction and valence bands. They have the required symmetry properties and lead here to a splitting of the degenerate exciton states or the linear Zeeman effect of excitons in the $\Gamma_6 \otimes \Gamma_8$ subspace (see Fig. 5.4).

Concerning electron-hole exchange-interaction terms, varying linearly with the strength of magnetic field $\boldsymbol{B}$, we see that two terms are in principle possible: Using Table 2.3, one has first to construct from the components of $\boldsymbol{B}$ and the Pauli-spin matrices $\boldsymbol{\sigma_e}$ new matrices pertaining to the conduction band that transform as (Γ_5, K^+) or (Γ_3, K^+). They can then be combined with the valence-band matrices (X_{v8}, Y_{v8}, Z_{v8}) and (U_{v8}, V_{v8}), which are given in Eq. (3.4.3). The resulting magnetic-field dependent exchange-interaction terms, transforming as (Γ_1, K^+), can be present in the exciton Hamiltonian. These terms are, however, very small since they involve not only the electron-hole exchange-interaction but also the spin-orbit coupling. They have not been studied further in literature. One can mention that other symmetry-breaking interactions, varying in higher order with $\boldsymbol{B}^n$ (with $n \geq 2$) can be constructed but, following the argumentation given above, terms involving the exchange interaction together with spin-orbit coupling can normally be neglected.

5.4.2 *Wave-Vector, Electric-Field, and Strain-Dependent Interactions*

Concerning the wave-vector $\boldsymbol{Q}$ dependent terms (transforming as (Γ_5, K^-)) or their power of nth order, noted $\boldsymbol{Q}^n$, we distinguish again two types: the first group is diagonal within the electron states and results directly from the conduction- and valence-band electrons and holes:

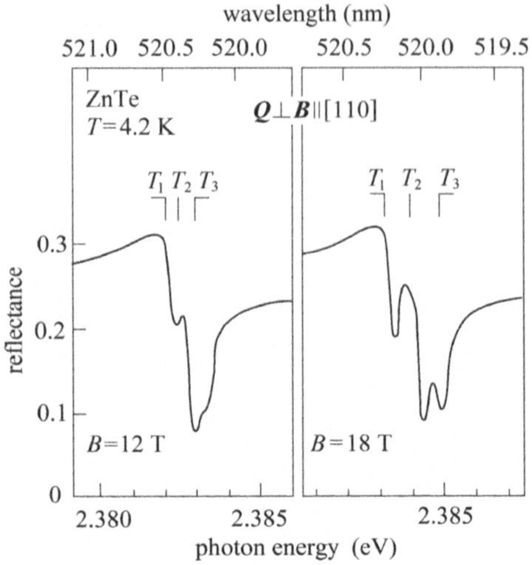

Fig. 5.4 Experimental normal-incidence magneto-reflectance of $\Gamma_6 \otimes \Gamma_8$-excitons in ZnTe at $T = 4.2$ K, for applied magnetic induction $\boldsymbol{B} \parallel [110]$ and incident light wave-vector $\boldsymbol{Q} \perp \boldsymbol{B}$ (Voigt configuration). Notice the splitting of the transverse Γ_5 ($J = 1$) exciton level into three components T_1, T_2, and T_3. (One could intuitively expect two split components only to occur; in this experimental configuration, however, a mixing between dipole-allowed ($J = 1$) and dipole-forbidden ($J = 2$) states resulted in the appearance of the third component.) The overall shift of the spectral pattern is due to a diamagnetic response of the exciton. Adapted after Ref. [11]

A $\boldsymbol{Q}$-linear term resulting from the valence-band hole has the same form as given in Eq. (3.4.6):

$$H_{Q1}^{ex8} = C_{Q1}^{ex8}[\{j_v^x, ((j_v^y)^2 - (j_v^z)^2)\}Q_x + \{j_v^y, ((j_v^z)^2 - (j_v^x)^2)\}Q_y + \{j_v^z, ((j_v^x)^2 - (j_v^y)^2)\}Q_z] \otimes 1_e. \tag{5.4.3}$$

In addition, a term quadratic in $\boldsymbol{Q}^2$ reads

$$\begin{aligned} H_{Q2}^{ex8} &= G_{1Q2}^{ex8} Q^2 1_{v8} \otimes 1_e, + \\ &+ G_{2Q2}^{ex8}[3(Q_x^2 - Q_y^2)((j_v^x)^2 - (j_v^y)^2) + (2Q_z^2 - Q_x^2 - Q_y^2)(2(j_v^z)^2 - (j_v^x)^2 - (j_v^y)^2)] \otimes 1_e + \\ &+ 2G_{3Q2}^{ex8}[Q_y Q_z\{j_v^y, j_v^z\} + Q_z Q_x\{j_v^z, j_v^x\} + Q_x Q_y\{j_v^x, j_v^y\}] \otimes 1_e. \end{aligned} \tag{5.4.4}$$

Although the terms H_{Q1}^{ex8} and H_{Q2}^{ex8} follow directly from the corresponding terms in the valence band, their numerical values in the exciton problem can be different from that of the electron or hole system. This is due to the fact that the wave-packets describing the exciton-envelope functions have to be constructed from a great number of electron and hole states. One also notes that the parameter G_{1Q2}^{ex8}, describing the isotropic-exciton translational-mass, depends explicitly on the value of the electron and hole effective masses.

A second group of terms describes the wave-vector $\boldsymbol{Q}$ dependence of the exchange interaction. In the subspace under consideration, eleven terms can be constructed, which are enumerated in Ref. [6]. As mentioned above, these terms are small and are usually not further considered in literature. The terms correspond to different orders of perturbation calculation, which is indicated by the power of $\boldsymbol{j}_{v8}$ and $\boldsymbol{\sigma}_e$ operators. Since they can, however, become important (they can e.g. explain the fact that longitudinal and transverse excitons with $J = 1$ may have different effective masses (see Ref. [6, 7]) we consider here terms linear in $\boldsymbol{j}_{v8}$ and $\boldsymbol{\sigma}_e$, which can be regarded as the most important ones. Then, the wave-vector dependence of the exchange interaction is given by

$$\begin{aligned} H_{Q2e}^{ex8} &= \delta_{1Q2}^{ex8} Q^2 (j_v^x \otimes \sigma_e^x + j_v^y \otimes \sigma_e^y + j_v^z \otimes \sigma_e^z) + \\ &+ \delta_{2Q2}^{ex8} [3(Q_x^2 - Q_y^2)((j_v^x \otimes \sigma_e^x - j_v^y \otimes \sigma_e^y) \\ &+ (2Q_z^2 - Q_x^2 - Q_y^2)(2 j_v^z \otimes \sigma_e^z - j_v^x \otimes \sigma_e^x - j_v^y \otimes \sigma_e^y)] + \\ &+ 2\delta_{3Q2}^{ex8} [Q_y Q_z [j_v^y, \sigma_e^z] + Q_z Q_x [j_v^z, \sigma_e^x] + Q_x Q_y [j_v^x, \sigma_e^y]], \end{aligned} \tag{5.4.5}$$

where

$$[j_v^x, \sigma_e^y] = (1/2)(j_v^x \otimes \sigma_e^y + j_v^y \otimes \sigma_e^x). \tag{5.4.6}$$

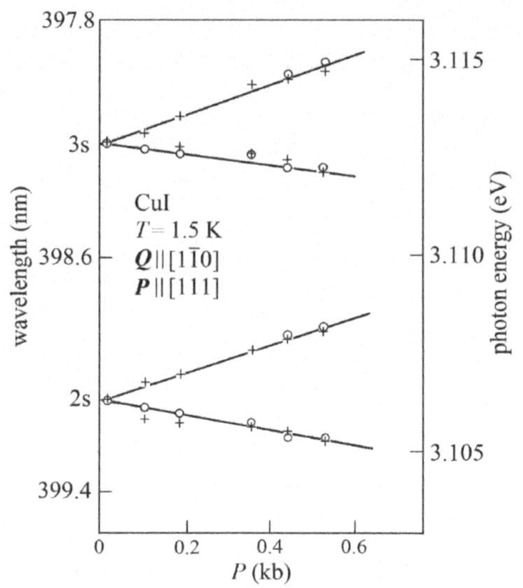

Fig. 5.5 Shift and splitting of the $n = 2, 3$ $\Gamma_6 \otimes \Gamma_8$-exciton reflectance lines in a CuI single crystal as a function of the applied uniaxial stress $\boldsymbol{P} \parallel [111]$. Wave vector of the incident light $\boldsymbol{Q} \parallel [1\bar{1}0]$, various symbols of the experimental points denote various light polarizations. It is interesting to note that because the higher exciton states $n = 2$ and $n = 3$ investigated here have large exciton radii, the electron and hole in the exciton are quite far apart. Therefore, the exchange interaction can be neglected and the exciton splitting simply reproduces the decomposition of the orbitally twofold degenerate Γ_8 valence band. According to Ref. [12]

The above discussion applies also to electric-field and strain-dependent interactions, which we will not discuss in more detail. (An example of strain-induced $\Gamma_6 \otimes \Gamma_8$-exciton splitting is shown in Fig. 5.5.) Compared to the wave-vector dependent symmetry-breaking effects one has only to pay attention to the fact that these physical quantities transform differently under time reversal than the wave-vector $\boldsymbol{Q}$. This implies that all even orders of the power development of the perturbation give rise to the same coupling scheme of the states, while the odd orders change in structure.

References

1. Cho, K.: Phys. Rev. B **14**, 4463 (1976)
2. Cho, K.: Excitons, Topics in Current Physics 14, Springer, Berlin, Heidelberg (1979)
3. Rahimpour Soleimani, H., Ostatnicky, T., Cronenberger, S., Gallart, M., Gilliot, P., Hönerlage, B.: J. Appl. Phys. **100**, 023705 (2006)
4. Dresselhaus, G.: Phys. Rev **100**, 580 (1955)
5. Luttinger, J.M.: Phys. Rev. **102**, 1030 (1956)
6. Hönerlage, B., Lévy, R., Grun, J.B., Klingshirn, C., Bohnert, K.: Phys. Rep. **124**, 161 (1985)
7. Mita, T., Satome, K., Ueta, M.: Solid State Commun. **33**, 1135 (1980)
8. Koda, T., Mitani, T., Murahashi, T.: Phys. Rev. Lett. **25**, 1495 (1970)
9. Nikitine, S., Biellmann, J., Deiss, J. L., Grosmann, M., Grun, J. B., Ringeissen, J., Schwab, C., Sieskind, M., Wursteisen, L.: In Report of the International Conference on The Physics of Semiconductors (Exeter, July 1962), The Institute of Physics and The Physical Society, London 1962, p. 431
10. Chelikowski, J.R., Cohen, M.L.: Phys. Rev. B **14**, 556 (1976)
11. Venghaus, H., Jusserand, B.: Phys. Rev. B **22**, 932 (1980)
12. Sauder, T., Daunois, A., Deiss, J.L., Merle, J.C.: Solid State Commun. **51**, 323 (1984)

Chapter 6
Invariant Expansion and Electron-Band Structure Effects in Wurtzite-Type Semiconductors

We are now interested in the invariant expansion of the electron, hole, and exciton states in direct semiconductors with wurtzite structure, i.e. with C_{6v} point-group symmetry. The electronic structure of many binary semiconductors with wurtzite structure is very similar to the case discussed above for zincblende structure. Often, materials can even realize both structures or can easily undergo phase transitions from one structure to the other if some parameter (as temperature or pressure) is varied, for instance CdS. Other important semiconductors crystallizing in wurtzite structure are GaN, InN and ZnO. It is, however, important to recall that the symmetry properties and the crystal structure of both systems are quite different, giving rise to their different physical properties.

The pile-up sequence in the crystal construction has been discussed in connection with Fig. 2.1 in the frame of a hard-sphere model. The pile-up sequence "ABABAB..." of the spheres in the "hexagonal close packed (hcp) structure" gives rise to an additional anisotropy of the crystal when compared to the fcc structure: it has a crystallographic axis, the so called "$\boldsymbol{c}$ -axis". In the following, we use Cartesian coordinates (x, y, z) and choose $z \parallel \boldsymbol{c}$ and $(x, y) \perp \boldsymbol{c}$.

While the electronic structure of atoms in the considered crystals with T_d and C_{6v} point-group symmetry is quite similar, the different crystal structures give rise to different electronic-band structures in the systems, i.e. a different organization of the electronic Bloch states. As we will see the main difference between the systems is due to an uniaxial crystal field, which appears in the hexagonal system. When compared to zincblende structure, this leads to an additional splitting of the valence band at the Γ-point and to some more important linear dispersion terms in the electronic-band structure.

Let us consider a semiconductor with wurtzite structure that has an empty conduction band (index "e"), which is made up from spin-degenerate atomic s-orbitals. When including the electron spin, the states are twofold degenerate and one has to describe them in the double-group representation. The symmetry-adapted conduction-band states then transform as Γ_7 at the Γ-point (see Refs. [1, 2]). This is in C_{6v} point-

© Springer International Publishing AG, part of Springer Nature 2018
B. Hönerlage and I. Pelant, *Symmetry and Symmetry-Breaking in Semiconductors*, Springer Tracts in Modern Physics 279,
https://doi.org/10.1007/978-3-319-94235-3_6

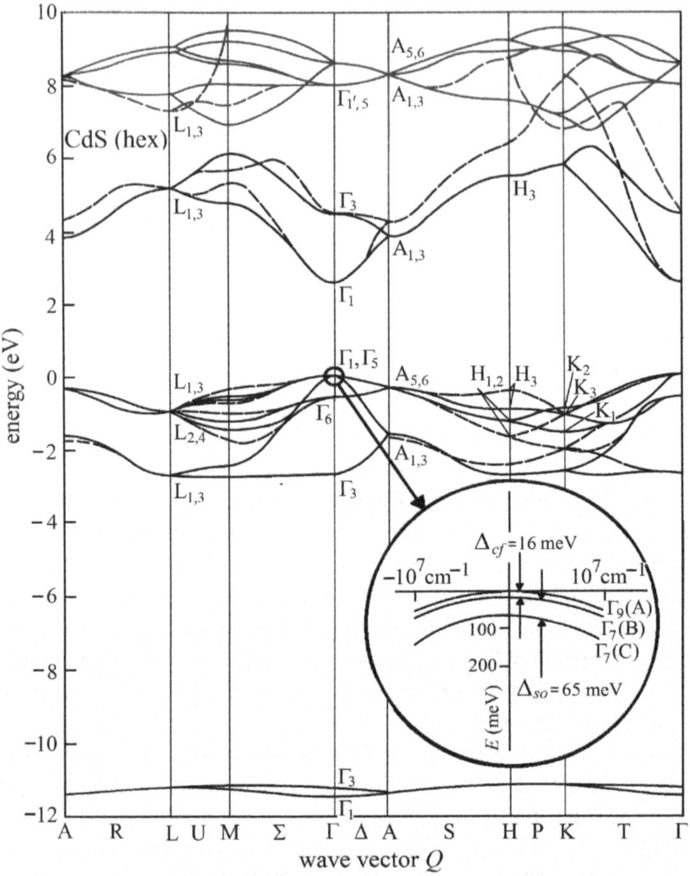

Fig. 6.1 The band structure of hexagonal CdS. We are mainly interested in the enlarged portion that displays schematically the resolved uppermost valence bands with Γ_9, Γ_7, and Γ_7-symmetries in the parabolic approximation close to the Γ-point. A uniaxial crystal field leads to a splitting Δ_{cf} of the valence bands as well as an isotropic spin-orbit coupling, characterized by the spin-orbit splitting Δ_{so}, which is also observed. Note: The valence band state of Γ_1-symmetry becomes, after including the electron spin, a $\Gamma_1 \otimes \Gamma_7 = \Gamma_7$ state in the double-group notation. Similarly, the states with Γ_5-symmetry develop into $\Gamma_5 \otimes \Gamma_7 = \Gamma_7 \oplus \Gamma_9$ states. (Figure adapted after Ref. [3])

group symmetry the double-group irreducible representation of 1/2 spins (while in T_d point-group symmetry it is called Γ_6; see Chap. 2).

Let us recall that the orbital part of the uppermost filled valence band states (index "v") are mostly originating from atomic p-orbitals with some admixture of atomic d-orbitals. These states are three-fold degenerate, described by an angular momentum $l = 1$. When including the electron spin, there are six degenerate valence-band states at the Γ-point, transforming pairwise as (Γ_7, Γ_7, and Γ_9), respectively. (See Figs. 1.4 and 6.1.)

Table 6.1 Notation of symmetry adapted operators or perturbation components in C_{6v} point-group symmetry. Their transformation properties are given in square brackets. (K^-, K^+) give the transformation properties with respect to Kramers' conjugation. From Refs. [1, 2, 4, 5]

$\Gamma_1 : (S)$	$[z]$ or $[l_x^2 + l_y^2]$ or $[l_z^2]$ or $[(\sigma_e^z)^2]$	$\|K^+\|$
$\Gamma_2 : (T)$	$[l_z]$ or $[\sigma_e^z]$	$\|K^-\|$
$\Gamma_3 : (U)$	$[(3x^2 - y^2)y]$	$\|K^+\|$
$\Gamma_4 : (V)$	$[(3y^2 - x^2)x]$	$\|K^+\|$
$\Gamma_5 : (X, Y)$	$[x, y]$	$\|K^+\|$
$\Gamma_5 : (X, Y)$	$[l_y, -l_x]$ or $[\sigma_e^y, -\sigma_e^x]$	$\|K^-\|$
$\Gamma_6 : (W, Z)$	$[(x^2 - y^2), 2xy]$	$\|K^+\|$

As we have discussed in the introduction the effective Hamiltonian H determines the energy E of the system, which is a real, scalar quantity. Therefore, H has to be a scalar operator, transforming as Γ_1 of the symmetry group under consideration. Since the Hamiltonian is supposed to be independent of time it transforms as K^+ under time reversal. All interaction terms that may exist in a crystal have to respect these symmetry properties. In the following we will closely follow the procedure that we have described above for zincblende-type crystals in order to construct an effective Hamiltonian of the system.

We analyze first the symmetry properties of perturbations under C_{6v} point-group symmetry. Neglecting the spin for the moment [1, 2] there exist six irreducible representations, into which all perturbation can be decomposed. Likewise Tables 2.1, 6.1 enumerates these irreducible representations in C_{6v} structures. They are labeled from Γ_1 to Γ_6. Then, Table 6.1 indicates the transformation properties of the different components using spatial basis functions or basis-operator components. According to their symmetry, the components are labeled by "S" to "Z". The angular-momentum ($[l_z]$ and $[l_y, -l_x]$) and the pseudo-spin operator-components ($[\sigma_e^z]$ and $[\sigma_e^y, -\sigma_e^x]$) transform as K^- under time reversal, all other functions and operator components as K^+.

In analogy with Tables 2.1 and 6.2 gives the transformation properties of several perturbations under C_{6v} point-group symmetry-operations. Following [1, 4] we consider for example the transformation properties of a magnetic field $\boldsymbol{B}$, a finite wave-vector $\boldsymbol{Q}$, of electric-field effects (proportional to $\boldsymbol{E}$ or $\boldsymbol{E}^2$), or strain components ϵ_{ij} (with $(i, j) = (x, y, z)$), and we give their symmetry with respect to Kramers' conjugation (K^-, K^+).

Table 6.3 gives the multiplication scheme for the components of Tables 6.1 and 6.2 for C_{6v} point-group symmetry. As shown before, when using it successively, symmetry-adapted perturbation-components of higher order can be generated. Combining them with symmetry-adapted matrix-operators the interaction terms of the effective Hamiltonian can be constructed. One should notice here that when generating matrix-operator components from the multiplication scheme, expressions, which are not yet symmetric or antisymmetric, have to be symmetrized to gener-

Table 6.2 Transformation properties of linear and quadratic perturbation components as magnetic field $\boldsymbol{B}$, wave-vector $\boldsymbol{Q}$ or electric field $\boldsymbol{E}$ effects, and strain tensor components ϵ_{ij} (with $(i, j) = (x, y, z)$) in C_{6v} point-group symmetry. (K^-, K^+) give the transformation properties with respect to Kramers' conjugation. From Refs. [1, 5]

	K^-	K^+	K^-	K^+	K^+	K^+	K^+
Γ_1 : S		B_z^2, $B_x^2 + B_y^2$	Q_z	Q_z^2, $Q_x^2 + Q_y^2$	E_z	E_z^2, $E_x^2 + E_y^2$	ϵ_{zz}, $\epsilon_{xx} + \epsilon_{yy}$
Γ_2 : T	B_z						
Γ_3 : U							
Γ_4 : V							
Γ_5 : X	B_y	$-B_z B_x$	Q_x	$Q_z Q_x$	E_x	$E_z E_x$	ϵ_{zx}
: Y	$-B_x$	$-B_y B_z$	Q_y	$Q_y Q_z$	E_y	$E_y E_z$	ϵ_{yz}
Γ_6 : W		$-B_x^2 + B_y^2$		$Q_x^2 - Q_y^2$		$E_x^2 - E_y^2$	$\epsilon_{xx} - \epsilon_{yy}$
: Z		$-2B_x B_y$		$2Q_x Q_y$		$2E_x E_y$	$2\epsilon_{xy}$

Table 6.3 Multiplication scheme for the components of Table 6.1 in C_{6v} point-group symmetry. RC: Resulting component. Concerning operator components, expressions, which are not yet symmetric or antisymmetric, they have to be symmetrized to obtain the correct product component. The "$XZ' - YW'$" and"$YZ' + XW'$" products lead to "resulting components (RC)" different from those given in Ref. [1]. From Refs. [1, 2]

RC	Product of components					
Γ_1 : S	SS'	TT'	UU'	VV'	$XX' + YY'$	$WW' + ZZ'$
Γ_2 : T	ST'		UV'		$XY' - YX'$	$WZ' - ZW'$
Γ_3 : U	SU'	TV'			$XZ' + YW'$	
Γ_4 : V	SV'	TU'			$YZ' - XW'$	
Γ_5 : X	SX'	TY'	UZ'	$-VW'$	$YZ' + XW'$	
: Y	SY'	$-TX'$	UW'	VZ'	$XZ' - YW'$	
Γ_6 : W	SW'	TZ'	$-UX'$	$-VX'$	$XX' - YY'$	$ZZ' - WW'$
: Z	SZ'	$-TW'$	UY'	VY'	$XY' + YX'$	$WZ' + ZW'$

ate the correct product component. In addition, the "$XZ' - YW'$" and"$YZ' + XW'$" products lead to "resulting components", which are different from those given in Ref. [1]. Finally, Table 6.3 is completed by the multiplication scheme of the irreducible representations given in Table 6.4 [1, 2, 4–6].

6.1 Spin-Degenerate Conduction-Bands in Semiconductors with C_{6v} Point-Group Symmetry

Let us first consider the one-electron band-structure of a spin-degenerate conduction-band in the vicinity of the Γ-point. Since all other atomic states are separated in energy from the conduction band, the electron is described by an effective Hamiltonian H_e, acting only in the subspace of the conduction-band states. Since this subspace is twofold degenerate we use the effective pseudo-spin operator $\boldsymbol{\sigma_e}$ (with $\sigma_e = 1/2$) in order to develop the effective Hamiltonian H_e. We choose again Pauli-spin matrices σ_e^i with $i = (x, y, z)$ as basis matrices:

$$\sigma_e^x = \begin{pmatrix} 0 & 1 \\ 1 & 0 \end{pmatrix}; \sigma_e^y = \begin{pmatrix} 0 & -i \\ i & 0 \end{pmatrix}; \sigma_e^z = \begin{pmatrix} 1 & 0 \\ 0 & -1 \end{pmatrix}. \tag{6.1.1}$$

These basis matrices transform spatially like the components of an angular-momentum operator and are K^- under time reversal. Because of the uniaxiality of wurtzite structure, the spatial transformation properties of (σ_e^x, σ_e^y) are different from that of σ_e^z. In systems with C_{6v} point-group symmetry (σ_e^x, σ_e^y) transform like the irreducible representation Γ_5, σ_e^z as Γ_2 [1, 2, 4–7].

The last pseudo-spin matrix used to span the basis of conduction-band states for the development of a Hamiltonian is the unit matrix 1_e, which can be obtained in systems with C_{6v} point-group symmetry by calculating

$$1_e = \begin{pmatrix} 1 & 0 \\ 0 & 1 \end{pmatrix} = (\sigma_e^z)^2, \tag{6.1.2}$$

which transforms as (Γ_1, K^+). As shown previously, 1_e may also be constructed from the Pauli matrices by calculating $(1/3)(\boldsymbol{\sigma_e})^2$ or $(1/2)[(\sigma_e^x)^2 + (\sigma_e^y)^2]$ (Concerning time-reversal symmetry one may recall that operators transforming as K^- under time reversal (as e.g. $\boldsymbol{\sigma_e}$) give rise to operators having K^+ symmetry when taken to even orders, as it is the case for the construction of the unit matrix 1_e.).

The four pseudo-spin matrices given in Eqs. (6.1.1) and (6.1.2) are linearly independent of each other and span a basis of interacting conduction-band states. Since the effective Hamiltonian H_e has to transform as (Γ_1, K^+) it has here to be proportional to 1_e. Like the case of conduction band in zincblende structure we find

$$H_e = a_e 1_e, \tag{6.1.3}$$

where a_e is a real number or function. It is the only form a Hamiltonian can take in its matrix form for the spin-degenerate conduction-band. Then H_e contains all interaction terms, which have the full crystal symmetry, i.e. which remain invariant under all symmetry operations that are compatible with the point group of the crystal. They are diagonal in the considered states and have the same energy value for both states. This means that these interactions do not couple the pseudo-Bloch functions

α_e and β_e, which are eigenstates of the Hamiltonian. The interaction terms have also to be invariant under time reversal. They determine completely the energy of the spin-degenerate electron-states in the conduction band at the center of the Brillouin zone.

Equation (6.1.3) shows that the spin degeneracy of the conduction-band electron-states can only be lifted through symmetry-breaking effects. These terms have to involve either the basis matrix σ_e^z, which transforms according to Table 6.1 like the irreducible representation Γ_2, or the pseudo-spin matrix-components (σ_e^x, σ_e^y), transforming as Γ_5. As mentioned above, these matrix operators transform according to K^- under time reversal.

Table 6.2 shows that a term

$$H_{B1} = a_{B1a} B_z \sigma_e^z + a_{B1b}(B_y \sigma_e^y + B_x \sigma_e^x) \tag{6.1.4}$$

may appear as a symmetry-breaking linear Zeeman effect. The two constants a_{B1a} and a_{B1b} are in principle independent of each other, and both terms are scalar quantities that show the correct behavior with respect to time reversal.

Quite similarly, Table 6.2 shows that also terms varying linearly with the wave-vector $\boldsymbol{Q}$ are possible. But since the transformation properties of the components of the magnetic field $\boldsymbol{B}$ and $\boldsymbol{Q}$ are different, the interaction terms rather read

$$H_{Q1} = a_{Q1}(Q_x \sigma_e^y - Q_y \sigma_e^x). \tag{6.1.5}$$

A linear term proportional to Q_z cannot exist since spatial symmetry would require the term to be multiplied by the matrix 1_e, which has, however, the wrong time-reversal symmetry when compared to that of Q_z.

As given in Table 6.2 the electric-field components (E_x, E_y) transform like (Γ_5, K^+) and E_z as (Γ_1, K^+). Because of its temporal symmetry properties, $\boldsymbol{E}^n$ ($\boldsymbol{E}$ taken to all possible orders n) have K^+ symmetry. Therefore, they can only be combined with the unit matrix 1_e. It follows immediately that an electric field cannot lead to a splitting of the spin-degenerate states. Thus, a linear Stark effect proportional to E_x or E_y or strain fields involving ϵ_{zx} or ϵ_{zy} have no effect on the energy of the conduction-band states, while a term

$$H_{Ez1} = a_{Ez1} E_z 1_e \tag{6.1.6}$$

is possible. It shifts the conduction-band states proportionally to the electric field component E_z linearly in energy as the crystal field does.

Concerning higher orders in $\boldsymbol{E}$, they can also lead to a Stark shift of the states. An example is the quadratic Stark effect, which results here in a perturbation of the form

$$H_{E2} = a_{E2a} E_z^2 1_e + a_{E2b}(E_x^2 + E_y^2) 1_e, \tag{6.1.7}$$

a_{E2a} and a_{E2b} being real constants. As it can be seen in Table 6.2, no other combinations of electric field components up to second order can exist. Figure 6.2 shows the

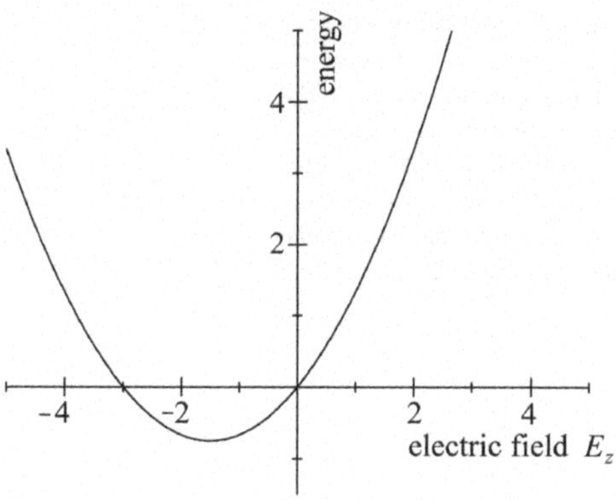

Fig. 6.2 Schematic energy variation of conduction-band electron-states if an electric field is applied in the z-direction, leading to a linear and quadratic Stark effect

energy variation if an electric field is applied in the z-direction: a linear and quadratic Stark effect is observed while the electric field components in the x- or y-direction lead only to a quadratic Stark effect.

Similar is the effect due to the strain components ϵ_{zz} or $(\epsilon_{xx} + \epsilon_{yy})$, giving rise to an interaction term of the form

$$H_{\epsilon 1} = a_{eplong}\epsilon_{zz}1_e + a_{eptrans}(\epsilon_{xx} + \epsilon_{yy})1_e, \tag{6.1.8}$$

which equally shift the states but do not lead to a splitting.

In the same way behave quadratic terms up to second order in the magnetic field $\boldsymbol{B}^2$ or wave-vector components $\boldsymbol{Q}^2$ since their spatial and temporal transformation properties are equal to that of $\boldsymbol{E}^2$. We find

$$H_{B2} = a_{B2a}B_z^2 1_e + a_{B2b}(B_x^2 + B_y^2)1_e$$
$$\text{and} \tag{6.1.9}$$
$$H_{Q2} = a_{Q2a}Q_z^2 1_e + a_{Q2b}(Q_x^2 + Q_y^2)1_e,$$

where a_{B2a} and a_{B2b} are parameters, characteristic for an anisotropic quadratic Zeeman effect, while a_{Q2a} and a_{Q2b} are the anisotropic effective mass parameters of the conduction-band electrons.

Figure 6.3 shows the energy variation if a magnetic field is applied in an arbitrary direction to a wurtzite-type semiconductor: a linear and quadratic Zeeman effect are observed and the degeneracy of the spin states is lifted because of the $\boldsymbol{B}$-linear interaction. The electron dispersion can have a similar form but, remembering Eq. (6.1.5), it presents a $\boldsymbol{Q}$-linear term only if the (Q_x, Q_y) components are $\neq 0$. Then, the dispersion of the spin-states are non-degenerate even in the absence of a magnetic field.

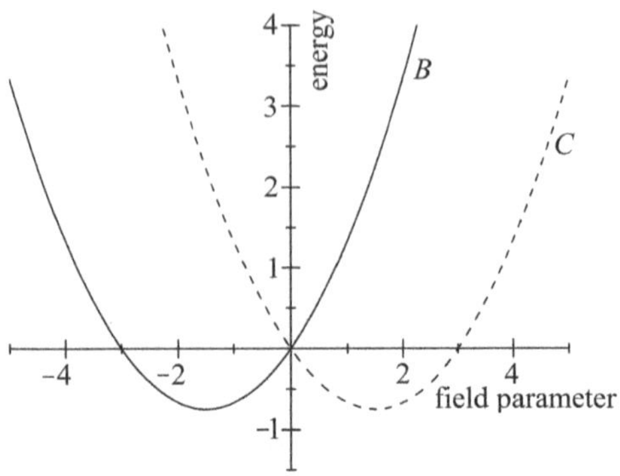

Fig. 6.3 Schematic energy variation of spin up (curve C) and spin down (curve B) conduction-band electron-states if a magnetic field is applied in an arbitrary direction, leading to a linear and quadratic Zeeman effect

6.2 Uniaxial Crystal Field Acting on a Three-Fold Degenerate Valence Band without Spin

Concerning the uppermost valence band of simple wurtzite-type semiconductors (C_{6v} point-group symmetry) let us first neglect (as in Sect. 3.1 for the zincblende structure) the electron spin. The valence-band states are mainly built from atomic p-orbitals. These orbitals $|m_j\rangle$ are eigenfunctions of the angular-momentum operator l and its z-component l_z with $l = 1$, m_j being the magnetic quantum number:

$$l_z|m_j\rangle = m_j|m_j\rangle \text{ with } (m_j = 1, 0, -1). \tag{6.2.1}$$

The energy eigenvalues e_{mj} are three-fold degenerate in systems with spherical symmetry. As indicated in Eq. (3.1.3) the matrix components of the angular-momentum operator are given by

$$l_x = (1/\sqrt{2})\begin{pmatrix} 0 & 1 & 0 \\ 1 & 0 & 1 \\ 0 & 1 & 0 \end{pmatrix}; l_y = (1/\sqrt{2})\begin{pmatrix} 0 & -i & 0 \\ i & 0 & -i \\ 0 & i & 0 \end{pmatrix}; l_z = \begin{pmatrix} 1 & 0 & 0 \\ 0 & 0 & 0 \\ 0 & 0 & -1 \end{pmatrix}. \tag{6.2.2}$$

According to Table 6.1, the operator component l_z transforms as the irreducible representation Γ_2 in systems with C_{6v} point-group symmetry, while (l_x, l_y) transform as the two-dimensional irreducible-representation Γ_5.

We will first construct the basis matrices, which are adapted to the crystal symmetry and span the three-dimensional angular-momentum subspace of the valence-band states. We can achieve this by forming products of the operator components given in Eq. (6.2.2). Using the multiplication scheme given in Table 6.4 we can form three different products of the irreducible representations:

$$\Gamma_2 \otimes \Gamma_2 = \Gamma_1, \quad \Gamma_5 \otimes \Gamma_5 = \Gamma_1 \oplus \Gamma_2 \oplus \Gamma_6, \text{ and } \Gamma_2 \otimes \Gamma_5 = \Gamma_5. \tag{6.2.3}$$

Table 6.4 Multiplication table for irreducible representations in crystals with C_{6v} point-group symmetry. From Refs. [1, 2]

	Γ_1	Γ_2	Γ_3	Γ_4	Γ_5	Γ_6	Γ_7	Γ_8	Γ_9
Γ_1	Γ_1	Γ_2	Γ_3	Γ_4	Γ_5	Γ_6	Γ_7	Γ_8	Γ_9
Γ_2		Γ_1	Γ_4	Γ_3	Γ_5	Γ_6	Γ_7	Γ_8	Γ_9
Γ_3			Γ_1	Γ_2	Γ_6	Γ_5	Γ_8	Γ_7	Γ_9
Γ_4				Γ_1	Γ_6	Γ_5	Γ_8	Γ_7	Γ_9
Γ_5					$\Gamma_1+\Gamma_2+\Gamma_6$	$\Gamma_3+\Gamma_4+\Gamma_5$	$\Gamma_7+\Gamma_9$	$\Gamma_8+\Gamma_9$	$\Gamma_7+\Gamma_8$
Γ_6						$\Gamma_1+\Gamma_2+\Gamma_6$	$\Gamma_8+\Gamma_9$	$\Gamma_7+\Gamma_9$	$\Gamma_7+\Gamma_8$
Γ_7							$\Gamma_1+\Gamma_2+\Gamma_5$	$\Gamma_3+\Gamma_4+\Gamma_6$	$\Gamma_5+\Gamma_6$
Γ_8								$\Gamma_1+\Gamma_2+\Gamma_5$	$\Gamma_5+\Gamma_6$
Γ_9									$\Gamma_1+\Gamma_2+\Gamma_3+\Gamma_4$

In this way we obtain seven additional matrices that may be used together with the three matrices in Eq. (6.2.2) to form a basis of our system, i.e. we have to choose now nine matrices, which are linearly independent and hermitian since they are used to develop the effective Hamiltonian. These basis matrices have to be adapted to the crystal symmetry. One can obtain such matrices or functions by using Table 6.3 that governs the product formation of symmetry-adapted components in crystals with C_{6v} point-group symmetry.

According to Tables 6.1 and 6.3 we construct the following components:

$$\begin{aligned} S = l_z^2;\ S_a = (l_x^2 + l_y^2) \Leftrightarrow \text{ transforming as} \Leftrightarrow (\Gamma_1, K^+) \\ T = l_z \Leftrightarrow (\Gamma_2, K^-) \\ U \Leftrightarrow (\Gamma_3) \text{ is not constructed for this basis} \\ V \Leftrightarrow (\Gamma_4) \text{ is not constructed for this basis} \\ (X, Y) = (l_y, -l_x) \Leftrightarrow \text{ transforming as} \Leftrightarrow (\Gamma_5, K^-) \\ (X_a, Y_a) = (-\{l_x l_z\}, -\{l_y l_z\}) \Leftrightarrow (\Gamma_5, K^+) \\ (W, Z) = (l_y^2 - l_x^2, -2\{l_x l_y\}) \Leftrightarrow (\Gamma_6, K^+), \end{aligned} \tag{6.2.4}$$

where the curly parentheses

$$\{l_y l_z\} = (1/2)(l_y l_z + l_z l_y) \tag{6.2.5}$$

indicates the symmetrized product of the two matrices l_y and l_z.

We see from Table 6.3 that products of components different from those in Eq. (6.2.4) could have been chosen to construct the basis of our subspace. They would, however, linearly depend on those given in Eq. (6.2.4) and differ from them probably by a phase factor (some multiplication by (-1) or $(\pm i)$). This has to be remembered when comparing results of calculations from different publications, e.g. approaches in Refs. [1, 2, 4–6]. Using the choice

$$T \to l_z \text{ and } (X, Y) \to (l_y, -l_x)$$

we obtain the following matrices together with their transformation properties:

$$\begin{gathered}
S = \begin{pmatrix} 1 & 0 & 0 \\ 0 & 0 & 0 \\ 0 & 0 & 1 \end{pmatrix}; S_a = \begin{pmatrix} 1 & 0 & 0 \\ 0 & 2 & 0 \\ 0 & 0 & 1 \end{pmatrix} \Leftrightarrow (\Gamma_1, K^+) \\
T = \begin{pmatrix} 1 & 0 & 0 \\ 0 & 0 & 0 \\ 0 & 0 & -1 \end{pmatrix} \Leftrightarrow (\Gamma_2, K^-) \\
X = (i/\sqrt{2}) \begin{pmatrix} 0 & -1 & 0 \\ 1 & 0 & -1 \\ 0 & 1 & 0 \end{pmatrix}; Y = (-1/\sqrt{2}) \begin{pmatrix} 0 & 1 & 0 \\ 1 & 0 & \\ 0 & 1 & 0 \end{pmatrix} \Leftrightarrow (\Gamma_5, K^-) \\
X_a = (1/(2\sqrt{2})) \begin{pmatrix} 0 & -1 & 0 \\ -1 & 0 & 1 \\ 0 & 1 & 0 \end{pmatrix}; Y_a = (i/(2\sqrt{2})) \begin{pmatrix} 0 & 1 & 0 \\ -1 & 0 & -1 \\ 0 & 1 & 0 \end{pmatrix} \Leftrightarrow (\Gamma_5, K^+) \\
W = \begin{pmatrix} 0 & 0 & -1 \\ 0 & 0 & 0 \\ -1 & 0 & 0 \end{pmatrix}; Z = \begin{pmatrix} 0 & 0 & i \\ 0 & 0 & 0 \\ -i & 0 & 0 \end{pmatrix} \Leftrightarrow (\Gamma_6, K^+).
\end{gathered} \tag{6.2.6}$$

One easily verifies that the matrices of Eq. (6.2.6) fulfill the requirements mentioned above and are characterized by a well-defined irreducible representation. As we have discussed in the introduction, since these matrices span the three-dimensional angular-momentum subspace of the valence-band states, they can be used to develop the effective Hamiltonian H^v of the system. In lowest order, this Hamiltonian does not consider any symmetry-breaking effect. It then contains all scalar interaction terms that remain unchanged under the symmetry operations of the crystal and are invariant under time reversal, i.e. interactions that transform as (Γ_1, K^+). In Eq. (6.2.4) the two operator-components $S = l_z^2$ and $S_a = (l_x^2 + l_y^2)$ fulfill this condition and H^v takes the form

$$H^v = a_{v1} S + a_{v2} S_a = a_{v1} l_z^2 + a_{v2}(l_x^2 + l_y^2), \tag{6.2.7}$$

a_{v1} and a_{v2} being real constants. The terms appearing in H^v are diagonal in the considered states, i.e. the interactions do not mix the pseudo-Bloch functions $|m_j\rangle$, which are eigenstates of the Hamiltonian H^v. The interaction terms shift, however, the energies of the states when passing from spherical symmetry to an environment with C_{6v} point-group symmetry. Equation (6.2.7) determines completely the energies of the valence-band states in the absence of any symmetry-breaking effect. The eigenvalues of H^v are given by

$$e_1 = e_{-1} = a_{v1} + a_{v2}$$
$$\text{and} \tag{6.2.8}$$
$$e_0 = 2a_{v2}.$$

Equation (6.2.8) shows that the uniaxial crystal field leads to an energy splitting between the eigenfunctions $|m_j\rangle$ with $(m_j = \pm 1)$ and $m_j = 0$. This so called "crystal-field splitting" Δ_{cf} is given by

$$\Delta_{cf} = a_{v2} - a_{v1}. \tag{6.2.9}$$

It vanishes if $a_{v1} = a_{v2}$ and we find again the energetic structure of crystals with zincblende structure. This situation is usually called to be "quasi-cubic". (Concerning Δ_{cf} see also Fig. 1.4, which will be discussed in more detail in Sect. 6.3 because of the simultaneous presence of crystal-field and spin-orbit coupling.)

The effective Hamiltonian H^v may be presented in a different form, in which the influence of the crystal field is better apparent. Using the matrix forms of S and S_a given in Eq. (6.2.6), one may use the form

$$H^v = a_{v1}(S + S_a) + (a_{v2} - a_{v1})S_a = 2a_{v1}\,1_v + \Delta_{cf}\,S_a = H_d^v + H_{cf}^v, \tag{6.2.10}$$

where 1_v stands for the unit matrix of the valence-band subspace when neglecting the electron spin. In analogy to Eq. (3.2.9) for zincblende-type semiconductors $H_d^v = 2a_{v1}1_v$ gives the direct term and $H_{cf}^v = \Delta_{cf}S_a$ describes the influence of the crystal field. Thus, H^v determines the valence-band energy at the Γ-point in wurtzite structure where the crystal-field splitting is separated from all other interactions.

We may now use the basis functions given in Eq. (6.2.4) together with various symmetry-breaking perturbations to construct the invariant expansion of the Hamiltonian H^v. Some simple examples of perturbations are given in Table 6.2 together with their transformation properties in crystals with C_{6v} point-group symmetry. As before, Table 6.3 is used to choose the correct product of the components according to their symmetry, which is given in Table 6.2. When combining e.g. the matrix components ($S = l_z^2$ and $S_a = (l_x^2 + l_y^2)$) of Eq. (6.2.4) with the perturbations (B_z^2 or $B_x^2 + B_y^2$); (Q_z^2 or $Q_x^2 + Q_y^2$); (E_z; E_z^2 or $E_x^2 + E_y^2$); (ϵ_{zz} or $\epsilon_{xx} + \epsilon_{yy}$) terms that may appear in the Hamiltonian H^v are obtained. This is due to the fact that all contributions to the products transform as (Γ_1, K^+).

A term linear in B_z of the form

$$H_{B1z}^v = a_{B1z}^v l_z B_z \tag{6.2.11}$$

may be present in the Hamiltonian while a term involving Q_z linearly has no influence on the energy of the valence band states. This is due to the fact that B_z and Q_z have different spatial transformation properties. On the other hand, the matrix components $(X, Y) = (l_y, -l_x)$ of Eq. (6.2.4) can be combined with the perturbation components $(B_y, -B_x)$ or (Q_x, Q_y) to contribute to the Hamiltonian in the form

$$H^{v}_{B1xy} = a^{v}_{B1xy}(l_y B_y + l_x B_x)$$
$$\text{or} \tag{6.2.12}$$
$$H^{v}_{Q1xy} = a^{v}_{Q1xy}(l_y Q_x - l_x Q_y).$$

All other combinations of perturbations from Table 6.2 with the matrix components (X, Y) are forbidden since they transform as K^+ under time reversal. In the same way, other symmetry adapted perturbation terms involving $(X_a, Y_a) = (-\{l_x l_z\}, -\{l_y l_z\})$ transforming as (Γ_5, K^+) or $(W, Z) = (l_y^2 - l_x^2, -2\{l_x l_y\})$ (that are of (Γ_6, K^+) symmetry) can be constructed.

6.3 Uniaxial-Crystal Field and Spin-Orbit Coupling in Semiconductor with C_{6v} Point-Group Symmetry: Valence Bands

Following closely the procedure described in Sect. 3.3 we develop the Hamiltonian describing fully electrons in the valence band of simple, wurtzite-type semiconductors as specified in the introduction. The invariant expansion has to respect the fact that the electrons in the valence band are spin 1/2 particles and have an orbital-angular momentum $\boldsymbol{l}_v = 1$. Again, we consider the spin degeneracy by introducing in addition to the orbital-angular momentum an effective-spin operator $\boldsymbol{\sigma}_v$, which operates only on the valence-band spin-states. The matrices describing this effective valence-electron spin transform as the effective spins of the conduction-band electrons, namely as (Γ_2, K^-) and (Γ_5, K^-) for the (σ_v^z) and (σ_v^x, σ_v^y) components, respectively. The direct product of the angular-momentum matrices given in Eq. (6.2.6) with the spin matrices span the subspace of dimension six, in which the valence-band states are defined. Again, the eigenstates of the product $\boldsymbol{l}_v \otimes \boldsymbol{\sigma}_v$ can be classified according to the total angular momentum $\boldsymbol{j}_v = \boldsymbol{l}_v \oplus \boldsymbol{\sigma}_v$ and its projection component j_z onto the z-axis of the crystal. Thus, states with a total angular momentum $\boldsymbol{j}_v = 3/2$ and $\boldsymbol{j}_v = 1/2$ are constructed. As given in Eq. (3.2.1) for the cubic zincblende structure, the six-fold valence-band electron-wave functions w_i^v take the form

$$w_i^v = a_{i,1}|1\rangle\alpha + a_{i,2}|1\rangle\beta + a_{i,3}|0\rangle\alpha + a_{i,4}|0\rangle\beta + a_{i,5}|-1\rangle\alpha + a_{i,6}|-1\rangle\beta \quad (i = 1, \ldots, 6), \tag{6.3.1}$$

where the eigenfunctions $|m_j\rangle$ of the momentum operator l_z have been introduced in Eq. (6.2.1) and α and β denote again the valence-band electron-spin up and down states, respectively.

Concerning the angular-momentum subspace, we start with the basis matrices (S; S_a; T; X, Y; X_a, Y_a; W, Z), which are obtained from the components (l_x, l_y, l_z) of the angular momentum and are given in Eqs. (6.2.4) and (6.2.6). The basis matrices transform as (Γ_1; Γ_1; Γ_2; Γ_5; Γ_5; and Γ_6), respectively. The valence-electron spin-matrices together with the unit matrix obtained from $(\sigma_v^z)^2$, which transforms as (Γ_1, K^+), are used as basis matrices to span the valence-band spin-subspace. They are defined in the same way as in Eqs. (6.1.1) and (6.1.2) for the conduction-band electron-states. Spin and angular-momentum subspaces are multiplied with each other, their direct product (Kronecker product) spanning the subspace of the valence-band states. As in Sect. 3.3, the matrices defined in the angular-momentum subspace given above are denoted in the following by capital Latin letters, those defined in the spin-subspace by a prime (′) and matrices defined in the product space by double quotation mark ("). According to Ref. [1] we denote in the following the valence-band spin-matrices by

$$\begin{aligned} (\sigma_v^z)^2 &\to S' \\ -\sigma_v^x &\to Y' \\ \sigma_v^y &\to X' \\ \sigma_v^z &\to T'. \end{aligned} \tag{6.3.2}$$

We thus obtain the thirty-six symmetry-adapted basis-matrices from the multiplication scheme in Table 6.4. They are defined through

$$\begin{aligned} (S; S_a; T; X, Y; X_a, Y_a; W, Z) \otimes S' \Leftrightarrow \text{transforming as} \Leftrightarrow (2(\Gamma_1, K^+); \Gamma_2, K^-; \Gamma_5, K^-; \Gamma_5, K^+; \Gamma_6, K^+) \\ (S; S_a; T; X, Y; X_a, Y_a; W, Z) \otimes (T') \Leftrightarrow (2(\Gamma_2, K^-); \Gamma_1, K^+; \Gamma_5, K^+; \Gamma_5, K^-; \Gamma_6, K^-) \\ (S; S_a) \otimes (X', Y') \Leftrightarrow (2(\Gamma_5, K^-)) \\ (T) \otimes (X', Y') \Leftrightarrow (\Gamma_5, K^+) \\ (X, Y) \otimes (X', Y') \Leftrightarrow (\Gamma_1 \oplus \Gamma_2 \oplus \Gamma_6, K^+) \\ (X_a, Y_a) \otimes (X', Y') \Leftrightarrow (\Gamma_1 \oplus \Gamma_2 \oplus \Gamma_6, K^-) \\ (W, Z) \otimes (X', Y') \Leftrightarrow (\Gamma_3 \oplus \Gamma_4 \oplus \Gamma_5, K^-). \end{aligned} \tag{6.3.3}$$

The symmetry-adapted basis-matrices corresponding to Eq. (6.3.3) can now be established using Table 6.3. Together with perturbation terms of Table 6.2 these matrices can be used to construct all possible contributions to a Hamiltonian, including symmetry-breaking effects. Their spatial and temporal transformation properties are noted according to the convention introduced in Sect. 3.3. Since different basis matrices may have the same spatial symmetry but arise from different components of the angular-momentum and spin operators, we indicate their origin by subscripts, where the first number indicates the irreducible representation to which the angular-momentum components belong and the second that of the spin component. For example the matrix X_{1a5}" characterizes the matrix X" (one of the components transforming as Γ_5 in the six-dimensional product space) obtained from the direct product of the angular-momentum matrix S_a transforming as Γ_1 with the spin component X' transforming as Γ_5. As given in Eq. (6.3.3) it has K^- time-reversal symmetry. For shortness we write in the following (as in Tables 2.3 and 6.3) the matrix product $S_{1a1}" = S_a \otimes S'$ in the form $S_{1a1}" = S_a S'$. We thus obtain the following thirty-six matrices from Eq. (6.3.3):

$$\begin{aligned}
&\text{Being } (K^+) \text{ under time reversal and resulting from } S' : \\
&S_{11}" = SS';\ S_{1a1}" = S_a S';\ X_{5a1}" = X_a S',\ Y_{5a1}" = Y_a S';\ W_{61}" = WS',\ Z_{61}" = ZS'; \\
&(\text{with } K^-) \text{ resulting from } S' : \\
&T_{21}" = TS';\ X_{51}" = XS',\ Y_{51}" = YS'; \\
&(\text{with } K^+) \text{ resulting from } T' : \\
&S_{22}" = TT';\ X_{52}" = YT',\ Y_{52}" = -XT'; \\
&(\text{with } K^-) \text{ resulting from } T' : \\
&T_{12}" = ST';\ T_{1a2}" = S_a T';\ X_{5a2}" = Y_a T',\ Y_{5a2}" = -X_a T';\ W_{62}" = ZT',\ Z_{62}" = -WT'; \\
&\text{resulting from } X'(\text{time-reversal symmetry not specified}) : \\
&Y_{25}" = -TX';\ X_{15}" = SX';\ X_{1a5}" = S_a X'; \\
&\text{resulting from } Y'(\text{time-reversal symmetry not specified}) : \\
&X_{25}" = TY';\ Y_{15}" = SY';\ Y_{1a5}" = S_a Y'; \\
&(\text{being } K^+) \text{ resulting from } (X, Y) \otimes (X', Y') : \\
&S_{55}" = XX' + YY';\ T_{55}" = XY' - YX';\ W_{55}" = XX' - YY',\ Z_{55}" = XY' + YX'; \\
&(\text{being } K^-) \text{ resulting from } (X_a, Y_a) \otimes (X', Y') : \\
&S_{5a5}" = X_a X' + Y_a Y';\ T_{5a5}" = X_a Y' - Y_a X';\ W_{5a5}" = X_a X' - Y_a Y',\ Z_{5a5}" = X_a Y' + Y_a X'; \\
&(\text{being } K^-) \text{ resulting from } (W, Z) \otimes (X', Y') : \\
&U_{65}" = ZX' + WY';\ V_{65}" = ZY' - WX';\ X_{65}" = WX' + ZY',\ Y_{65}" = -WY' + ZX'.
\end{aligned} \tag{6.3.4}$$

Let us now discuss the Hamiltonian H^v of the valence band in wurtzite-type semiconductors in detail. We start with those terms, which have the full crystal symmetry at the Γ-point. They are characterized by the fact that only scalar values multiply the matrices of Eq. (6.3.4). Inspection shows immediately that four matrices of the product space are compatible with interaction terms that may appear in a Hamiltonian, i.e. products transforming as (Γ_1, K^+). The first two are given by

$$S_{11}" = S \otimes S' \text{ and } S_{1a1}" = S_a \otimes S'. \tag{6.3.5}$$

They are independent of the valence-band spin-state and have been discussed in the foregoing chapter as giving rise to the energy of the valence band states and the crystal-field splitting. Two further terms

$$S_{22}" = T \otimes T' \text{ together with } S_{55}" = X \otimes X' + Y \otimes Y' \tag{6.3.6}$$

depend on the spin state.

Formally, the Hamiltonian H^v respecting the full point-group symmetry can therefore be written as

$$\begin{aligned}
H^v &= H^v_{cf1} + H^v_{cf2} + H^v_{soz} + H^v_{soxy} = \\
&= a^v_{cf1} S_{11}" + a^v_{cf2} S_{1a1}" + a^v_{soz} S_{22}" + a^v_{soxy} S_{55}",
\end{aligned} \tag{6.3.7}$$

where a^v_{cf1}, a^v_{cf2}, a^v_{soz}, and a^v_{soxy} are real constants with arbitrary values. (We use here the terms a^v_{cf1} and a^v_{cf2} instead of a_{v1} and a_{v2} of the foregoing section in order to distinguish clearly between interactions originating from spin-orbit coupling and crystal-field effects.) The first and the second contributions in Eq. (6.3.7) give the energy of the valence-band states at the Γ-point (see also Eq. (6.2.7)), which are no longer degenerate in energy. In the preceding section we have identified the splitting as being due to the uniaxial crystal field. It splits the states into two different blocks, characterized by the orbital-angular momentum according to ($m_j = \pm 1$) and ($m_j = 0$). In the product space including the spin, these angular-momentum states become each twofold degenerate.

As shown in Eq. (6.3.6) two additional contributions that depend on the spin and the angular momentum of the valence-band states may exist in a Hamiltonian of systems with wurtzite structure. Since they depend on the spin state, by analogy to Eq. (3.2.10), they are due to the spin-orbit interaction. The latter may also be influenced by the axial anisotropy of the crystal field, i.e. the spin-orbit interaction can be different in the z-direction and perpendicular to it. This may result in two different values of the spin-orbit coupling-constants a^v_{soz} and a^v_{soxy}.

In order to discuss further the different terms in Eq. (6.3.7) we calculate the basis matrices indicated in Eqs. (6.3.5) and (6.3.6). We obtain the following set of matrices:

$$S_{11}" = \begin{pmatrix} 1&0&0&0&0&0\\ 0&1&0&0&0&0\\ 0&0&0&0&0&0\\ 0&0&0&0&0&0\\ 0&0&0&0&1&0\\ 0&0&0&0&0&1 \end{pmatrix}; \; S_{1a1}" = \begin{pmatrix} 1&0&0&0&0&0\\ 0&1&0&0&0&0\\ 0&0&2&0&0&0\\ 0&0&0&2&0&0\\ 0&0&0&0&1&0\\ 0&0&0&0&0&1 \end{pmatrix} \quad \text{and} \tag{6.3.8}$$

$$S_{22}" = \begin{pmatrix} 1&0&0&0&0&0\\ 0&-1&0&0&0&0\\ 0&0&0&0&0&0\\ 0&0&0&0&0&0\\ 0&0&0&0&-1&0\\ 0&0&0&0&0&1 \end{pmatrix}; \; S_{55}" = \begin{pmatrix} 0&0&0&0&0&0\\ 0&0&\sqrt{2}&0&0&0\\ 0&\sqrt{2}&0&0&0&0\\ 0&0&0&0&\sqrt{2}&0\\ 0&0&0&\sqrt{2}&0&0\\ 0&0&0&0&0&0 \end{pmatrix}.$$

Let us first consider the first and second term in Eq. (6.3.7), namely, $H^v_{cf1} + H^v_{cf2} = a^v_{cf1} S_{11}" + a^v_{cf2} S_{1a1}"$. If $a^v_{cf1} = a^v_{cf2}$ we obtain for their sum ($H^v_{cf1} + H^v_{cf2}$)

$$H^v_{cf1} + H^v_{cf2} = 2a^v_{cf1} 1_v. \tag{6.3.9}$$

As discussed in the foregoing section this expression is proportional to the unit matrix 1_v of the valence-band subspace and we see that the crystal-field splitting vanishes. Like the direct term H^v_d in Eq. (3.2.9) for zincblende-type semiconductors, Eq. (6.3.9) describes also the valence-band energy at the Γ-point in wurtzite structure.

Similarly to Eqs. (6.2.10), (6.3.7) can be simplified writing it as

$$\begin{aligned} H^{v} &= H^{v}_{d} + H^{v}_{cf} + H^{v}_{soz} + H^{v}_{soxy} = \\ &= 2a^{v}_{cf2} 1_{v} + a^{v}_{cf} S_{11}\text{''} + a^{v}_{soz} S_{22}\text{''} + a^{v}_{soxy} S_{55}\text{''}, \end{aligned} \tag{6.3.10}$$

where $a^{v}_{cf} = a^{v}_{cf1} - a^{v}_{cf2}$ denotes the crystal-field coupling-parameters. Then, the direct term $H^{v}_{d} = 2a^{v}_{cf2} 1_{v}$ can be used to fix the zero of energy chosen at the Γ-point inside the valence band in the absence of all fine-structure interactions that lift the degeneracy. As a consequence, this term can be neglected in the following discussion and the crystal-field interaction reduces to the Hamiltonian $H^{v}_{cf} = a^{v}_{cf} S_{11}$". We will now discuss several limiting cases:

- $a^{v}_{cf} = 0,\ a^{v}_{soz} = a^{v}_{soxy} \neq 0$

If $a^{v}_{cf} = 0$ and $a^{v}_{soz} = a^{v}_{soxy}$, i.e. if the spin-orbit coupling is isotropic in the absence of a crystal-field splitting, we obtain for the sum $(H^{v}_{soz} + H^{v}_{soxy})$ of the third and fourth term of Eq. (6.3.7) or Eq. (6.3.10) the matrix form

$$H^{v}_{soz} + H^{v}_{soxy} = a^{v}_{soz} \begin{pmatrix} 1 & 0 & 0 & 0 & 0 & 0 \\ 0 & -1 & \sqrt{2} & 0 & 0 & 0 \\ 0 & \sqrt{2} & 0 & 0 & 0 & 0 \\ 0 & 0 & 0 & 0 & \sqrt{2} & 0 \\ 0 & 0 & 0 & \sqrt{2} & -1 & 0 \\ 0 & 0 & 0 & 0 & 0 & 1 \end{pmatrix}. \tag{6.3.11}$$

This expression reproduces exactly the spin-orbit interaction of the valence-band in zincblende-type semiconductors H^{v}_{so} given in Eq. (3.2.10). Diagonalization of the matrix in Eq. (6.3.11) leads us to the eigenvalues and eigenvectors of the Hamiltonian as given in Eqs. (3.2.11) and (3.2.12). In this respect the conditions $a^{v}_{cf1} = a^{v}_{cf2}$ and $a^{v}_{soz} = a^{v}_{soxy}$ used to derive Eq. (6.3.11) define the "quasi-cubic limit" of wurtzite-structure crystals. In this limit the wave functions transforming as $(\boldsymbol{x}, \boldsymbol{y}, \boldsymbol{z}\ )$ are equivalent and one can carry one wave function into the others by a simple rotation of the coordinate system.

It is important to notice that the "quasi-cubic limit" does not respect, however, the fact that the cubic axes $(\boldsymbol{x}, \boldsymbol{y}, \boldsymbol{z})$ are defined through the Cartesian coordinate system in zincblende structure while in wurtzite structure the cubic [111]-crystal direction is taken as $\boldsymbol{z}$-axis, and the $(\boldsymbol{x}, \boldsymbol{y})$-axis are perpendicular to it but they are not further specified. This means that the symmetry properties, i.e. the cylindrical (uniaxial) symmetry of wurtzite-type crystals, remain valid and only the eigenvalues and eigenfunctions of the states are the same in both crystal systems. Or, to put it differently, in order to simulate the wurtzite structure when starting from the cubic zincblende structure, stress may be applied in the [111]-crystal direction. This leads to a symmetry breaking and a crystal-field splitting of the valence band with Γ_8 symmetry. This splitting vanishes of course with vanishing stress, defining thus again the "quasi-cubic limit" of wurtzite structure. This discussion indicates that this limit may be realized despite the different point-group symmetries of the structures and the tetrahedral arrangement of atoms in zincblende-type crystals.

- $a_{cf}^{v} \neq 0$, $a_{soz}^{v} \neq 0$, and $a_{soxy}^{v} = 0$

Another interesting limit of Eq. (6.3.10) is characterized by $a_{cf}^{v} \neq 0$ and $a_{soxy}^{v} = 0$. We see, via comparing Eq. (6.3.10) with Eq. (6.3.8), that the terms H_{cf}^{v} and H_{soz}^{v}, which characterize the crystal field and a part of the spin-orbit coupling correspond to diagonal matrices, i.e. when looked upon separately, the states $|m_j\rangle\alpha$ and $|m_j\rangle\beta$, with $j = (-1, 0, 1)$ defined in Eq. (6.3.1) are eigenstates of this Hamiltonian. The states are two-fold degenerate in energy e_{ij} by pairs obeying to

$$\begin{aligned} E_1 = e_{16} &= (a_{cf}^{v} + a_{soz}^{v}) \text{ for } (|1\rangle\alpha \text{ and } |-1\rangle\beta) \text{ transforming as } \Gamma_9 \\ E_2 = e_{25} &= (a_{cf}^{v} - a_{soz}^{v}) \text{ for } (|1\rangle\beta \text{ and } |-1\rangle\alpha) \text{ transforming as } \Gamma_7 \\ E_3 = e_{34} &= 0 \text{ for } (|0\rangle\alpha \text{ and } |0\rangle\beta) \text{ transforming as } \Gamma_7. \end{aligned} \tag{6.3.12}$$

This result can be inferred immediately by scrutinizing the sum of the diagonal matrices $a_{cf}^{v} S_{11}$" $+ a_{soz}^{v} S_{22}$". For instance the same eigenvalue $(a_{cf}^{v} + a_{soz}^{v})$ corresponds to eigenstates $|1\rangle\alpha$ (upper left corner) and $|-1\rangle\beta$ (lower right corner); this is then the energy e_{16}. One can proceed similarly for e_{25} and e_{34}.

The states $|1\rangle\alpha$ and $|-1\rangle\beta$ transform according to the irreducible representation Γ_9 in C_{6v} symmetry, the pair $|1\rangle\beta$ and $|-1\rangle\alpha$ according to Γ_7, and finally $|0\rangle\alpha$ and $|0\rangle\beta$ transform also as Γ_7.

When comparing Eqs. (6.3.12) to (3.2.13) we see that the states $w_1^{v} = |1\rangle\alpha$ and $w_6^{v} = |-1\rangle\beta$ with energy e_{16} correspond directly to the states $|3/2, 3/2\rangle$ and $|3/2, -3/2\rangle$ in Eq. (3.2.13), respectively. In zincblende structure this doublet of states is degenerate with the pair of states $|3/2, \pm 1/2\rangle$ of Eq. (3.2.13). Together, they are transforming as Γ_8 in T_d point-group symmetry. In C_{6v} symmetry, the $|3/2, \pm 1/2\rangle$ states form a doublet of states with energy e_{25}, transforming as Γ_7. They are separated in energy from the other states depending on the uniaxial crystal-field a_{cf}^{v} and the z-component of the spin-orbit coupling-constant a_{soz}^{v}. Last but not least the states $|1/2, \pm 1/2\rangle$ with energy e_{34} form a second doublet of states transforming as Γ_7.

One should notice here that the energetic order of the three doublets with Γ_9, Γ_7, and Γ_7 symmetry depends on the relative strengths and signs of a_{cf}^{v} and a_{soz}^{v}. Historically, the valence bands in wurtzite structure were labeled "A", "B", and "C" according to their energy, the "A" band having the highest energy value. In most of the simple wurtzite-type semiconductors (as e.g. in CdS) the "A" band is the one with Γ_9 symmetry, but there are exceptions: Most probably in ZnO under normal conditions one of the Γ_7 bands has the highest energy and is labeled "A", followed by the band "B", which corresponds to the doublet with Γ_9 symmetry. Since the symmetry of the states influences e.g. the selection rules of optical transitions, it is more relevant to classify the states according to their symmetry than to their energetic order. Therefore, when referring to the "A", "B", and "C" valence bands in the following, we will suppose that the "A" band has Γ_9 symmetry and "B" and "C" are doublets with Γ_7 symmetry.

- $a^v_{cf} \neq 0$, $a^v_{soz} \neq 0$, and $a^v_{soxy} \neq 0$

Concerning the spin-orbit coupling, let us now consider in addition to $H^v_{soz} \neq 0$ the interaction term H^v_{soxy} that was neglected in the preceding case. We see from the form of the S_{55}" matrix that the states with Γ_9 symmetry

$$w^v_1 = |1\rangle\alpha \text{ and } w^v_6 = |-1\rangle\beta \tag{6.3.13}$$

are not affected by H^v_{soxy} and do not couple to other valence-band states. They have still the energy $e_{16} = (a^v_{cf} + a^v_{soz})$. On the other hand, other states characterized by the same j_z components become coupled, i.e. as indicated by the matrix S_{55}" the state

$$\begin{aligned} |1\rangle\beta &\text{ couples to } |0\rangle\alpha \\ &\text{and} \\ |0\rangle\beta &\text{ couples to } |-1\rangle\alpha \end{aligned} \tag{6.3.14}$$

due to the planar spin-orbit coupling H^v_{soxy}.

Using the notation introduced in Refs. [6, 8] the full Hamiltonian H^v describing the valence band at the Γ-point takes in this general case ($a^v_{cf} \neq 0$, $a^v_{soz} \neq 0$, and $a^v_{soxy} \neq 0$) the form

$$H^v = H^v_{cf} + H^v_{soz} + H^v_{soxy} = \begin{pmatrix} F & 0 & 0 & 0 & 0 & 0 \\ 0 & G & \Delta & 0 & 0 & 0 \\ 0 & \Delta & 0 & 0 & 0 & 0 \\ 0 & 0 & 0 & 0 & \Delta & 0 \\ 0 & 0 & 0 & \Delta & G & 0 \\ 0 & 0 & 0 & 0 & 0 & F \end{pmatrix}, \tag{6.3.15}$$

where

$$\begin{aligned} F &= (a^v_{cf} + a^v_{soz}) \\ G &= (a^v_{cf} - a^v_{soz}) \\ \Delta &= \sqrt{2} a^v_{soxy}. \end{aligned} \tag{6.3.16}$$

After diagonalizing Eq. (6.3.15) we obtain the eigenvalues of the three doublets to

$$\begin{aligned} E_1 &= (a^v_{cf} + a^v_{soz}) \\ &\text{and} \\ E_{2,3} &= \{(a^v_{cf} - a^v_{soz}) \pm \sqrt{(a^v_{cf} - a^v_{soz})^2 + 8(a^v_{soxy})^2}\}/2, \end{aligned} \tag{6.3.17}$$

which give the energies of the valence sub-bands in the presence of the planar spin-orbit coupling and of the crystal field. The normalized energies $E_i/a^v_{cf}, i = (1, 2, 3)$ are plotted against $A = a^v_{soz}/a^v_{cf}$ in Fig. 6.4a–c for different factors a^v_{soxy}/a^v_{soz} that measure the anisotropy of the system. An anti-crossing effect of the states with Γ_7

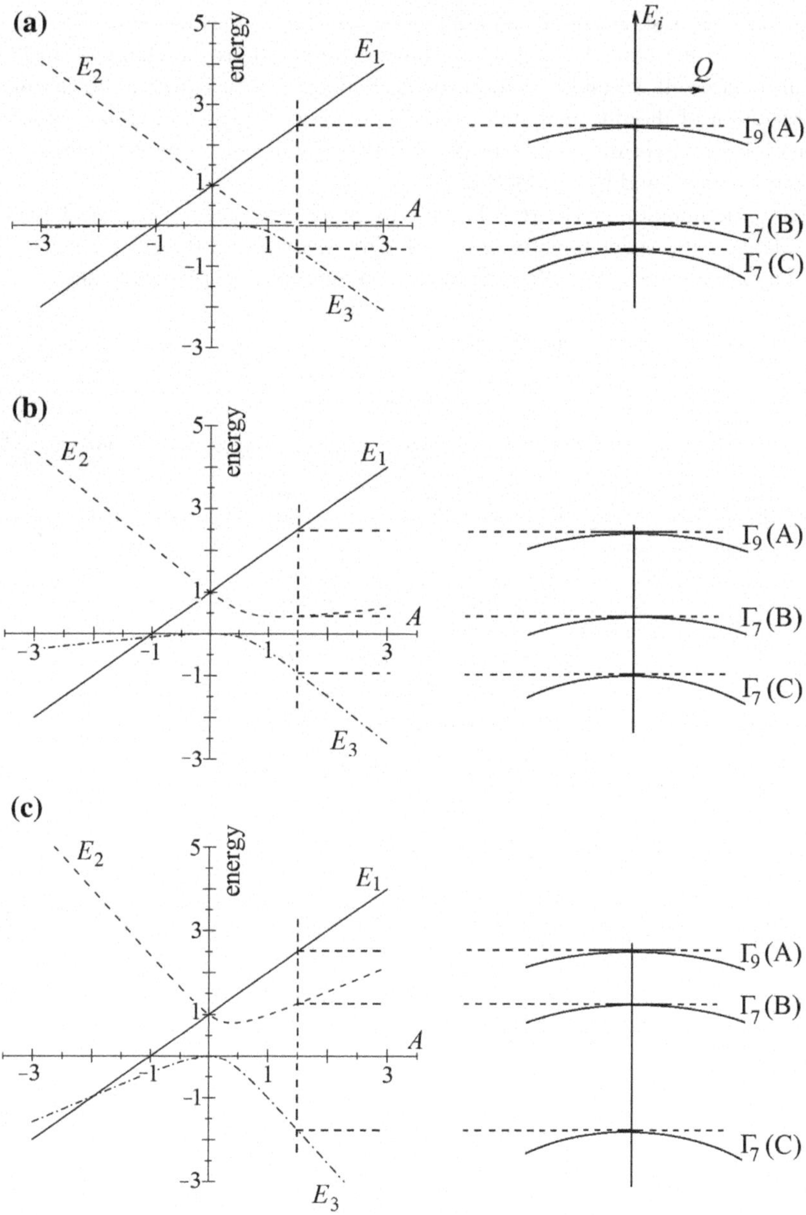

Fig. 6.4 (**a**)–(**c**) Left panel: Normalized energies E_i/a^v_{cf}, $i = (1, 2, 3)$ in the presence of the crystal field ($a^v_{cf} \neq 0$) and of an anisotropic spin-orbit coupling (**a**) $a^v_{soxy} = 0.1a^v_{soz}$, (**b**) $a^v_{soxy} = 0.3a^v_{soz}$ and (**c**) $a^v_{soxy} = 0.7a^v_{soz}$, plotted in dependence of $A = a^v_{soz}/a^v_{cf}$ as given by solution of Eq. (6.3.17). See text. Right panel displays schematically the three corresponding valence-band dispersions in the vicinity of the Brillouin zone center for $A = 1.5$. It can be seen how the energy separation between the Γ_7 bands increases with increasing planar spin-orbit coupling (i.e. increasing a^v_{soxy}), while the Γ_9 states are not influenced

symmetry clearly shows up. Its amount increases with increasing value of the factor a^v_{soxy}/a^v_{soz}. In the case if a^v_{cf} and a^v_{soz} have different signs the energetic order of the sub-bands with Γ_7 and Γ_9 symmetry is changed. In consequence the amount of hybridization of the different states with Γ_7 symmetry is also modified. One also remarks that no hybridization takes place between states of different symmetry, i.e. the states with Γ_9 and Γ_7 symmetry.

Using the notation of Eq. (6.3.1) the remaining eigenstates of Eq. (6.3.15) (the "A" valence-band states transforming as Γ_9 are given in Eq. (6.3.13)) transform as Γ_7. They are given by the following linear combinations of spin-orbit states:

$$w^v_2 = a_{2,2}|1\rangle\beta + a_{2,3}|0\rangle\alpha \text{ with energy } E_2 = \left\{(a^v_{cf} - a^v_{soz}) + \sqrt{(a^v_{cf} - a^v_{soz})^2 + 8(a^v_{soxy})^2}\right\}/2$$

and

$$w^v_3 = a_{3,2}|1\rangle\beta + a_{3,3}|0\rangle\alpha \text{ with energy } E_3 = \left\{(a^v_{cf} - a^v_{soz}) - \sqrt{(a^v_{cf} - a^v_{soz})^2 + 8(a^v_{soxy})^2}\right\}/2, \tag{6.3.18}$$

where the corresponding eigenvalues have been added. The coefficients $a_{2,2}$, $a_{2,3}$, $a_{3,2}$, $a_{3,3}$ have the values:

$$\begin{aligned} a_{2,2} &= \sqrt{2}a^v_{soxy}/\sqrt{(E_3)^2 + 2(a^v_{soxy})^2} \\ a_{2,3} &= (-E_3)/\sqrt{(E_3)^2 + 2(a^v_{soxy})^2} \\ a_{3,2} &= \sqrt{2}a^v_{soxy}/\sqrt{(E_2)^2 + 2(a^v_{soxy})^2} \\ a_{3,3} &= (-E_2)/\sqrt{(E_2)^2 + 2(a^v_{soxy})^2} \end{aligned} \tag{6.3.19}$$

For the valence band states w^v_4 and w^v_5 we obtain similarly

$$w^v_4 = a_{4,4}|0\rangle\beta + a_{4,5}|-1\rangle\alpha \text{ with energy } E_2 = \left\{(a^v_{cf} - a^v_{soz}) + \sqrt{(a^v_{cf} - a^v_{soz})^2 + 8(a^v_{soxy})^2}\right\}/2$$

and

$$w^v_5 = a_{5,4}|0\rangle\beta + a_{5,5}|-1\rangle\alpha \text{ with energy } E_3 = \left\{(a^v_{cf} - a^v_{soz}) - \sqrt{(a^v_{cf} - a^v_{soz})^2 + 8(a^v_{soxy})^2}\right\}/2 \tag{6.3.20}$$

with

$$\begin{aligned} a_{4,4} &= \sqrt{2}a^v_{soxy}/\sqrt{(E_2)^2 + 2(a^v_{soxy})^2} \\ a_{4,5} &= E_2/\sqrt{(E_2)^2 + 2(a^v_{soxy})^2} \\ a_{5,4} &= \sqrt{2}a^v_{soxy}/\sqrt{(E_3)^2 + 2(a^v_{soxy})^2} \\ a_{5,5} &= E_3/\sqrt{(E_3)^2 + 2(a^v_{soxy})^2}. \end{aligned} \tag{6.3.21}$$

One can thus see that the states w^v_2, w^v_4 and w^v_3, w^v_5 are degenerate in energy. We see that these coupled states that diagonalize Eq. (6.3.15) (involving the crystal-field

a^v_{cf} and the anisotropic spin-orbit interaction through a^v_{soz} and a^v_{soxy}) depend on four parameters. Introducing

$$a_{2,2} = a_{5,4};\ -a_{2,3} = a_{5,5} \quad \text{and} \quad a_{3,2} = a_{4,4};\ -a_{3,3} = a_{4,5}, \tag{6.3.22}$$

the "B" band states (having the energy $E_2 > E_3$) are given by

$$w^v_2 = a_{2,3}|0\rangle\alpha + a_{2,2}|1\rangle\beta \quad \text{and} \quad w^v_4 = -a_{3,3}|-1\rangle\alpha + a_{3,2}|0\rangle\beta. \tag{6.3.23}$$

In the same way the second doublet, the "C" valence-band states with energy E_3 are given by

$$w^v_3 = a_{3,3}|0\rangle\alpha + a_{3,2}|1\rangle\beta \quad \text{and} \quad w^v_5 = -a_{2,3}|-1\rangle\alpha + a_{2,2}|0\rangle\beta. \tag{6.3.24}$$

Equations (6.3.13), (6.3.17), (6.3.23) and (6.3.24) determine completely the ("A", "B", and "C") valence-band wave-functions and their energies in the considered wurtzite-type semiconductors in the presence of a crystal field and an anisotropic spin-orbit coupling. It is, however, important to notice that (depending on the value of the coefficients $a_{2,2}$, $a_{2,3}$, $a_{3,2}$, and $a_{3,3}$) the eigenfunctions of the "B" and "C" valence-band states are not necessarily orthogonal to each other. This indicates that the simultaneous presence of a crystal field and the spin-orbit coupling mixes the different spin-orbitals.

- $a^v_{cf} \neq 0, a^v_{soz} = a^v_{soxy} \neq 0$ (i.e. wurtzite valence bands in the so-called quasi-cubic approximation)

Let us discuss the special case $a^v_{cf} \neq 0$, $a^v_{soz} = a^v_{soxy} \neq 0$ in detail. After diagonalizing Eq. (6.3.15) we obtain the eigenvalues of the three doublets

$$E_1 = (a^v_{cf} + a^v_{soz}) \quad \text{and} \quad E_{2,3} = \{(a^v_{cf} - a^v_{soz}) \pm \sqrt{(a^v_{cf} - a^v_{soz})^2 + 8(a^v_{soz})^2}\}/2, \tag{6.3.25}$$

which give the energies of the valence sub-bands in the presence of the crystal field and of an isotropic spin-orbit coupling. Figure 6.5 shows the resulting energy E_i/a^v_{cf} as a function of $A = a^v_{soz}/a^v_{cf}$.

The situation ($a^v_{cf} \neq 0$, $a^v_{soz} = a^v_{soxy} \neq 0$) is often realized in real crystals since a finite crystal field modifies only slightly the spin-orbit coupling, which is usually

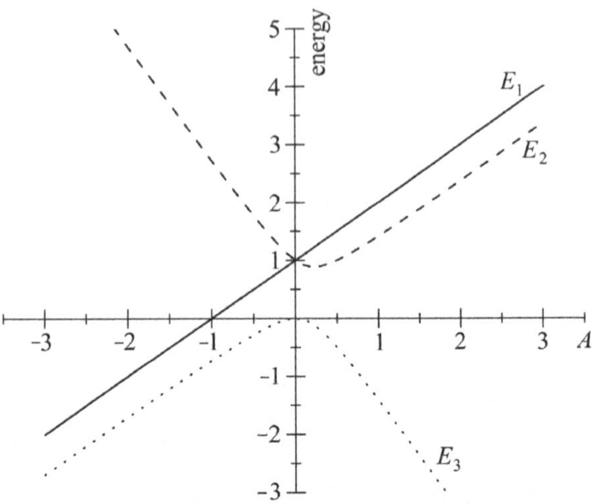

Fig. 6.5 Normalized energies E_i/a^v_{cf}, $i = (1, 2, 3)$ in the presence of the crystal field and of an isotropic spin-orbit coupling $a^v_{soz} = a^v_{soxy} \neq 0$ as a function of $A = a^v_{soz}/a^v_{cf}$ as given by the solution of Eq. (6.3.17). E_1 denotes the energy of the exciton states with Γ_9-symmetry, E_2 and E_3 are of Γ_7-symmetry

isotropic in bulk semiconductors. Therefore it is a good approximation [1, 9] to assume that $a^v_{soz} = a^v_{soxy}$ even in uniaxial semiconductors.

In the literature (c.f. Refs. [1, 9]) an isotropic spin-orbit coupling has been considered within an approximation called the "quasi-cubic approximation", which we will now discuss. It consists in diagonalizing the spin-orbit interaction $\sim \boldsymbol{l} \cdot \boldsymbol{\sigma}$ and a small effective stress along the cubic z axis. For small (and almost vanishing) crystal-field splittings the wave-function admixtures between the "B" and "C" sub-bands are supposed to be negligible and the states remain orthogonal to each other at finite field strengths. (As shown in Fig. 6.4a–c the admixtures between the "B" and "C" sub-bands lead to a hybridization of their states, which is not considered in the "quasi-cubic approximation").

Applying this approximation the valence-band wave-functions are characterized by only two different mixing parameters called "γ" and "δ". If the crystal-field splitting vanishes they reproduce the valence-band wave-functions of zincblende-type crystals. This approximation is only valid if the crystal-field splitting is small compared to the energy gap and if the spin-orbit coupling is isotropic.

In the literature (see Ref. [1]) the valence-band states are defined by the functions $\Phi_i, i = 1, \ldots 6$. First, the A-band states, which have Γ_9 symmetry read

$$\Phi_5 = -|1\rangle\alpha(\text{corresponding } w^v_1 \text{ introduced above})$$
$$\text{and} \qquad (6.3.26)$$
$$\Phi_6 = |-1\rangle\beta(\text{corresponding } w^v_6 \text{ introduced above}).$$

They are not affected by the planar spin-orbit coupling.

The "A" band is followed by the two doublets of states transforming as Γ_7. First, the "B" band states are characterized by

$$\Phi_1 = \delta|0\rangle\alpha - \gamma|1\rangle\beta(\text{corresponding } \approx w_2^v)$$
$$\text{and} \tag{6.3.27}$$
$$\Phi_2 = \gamma|-1\rangle\alpha - \delta|0\rangle\beta(\text{corresponding } \approx w_4^v)$$

and the second doublet, the "C" valence-band states, are given by

$$\Phi_3 = -\gamma|0\rangle\alpha - \delta|1\rangle\beta(\text{corresponding } \approx w_3^v)$$
$$\text{and} \tag{6.3.28}$$
$$\Phi_4 = \delta|-1\rangle\alpha + \gamma|0\rangle\beta(\text{corresponding } \approx w_5^v).$$

The parameters γ and δ are defined through energy differences as

$$\begin{gathered} \gamma = \sqrt{2/(2+B_0^2)} \\ \delta = B_0/\sqrt{(2+B_0^2)} \\ \text{obeying } \gamma^2+\delta^2 = 1 \\ \text{and} \\ B_0 = -2 + 3\Delta E/\Delta_{so}, \end{gathered} \tag{6.3.29}$$

where $\Delta E = E_1 - E_2$ is the energy difference between the valence bands as given in Eq. (6.3.25) and Δ_{so} denotes the spin-orbit splitting in the quasi-cubic limit as given in Eq. (3.2.15).

If $a_{cf}^v = 0$, the crystal field vanishes and the states Φ_5 and Φ_6, which have Γ_9 symmetry and Φ_1 and Φ_2 with Γ_7 symmetry become degenerate. This situation is called the "quasi-cubic limit". In this case, as one can verify using a simple algebra, the wave functions are determined by $\gamma = \sqrt{1/3}$ and $\delta = -\sqrt{2/3}$.

In the quasi-cubic limit the three cubic axes $(\boldsymbol{x}, \boldsymbol{y}, \boldsymbol{z})$ are equivalent. Then the spin-orbit coupling is the same for the atomic p-orbitals, which are considered. This leads here to the fact that $a_{soz}^v = a_{soxy}^v$, and the splitting energy $\Delta_{so} = 3a_{soz}^v = 3a_{soxy}^v$ corresponds to the spin-orbit splitting observed in zincblende structure.

Let us discuss in the following our wave functions w_i^v in the quasi-cubic limit, i.e. determine their form in the limit that $a_{cf}^v \to 0$ and $a_{soz}^v = a_{soxy}^v$. We note the resulting approximated functions by q_i^v with $i = (1, ..., 6)$. We will discuss here in terms of the crystal-field coupling parameter a_{cf}^v, which we have introduced in Eq. (6.3.10), rather than in terms of the energy difference ΔE introduced in Ref. [1]. As in Ref. [1] we find from Eq. (6.3.25) in the case $a_{cf}^v = 0$

$$\begin{gathered} E_2 = a_{soz}^v \text{ and } E_3 = -2a_{soz}^v, \\ \text{leading with Eq. (6.3.19) and } a_{soz}^v = a_{soxy}^v \text{ to} \\ a_{2,3} = a_{3,2} \text{ and } a_{2,2} = -a_{3,3}. \end{gathered} \tag{6.3.30}$$

The coefficients in Eqs. (6.3.19) and (6.3.21) can now be determined for the limit that $a^v_{cf} \rightarrow 0$. Starting with $a_{2,2}$ we have:

$$a_{2,2} = \sqrt{2}/\sqrt{2 + (E_3/a^v_{soz})^2}. \tag{6.3.31}$$

Introducing $\epsilon = a^v_{cf}/a^v_{soz}$, we now obtain for the term (E_3/a^v_{soz}) by a first order power development of ϵ

$$\begin{aligned} E_3/a^v_{soz} &= [-1 + \epsilon - \sqrt{(-1+\epsilon)^2 + 8}]/2 = \\ &= [-1 + \epsilon - 3\sqrt{1 - (2/9)\epsilon}]/2 = [-2 + (2/3)\epsilon]. \end{aligned} \tag{6.3.32}$$

Introducing in analogy with Eq. (6.3.29)

$$B_1 = -2 + (2/3)(a^v_{cf}/a^v_{soz}) \tag{6.3.33}$$

one obtains:

$$\begin{aligned} \gamma = a_{2,2} &= \sqrt{2/(2 + B_1^2)} \\ &\text{and equally} \\ \delta = a_{2,3} &= B_1/\sqrt{(2 + B_1^2)}, \end{aligned} \tag{6.3.34}$$

where $a_{2,3}$ can be calculated in the same way as $a_{2,2}$ or by simply using the fact that the wave functions are normalized leading to $(\gamma^2 + \delta^2) = 1$. The other coefficients in Eqs. (6.3.22) and (6.3.30) can be obtained in a similar way.

Figure 6.6 shows the mixing coefficients γ and δ of the Γ_7 states in the presence of a crystal field and an isotropic spin-orbit coupling in dependence on the parameter $E_{cf}/E_{so} = (2/3)(a^v_{cf}/a^v_{soz})$, which measures the relative strength of the crystal-field interaction compared to the spin-orbit coupling.

Since the "A" band states with Γ_9 symmetry are not affected by the planar spin-orbit coupling we find for the valence-band wave-functions also in the frame of the quasi-cubic approximation:

$$w^v_1 = q^v_1 = |1\rangle\alpha \text{ and } w^v_6 = q^v_6 = |-1\rangle\beta. \tag{6.3.35}$$

Together with Eq. (6.3.22) we see that the set of wave functions w^v_i given in Eqs. (6.3.23) and (6.3.24) can be approximated by introducing the two mixing parameters γ and δ defined in Eq. (6.3.34). The "B" band states are given by

$$\begin{aligned} q^v_2 &= \delta|0\rangle\alpha + \gamma|1\rangle\beta \\ &\text{and} \\ q^v_4 &= \gamma|-1\rangle\alpha + \delta|0\rangle\beta \end{aligned} \tag{6.3.36}$$

and in the same way the "C" valence-band states

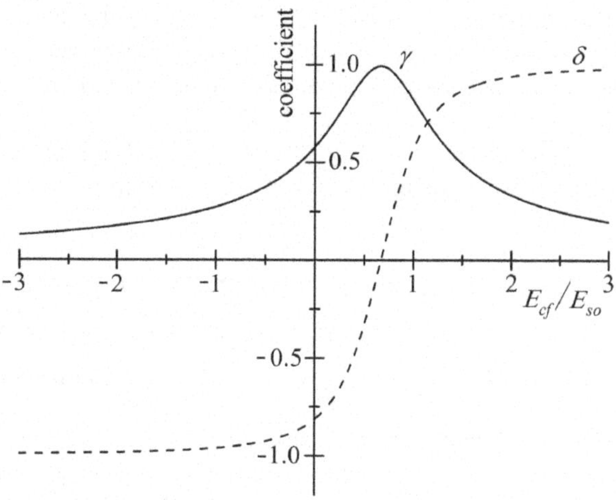

Fig. 6.6 Mixing coefficients γ and δ of the Γ_7 states in the quasi-cubic approximation in the presence of a crystal field and an isotropic spin-orbit coupling as a function of E_{cf}/E_{so}. In the quasi-cubic limit (i.e. $E_{cf} = 0$) $\gamma = \sqrt{1/3}$ and $\delta = -\sqrt{2/3}$. See text and Eq. (6.3.34)

$$q_3^v = -\gamma|0\rangle\alpha + \delta|1\rangle\beta \quad \text{and} \quad q_5^v = -\delta|-1\rangle\alpha + \gamma|0\rangle\beta. \tag{6.3.37}$$

Differences in the notations between the functions Φ_i and our functions q_i^v given above are due to not important phase factors in the definition of the wave functions. Using our definitions we keep the same phase-factor of the wave functions w_i^v and q_i^v. The advantage to use the functions q_i^v instead of the w_i^v for the valence band states lies in the fact that the different sub-band states are mutually orthogonal to each other. This facilitates the discussion of selection rules for optical transitions. The quasi-cubic approximation and the resulting notation for the wave functions will be explicitly used in Chap. 7 concerning excitons in wurtzite-type semiconductors.

- Symmetry-breaking interactions in the valence bands of crystals with C_{6v} point-group symmetry

Let us now discuss some symmetry-breaking interactions and consider first the dispersion of the valence bands. As we have demonstrated in preceding chapters, we develop the dispersion into a power series of the wave-vector $\boldsymbol{Q}$. We then consider symmetry-adapted combinations of the wave-vector components and construct, using the basis matrices given in Eq. (6.3.4), those terms, which have the correct transformation properties of a Hamiltonian. Since the number of basis matrices is finite (here 36), the number of symmetry-breaking interaction terms for a given order of the development is finite.

Table 6.2 indicates that the wave-vector component Q_z transforms as (Γ_1, K^-) and (Q_x, Q_y) as (Γ_5, K^-). Concerning terms varying linearly with the Q_z component, Eq. (6.3.4) shows that the only term transforming as (Γ_1, K^+) that can be constructed is the interaction

$$H_{Q1z} = a_{Q1z} Q_z S_{5a5}\text{''}. \tag{6.3.38}$$

The matrix S_{5a5}" couples the same states as the matrix S_{55}" given in Eq. (6.3.8) that describes the planar form of the spin-orbit interaction, i.e. it mixes further the states of the Γ_7 sub-bands. This mixing increases with increasing wave-vector component Q_z.

Concerning Q_x and Q_y, there are five different terms possible that vary linearly with the wave-vector components and transform as (Γ_1, K^+). These terms are given by:

$$\begin{aligned} H_{Q1xy} = a_{Q1xy1}(Q_x X_{51}" + Q_y Y_{51}") + a_{Q1xy2}(Q_x X_{5a2}" + Q_y Y_{5a2}")+ \\ +a_{Q1xy3}(Q_x X_{15}" + Q_y Y_{15}") + a_{Q1xy4}(Q_x X_{1a5}" + Q_y Y_{1a5}")+ \\ +a_{Q1xy5}(Q_x X_{65}" + Q_y Y_{65}"). \end{aligned} \quad (6.3.39)$$

These $\boldsymbol{Q}$-linear terms are also given in Refs. [1, 6]. Their mixing scheme is very complex since they couple different valence-band states of Γ_7 symmetry with each other. It is interesting to notice that all the different $\boldsymbol{Q}$-linear terms do not act on the Γ_9 sub-bands, which are not coupled with each other nor with the Γ_7 sub-bands, i.e. the Γ_9 sub-bands cannot carry a $\boldsymbol{Q}$-linear term in wurtzite-type semiconductors. This can be directly seen from the form of the matrices X_{ij}" and Y_{ij}" and Eq. (6.3.15): In Eq. (6.3.15) the Γ_9-states are diagonal (not coupled to other states) and remain unchanged after having diagonalized the spin-orbit coupling.

Another point is helpful when considering Eq. (6.3.39): The coefficient labeled a_{Q1xy1} should give rise to the most important $\boldsymbol{Q}$-linear term in wurtzite-type semiconductors since it is due to the mixing of the wave-vector with the angular momentum of the state. Including the electron spin does not modify this interaction. The terms proportional to a_{Q1xy2} and a_{Q1xy5} appear only when the spin-orbit interaction is added. They vanish with vanishing spin-orbit coupling. The terms proportional to a_{Q1xy3} and a_{Q1xy4} have their origin in the direct interaction of the wave-vector with the electron spin. But this coupling appears only if the spin states are mixed with the orbital momentum due to the spin-orbit coupling.

We consider that it is beyond the scope of this book to develop all terms that are of second or higher order in $\boldsymbol{Q}$. In second order one usually restricts to the common effective-mass parameters, which are well-known in wurtzite-type crystals. But the basis matrices enumerated in Eq. (6.3.4) show (together with the terms quadratic in $\boldsymbol{Q}^2$ given in Table 6.2) that nine different effective-mass terms can be constructed. They determine the warping and splitting of the valence-band dispersion. Symmetry breaking terms due to an externally applied electric field $\boldsymbol{E}$ or to strain can be derived in the same way from Table 6.2 together with Eq. (6.3.4).

Concerning the symmetry-breaking interaction-terms varying linearly with the magnetic field $\boldsymbol{B}$, Table 6.2 indicates that the B_z component transforms as (Γ_2, K^-), while $(B_y, -B_x)$ have the same transformation properties as (Q_x, Q_y), respectively. Therefore the discussion of Eq. (6.3.38) is also valid for the magnetic-field components $(B_y, -B_x)$. The linear B_z component, however, must be constructed explicitly using Eq. (6.3.4). It gives rise to an interaction Hamiltonian consisting of four terms that is given as

$$H_{B1z} = a_{B1z1} B_z T_{21}\text{''} + a_{B1z2} B_z T_{12}\text{''} + a_{B1z3} B_z T_{1a2}\text{''} + a_{B1z4} B_z T_{5a5}\text{''}. \quad (6.3.40)$$

The corresponding matrices have a similar structure as those given in Eq. (6.3.8). The coefficients $a_{B1z1} B_z$ to $a_{B1z3} B_z$ are diagonal and only split the states of the different sub-bands. The last term, proportional to $a_{B1z4} B_z$, mixes, however, also states from the different sub-bands with Γ_7 symmetry.

References

1. Cho, K.: Phys. Rev. B **14**, 4463 (1976)
2. Koster, J.F., Dimmock, J.O., Wheeler, R.G., Statz, H.: Properties of the Thirty-two Point Groups. MIT Press, Cambridge, Mass. (1963)
3. Landolt-Börnstein: New series, III/22: Semiconductors, Subvolume a: Intrinsic Properties of Group IV Elements and III-V, II-VI, and I-VII Compounds. Madelung, O., Schulz, M. (eds.), p. 400. Springer, Berlin (1987)
4. Hönerlage, B., Lévy, R., Grun, J.B., Klingshirn, C., Bohnert, K.: Phys. Rep. **124**, 161 (1985)
5. Cho, K.: Excitons, Topics in Current Physics 14, Springer, Berlin, Heidelberg (1979)
6. Bir, G.L., Pikus, G.E.: Symmetry and Strain-Induced Effects in Semiconductors. Wiley, New York, Toronto (1974)
7. Ludwig, W., Falter, C.: Symmetries in Physics : Group Theory Applied to Physical Problems, Springer Series in Solid State Sciences 64. Springer, Berlin, Heidelberg (1988)
8. Suzuki, M., Uenoyama, T., Yanase, A.: Phys. Rev. B **52**, 8132 (1995)
9. Hopfield, J.J.: J. Phys. Chem. Solids **15**, 97 (1960)

Chapter 7
Excitons in Wurtzite-Type Semiconductors

7.1 Exciton Ground-State Energy

As in Chap. 4, let us first consider the symmetry of exciton states at the Γ-point [1, 2]. One should mention at this stage that the total symmetry of exciton states Γ_{ex} is given by the symmetry of the envelope function (labeled Γ_{env}) multiplied by the symmetry of the valence-band states "w" (Γ_w) and multiplied by the symmetry of the conduction-band states (in our case Γ_7). Since the envelope function of the exciton ground state considered in the following has spherical symmetry (i.e. it has Γ_1-symmetry) one obtains:

$$\Gamma_{ex} = \Gamma_{env} \otimes \Gamma_w \otimes \Gamma_7 = \Gamma_1 \otimes \Gamma_w \otimes \Gamma_7 = \Gamma_w \otimes \Gamma_7. \tag{7.1.1}$$

We will again use the following convention: Matrices indicated by a subscript "e" are defined in the conduction-band electron-subspace. In the pseudo-spin formalism they are given by $\boldsymbol{\sigma_e} = (\sigma_e^x, \sigma_e^y, \sigma_e^z)$ and 1_e and are labeled according to Table 6.1. One thus obtains:

$$\begin{aligned} 1_e = (\sigma_e^z)^2 &\to S_e \\ \sigma_e^y &\to X_e \\ -\sigma_e^x &\to Y_e \\ \sigma_e^z &\to T_e. \end{aligned} \tag{7.1.2}$$

As mentioned before, S_e transforms according to (Γ_1, K^+). The terms (X_e, Y_e) represent the pseudo-spin matrices that transform as (Γ_5, K^-), and T_e transforms as (Γ_2, K^-), respectively. In order to construct the interaction Hamiltonian of exciton states, we calculate again the Kronecker product of the thirty-six valence-band matrices (which are given in Eq. (6.3.4) and indicated by a double quotation mark (")) with the conduction-band electron-matrices defined through Eq. (7.1.2).

Since conduction- and valence-band electron-states are characterized by spin- and orbital-angular momentum, one has to use the double-group representation also for

© Springer International Publishing AG, part of Springer Nature 2018

B. Hönerlage and I. Pelant, *Symmetry and Symmetry-Breaking in Semiconductors*, Springer Tracts in Modern Physics 279,
https://doi.org/10.1007/978-3-319-94235-3_7

the characterization of exciton states. As mentioned above, conduction-band states are made up from s-type atomic orbitals and are only spin degenerate. They transform as Γ_7. Similarly to excitons in zincblende structure, the uppermost filled valence-band states in wurtzite structure have a dimension of six at the Γ-point. They are usually made up from atomic p-orbitals with some admixtures of d-orbitals. As discussed above, their degeneracy is partly lifted by crystal-field and spin-orbit interactions. This results into three sets of doublet states, one transforming as Γ_9 and the two others as Γ_7.

After coupling of electrons and holes one obtains also three exciton series, labeled "A", "B", and "C" that have different energies. (Historically, in different substances with wurtzite structure, excitons with the lowest ground-state energy are labeled "A"-excitons. On the contrary, we will classify here the exciton series according to their transformation properties. Thus, we identify the "A" exciton series with the one where the valence-band states transform as Γ_9. As we will see this symmetry-dependent attribution is more interesting than the one relying on the energy since the symmetry of states determines the selection rules for optical transitions that are important in exciton physics.)

Introducing the symmetries of the valence-band states into Eq. (7.1.1), the transformation properties of the exciton ground states are given by:

$$\begin{aligned} \text{For the ``A'' exciton series} &\rightarrow \Gamma_1 \otimes \Gamma_9 \otimes \Gamma_7 = \Gamma_5 \oplus \Gamma_6 \\ \text{and for the ``B'' and ``C'' series} &\rightarrow \Gamma_1 \otimes \Gamma_7 \otimes \Gamma_7 = \Gamma_1 \oplus \Gamma_2 \oplus \Gamma_5. \end{aligned} \tag{7.1.3}$$

This product space, in which the exciton ground state is defined, is of dimension twelve, and all direct, exchange and symmetry-breaking perturbations can now be formulated in this exciton subspace.

According to the transformation properties of the valence-band matrices given in Eqs. (6.3.3) and (6.3.4), we see that the Hamiltonian H^{ex} describing the exciton ground state at the Γ-point may have four different contributions transforming as (Γ_1, K^+):

$$H^{ex} = (H_d^{ex} + H_{cf}^{ex}) + H_{so}^{ex} + H_{ech}^{ex}. \tag{7.1.4}$$

Here the first and the second term are not explicitly separated into the direct electron-hole interaction and a crystal-field term, appearing in the exciton problem due to the valence band:

$$(H_d^{ex} + H_{cf}^{ex}) = (a_d^{ex} S_{11}\text{''} + a_{cf}^{ex} S_{1a1}\text{''}) \otimes S_e. \tag{7.1.5}$$

The third term is due to the spin-orbit interaction of the valence band, which can again be anisotropic and have different values along the $\boldsymbol{c}$-direction and perpendicular to it, i.e. in the $(\boldsymbol{x}, \boldsymbol{y})$-plane. The spin-orbit coupling Hamiltonian then reads:

$$H_{so}^{ex} = (a_{soz}^{ex} S_{22}\text{''} + a_{soxy}^{ex} S_{55}\text{''}) \otimes S_e. \tag{7.1.6}$$

The electron spin, and therefore the indiscernibility of electrons does not play any role in the terms given above. They all involve only the unit matrix of the conduction-band electron-subspace. Therefore, at this stage, the exciton energies are given by the energy difference between the conduction band and the "A", "B", and "C" valence bands, reduced by the exciton binding energy, which can be determined by calculating the eigenvalue of the direct term H_d^{ex} in the Hamiltonian.

The last term in Eq. (7.1.4), H_{ech}^{ex}, has a different structure. It concerns the exchange interaction between a conduction-band electron and a hole in the valence band. It affects the exciton fine structure and modifies not only the direct electron-hole interaction but can also influence the crystal-field and the spin-orbit coupling. Analogously to zincblende-type semiconductors different exchange-interaction terms can be also constructed for wurtzite-type material, but, since the material is uniaxial, the exchange interaction can also be anisotropic. We will discuss these terms in a separate section.

The terms given in Eq. (7.1.4) determine now completely the energy-fine structure of the twelve-fold degenerate exciton ground state in wurtzite-type semiconductors, which is shown in Fig. 7.1. Δ_B indicates the exciton-binding energy determined with respect to the energy minimum of the electron-hole continuum of states. The degeneracy of the exciton ground state is partly lifted by crystal-field and spin-orbit interactions into the three exciton ground states labeled "A", "B", and "C". Their energy is further split by the different electron-hole exchange interactions. One should notice that exciton states with Γ_1 and Γ_5 symmetry are dipole active states.

7.2 Construction of Exciton States in Wurtzite-Type Semiconductors

As we have discussed in connection with zincblende structure, the exciton states may be constructed from the valence-band states, for which the crystal-field and the spin-orbit interaction are diagonalized. They are associated with the conduction-band states, whose interaction matrices are also diagonal with respect to these interactions. We thus build e.g. the "A"-exciton states from the valence-band states given in Eq. (6.3.13):

$$w_1^v = |1\rangle\alpha,\ w_6^v = |-1\rangle\beta \text{ and the electron states } (\alpha_e, \beta_e).$$

As discussed above (Eq. (7.1.3)) the symmetry-adapted wave functions of the exciton ground state are transforming according to:

$$\Gamma_5 \oplus \Gamma_6 \leftrightarrow \text{ for the "A"-exciton series}$$

and

$$\Gamma_1 \oplus \Gamma_2 \oplus \Gamma_5 \leftrightarrow \text{ for the "B"- and "C"-exciton series.}$$

As indicated in Table 6.1 the Γ_5 and Γ_1 states transforming as the space coordinates (x, y, z) are states, which posses a dipole-moment, while the other states are spin-triplet states, which are not dipole-active. Since we want to discuss optical transitions induced by linearly polarized light in the following we are especially interested in the states (x, y, z). (Circular or elliptically polarized light can be decomposed into two linearly polarized light fields with a fixed phase difference and thus they can also be treated within this approach.) For the different exciton series "i"$(i \in (A, B, C))$ these states are noted $|x\rangle_i^{ex}$, $|y\rangle_i^{ex}$, and $|z\rangle_i^{ex}$, respectively. The remaining triplet states are labeled $|t\rangle_i^{ex}$.

Let us now introduce conduction- and valence-band product-states. It is worthwhile to notice again that electronic transitions from valence- to conduction-band states that are induced by an electromagnetic radiation field leave the spin state of the transiting electron unchanged. Therefore, dipole-active product-states must involve in our notation spin configurations of the type $\alpha\alpha_e$ or $\beta\beta_e$.

As we will see, "A"-exciton wave functions having a linear polarization $|x\rangle_A^{ex}$, $|y\rangle_A^{ex}$ (i.e. transforming as x or y, respectively) can be be formed from the valence-band states given in Eq. (6.3.13) by $(w_1^v = |1\rangle\alpha,\ w_6^v = |-1\rangle\beta)$ and the electron states (α_e, β_e). Concerning "B"- and "C"-exciton wave functions this is in general not possible since the valence-band states w_2^v to w_5^v given in Eqs. (6.3.23) and (6.3.24) involve four different parameters, which are not independent: For the coefficients a_{ij} in Eq. (6.3.23) and (6.3.24) relations $a_{23} = a_{32} = \delta$ and $a_{22} = -a_{33} = \gamma$ hold, as we demonstrated in the preceding chapter. It means that the only combination of wave functions from Eq. (6.3.23) and (6.3.24), which could result in completely linearly polarized "B"- or "C"-exciton states, is of the type $(|1\rangle - |-1\rangle)$. Glancing at Eq. (3.1.2) implies, however, that in this way a linear polarization $|y\rangle_B^{ex}$ or $|y\rangle_C^{ex}$ cannot be achieved, since

$$a|1\rangle - b|-1\rangle \neq c|y\rangle \text{ if} a \neq b \text{ but arbitrary.}$$

This indicates that "B"- and "C"-exciton states states possess a remaining circular polarization contribution, which cannot be eliminated. In consequence, the resulting selection rules for transitions with linearly polarized light are not strict. Therefore we will construct in this section the exciton wave functions in the framework of the quasi-cubic approximation, i.e. we will use for their construction the only approximately valid valence-band wave-functions q_i^v given in Eqs. (6.3.35)–(6.3.37) instead of the functions w_i^v with $i = 1, \ldots, 6$. As discussed above this approximation is widely applicable. In addition, even if the wave functions w_i^v have to be used for a specific application, our approach gives at least an indication of the selection rules for the considered exciton transitions. Attention has to be payed, however, that the "true" states are different from the ones formed within this approximation. (Or, to put it differently, the "true" states can carry a small dipole-moment (i.e. be dipole active) while the approximated states are pure triplet states.)

It is evident that dipole-active transitions are easier to understand in the electron–defect-electron representation than in the electron-hole formulation. We will therefore start from the following product-states for the construction of the "A"-exciton states. Then, we will only at the end of this chapter transform the exciton-wave functions to the electron-hole representation. Let us introduce:

$$\begin{aligned} P_1^{ex} &= |1\rangle\alpha\beta_e = q_1^v\beta_e \\ P_2^{ex} &= |1\rangle\alpha\alpha_e = q_1^v\alpha_e \\ P_3^{ex} &= |-1\rangle\beta\beta_e = q_6^v\beta_e \\ P_4^{ex} &= |-1\rangle\beta\alpha_e = q_6^v\alpha_e. \end{aligned} \tag{7.2.1}$$

We see that the states P_2^{ex} and P_3^{ex} are dipole-active states, because they do not flip the spin of the excited electron. We thus decompose (by analogy to Eq. (4.2.4) in zincblende structure) the "A"-exciton states in wurtzite-type material into a linear combination of proper basis-product states. The same result can be obtained by direct but tedious diagonalization of the exciton Hamiltonian in the quasi-cubic approximation. The product states read

$$\begin{aligned} |x\rangle_A^{ex} &= (q_6^v\beta_e - q_1^v\alpha_e)/\sqrt{2} = (|-1\rangle\beta\beta_e - |1\rangle\alpha\alpha_e)/\sqrt{2} = (P_3^{ex} - P_2^{ex})/\sqrt{2} \\ |y\rangle_A^{ex} &= i(q_6^v\beta_e + q_1^v\alpha_e)/\sqrt{2} = i(|-1\rangle\beta\beta_e + |1\rangle\alpha\alpha_e)/\sqrt{2} = i(P_3^{ex} + P_2^{ex})/\sqrt{2} \\ |t_1\rangle_A^{ex} &= (q_6^v\alpha_e - q_1^v\beta_e)/\sqrt{2} = (|-1\rangle\beta\alpha_e - |1\rangle\alpha\beta_e)/\sqrt{2} = (P_4^{ex} - P_1^{ex})/\sqrt{2} \\ |t_2\rangle_A^{ex} &= -i(q_6^v\alpha_e + q_1^v\beta_e)/\sqrt{2} = -i(|-1\rangle\beta\alpha_e + |1\rangle\alpha\beta_e)/\sqrt{2} = -i(P_1^{ex} + P_4^{ex})/\sqrt{2}. \end{aligned} \tag{7.2.2}$$

In order to describe the "B"-exciton states we introduce product states that are constructed from the valence-band states $q_2^v = \delta|0\rangle\alpha + \gamma|1\rangle\beta$ and $q_4^v = \delta|0\rangle\beta + \gamma|-1\rangle\alpha$ given in Eq. (6.3.36). Using the notation

$$\begin{aligned} P_5^{ex} &= q_2^v\beta_e = (\delta|0\rangle\alpha + \gamma|1\rangle\beta)\beta_e \\ P_6^{ex} &= q_2^v\alpha_e = (\delta|0\rangle\alpha + \gamma|1\rangle\beta)\alpha_e \\ P_7^{ex} &= q_4^v\beta_e = (\delta|0\rangle\beta + \gamma|-1\rangle\alpha)\beta_e \\ P_8^{ex} &= q_4^v\alpha_e = (\delta|0\rangle\beta + \gamma|-1\rangle\alpha)\alpha_e \end{aligned} \tag{7.2.3}$$

one obtains for the "B"-exciton states:

$$\begin{aligned} |x\rangle_B^{ex} &= (-q_2^v\beta_e + q_4^v\alpha_e)/\sqrt{2} = (-(\delta|0\rangle\alpha + \gamma|1\rangle\beta)\beta_e + (\delta|0\rangle\beta + \gamma|-1\rangle\alpha)\alpha_e)/\sqrt{2} = \\ &= (-P_5^{ex} + P_8^{ex})/\sqrt{2} \\ |y\rangle_B^{ex} &= i(q_2^v\beta_e + q_4^v\alpha_e)/\sqrt{2} = i((\delta|0\rangle\alpha + \gamma|1\rangle\beta)\beta_e + (\delta|0\rangle\beta + \gamma|-1\rangle\alpha)\alpha_e)/\sqrt{2} = \\ &= i(P_5^{ex} + P_8^{ex})/\sqrt{2} \\ |z\rangle_B^{ex} &= (q_2^v\alpha_e + q_4^v\beta_e)/\sqrt{2} = ((\delta|0\rangle\alpha + \gamma|1\rangle\beta)\alpha_e + (\delta|0\rangle\beta + \gamma|-1\rangle\alpha)\beta_e)/\sqrt{2} = \\ &= (P_6^{ex} + P_7^{ex})/\sqrt{2} \\ |t\rangle_B^{ex} &= -i(q_2^v\alpha_e - q_4^v\beta_e)/\sqrt{2} = -i((\delta|0\rangle\alpha + \gamma|1\rangle\beta)\alpha_e - (\delta|0\rangle\beta + \gamma|-1\rangle\alpha)\beta_e)/\sqrt{2} = \\ &= -i(P_6^{ex} - P_7^{ex})/\sqrt{2}. \end{aligned} \tag{7.2.4}$$

To construct the last quartet of states (the "C"-exciton states) we start with $q_3^v = -\gamma|0\rangle\alpha + \delta|1\rangle\beta$ and $q_5^v = \gamma|0\rangle\beta - \delta|-1\rangle\alpha$ from Eq. (6.3.37) and obtain

$$\begin{aligned} P_9^{ex} &= q_3^v\beta_e = (-\gamma|0\rangle\alpha + \delta|1\rangle\beta)\beta_e \\ P_{10}^{ex} &= q_3^v\alpha_e = (-\gamma|0\rangle\alpha + \delta|1\rangle\beta)\alpha_e \\ P_{11}^{ex} &= q_5^v\beta_e = (\gamma|0\rangle\beta - \delta|-1\rangle\alpha)\beta_e \\ P_{12}^{ex} &= q_5^v\alpha_e = (\gamma|0\rangle\beta - \delta|-1\rangle\alpha)\alpha_e. \end{aligned} \tag{7.2.5}$$

One then obtains for the "C"-exciton states:

$$\begin{aligned} |x\rangle_C^{ex} &= (-q_3^v\beta_e - q_5^v\alpha_e)/\sqrt{2} = (-(-\gamma|0\rangle\alpha + \delta|1\rangle\beta)\beta_e - (\gamma|0\rangle\beta - \delta|-1\rangle\alpha)\alpha_e)/\sqrt{2} = \\ &= (-P_{12}^{ex} - P_9^{ex})/\sqrt{2} \\ |y\rangle_C^{ex} &= i(q_3^v\beta_e - q_5^v\alpha_e)/(\sqrt{2}) = i((-\gamma|0\rangle\alpha + \delta|1\rangle\beta)\beta_e - (\gamma|0\rangle\beta - \delta|-1\rangle\alpha)\alpha_e)/(\sqrt{2}) = \\ &\quad i(-P_{12}^{ex} + P_9^{ex})/\sqrt{2} \\ |z\rangle_C^{ex} &= (-q_3^v\alpha_e + q_5^v\beta_e)/\sqrt{2} = (-(-\gamma|0\rangle\alpha + \delta|1\rangle\beta)\alpha_e + (\gamma|0\rangle\beta - \delta|-1\rangle\alpha)\beta_e)/\sqrt{2} = \\ &= (-P_{10}^{ex} + P_{11}^{ex})/\sqrt{2} \\ |t\rangle_C^{ex} &= -i(q_3^v\alpha_e + q_5^v\beta_e)/\sqrt{2} = -i((-\gamma|0\rangle\alpha + \delta|1\rangle\beta)\alpha_e + (\gamma|0\rangle\beta - \delta|-1\rangle\alpha)\beta_e)/\sqrt{2} = \\ &= -i(P_{10}^{ex} + P_{11}^{ex})/\sqrt{2}. \end{aligned} \tag{7.2.6}$$

Using the exciton basis functions of Eqs. (7.2.2), (7.2.4), and (7.2.6), i.e. the states ($|x\rangle_A^{ex}$ to $|t_2\rangle_A^{ex}$), ($|x\rangle_B^{ex}$ to $|t\rangle_B^{ex}$), and ($|x\rangle_C^{ex}$ to $|t\rangle_C^{ex}$), one obtains three exciton blocks. In these exciton blocks the direct electron-hole interaction, the crystal-field interaction, and the spin-orbit coupling are diagonalized. The blocks are, however, separated in energy by the spin-orbit and crystal-field interaction, acting on the valence-band states.

It is interesting to notice the difference between the "A"-exciton states on one side and the "B"- and "C"-exciton states on the other side: in the "B"- and "C"-exciton series, where the ground-state exciton-wave function transforms as $(\Gamma_1 \oplus \Gamma_2 \oplus \Gamma_5)$, not only the dipole active ($|x\rangle^{ex}$, $|y\rangle^{ex}$) exciton states transforming as Γ_5 are present, but also the $|z\rangle^{ex}$-exciton state (having Γ_1-symmetry).

Let us now discuss the dipole moment of the exciton states in more detail. Let us consider linearly polarized light, (ξ, η, ζ) being the projection of the polarization unit vector onto the cubic axes $(\boldsymbol{x}, \boldsymbol{y}, z)$. (Here we remember that the z-axis is chosen parallel to the crystallographic $\boldsymbol{c}$-axis and the $(\boldsymbol{x}, \boldsymbol{y})$-plane is perpendicular to it.) After summing over the spin states of the conduction-band electron and the valence-band defect-electron states, we obtain the dipole moments of the exciton-states to (see Ref. [1]):

$$\begin{aligned}
&\text{for the "A"-exciton states}\\
&(|x\rangle_A^{ex}, |y\rangle_A^{ex}) = (\xi, \eta)M_0\\
&(|t_1\rangle_A^{ex}, |t_2\rangle_A^{ex}) = 0,\\
&\text{for the "B"-exciton states}\\
&(|x\rangle_B^{ex}, |y\rangle_B^{ex}, |z\rangle_B^{ex}) = (\gamma\xi, \gamma\eta, \sqrt{2}\delta\zeta)M_0\\
&(|t\rangle_B^{ex}) = 0,\\
&\text{and for the "C"-exciton states}\\
&(|x\rangle_C^{ex}, |y\rangle_C^{ex}, |z\rangle_C^{ex}) = (\delta\xi, \delta\eta, \sqrt{2}\gamma\zeta)M_0\\
&(|t\rangle_C^{ex}) = 0,
\end{aligned} \tag{7.2.7}$$

where $M_0 = \langle s|ex|x\rangle$ indicates the transition-matrix element from the crystal ground state labeled $\langle s|$ to the exciton state $|x\rangle$ induced by the dipole-moment operator ex. The exciton states labeled $(|x\rangle_i^{ex}, |y\rangle_i^{ex}, |z\rangle_i^{ex}$ $(i \in (A, B, C)))$ are dipole active and their dipole moment depends on the orientation of the polarization vector with respect to the crystallographic axis.

It is interesting to notice that the dipole moment of the "B"- and "C"-exciton states depends (through the coefficients δ and γ) in the quasi-cubic approximation on the crystal-field and spin-orbit coupling while that of the "A"-exciton states is independent of it and depends only (as it is also the case for zincblende structure) on the orientation of the dipole. As discussed above, this is due to the different admixtures of the spin-orbitals of the valence band to the singlet-exciton states in the case of the "B"- and "C"-excitons. The triplet states $(|t\rangle_i^{ex})$ carry no dipole moment in the quasi-cubic approximation.

As described in Chap. 4 and going back to Eqs. (6.3.35)–(6.3.37), we now formulate exciton states in wurtzite type crystals in the frame of the product space of electron and hole states. Therefore, we have to introduce hole states as Kramers' conjugated states $|\psi_i^h\rangle$ of the defect-electron valence-band states q_j^v considered up to now. We introduce again the index "h" in the wave-functions in order to indicate that we are dealing with the "hole" representation for the valence-band states.

The "A"-band states (q_1^v, q_6^v) that transform according to the irreducible representation Γ_9 become:

$$\begin{aligned}
q_1^v = |1\rangle\alpha = -(x+iy)\alpha/\sqrt{2} \Rightarrow\\
\Rightarrow -(x-iy)\beta_h/\sqrt{2} = -|-1\rangle\beta_h = -|\psi_1^h\rangle\\
q_6^v = |-1\rangle\beta = (x-iy)\beta/\sqrt{2} \Rightarrow\\
\Rightarrow -(x+iy)\alpha_h/\sqrt{2} = |1\rangle\alpha_h = |\psi_2^h\rangle.
\end{aligned}$$

The "B"-band states (q_2^v, q_4^v) that transform as Γ_7 are:

$$\begin{aligned}
q_2^v = \gamma|1\rangle\beta + \delta|0\rangle\alpha = \delta z\alpha - \gamma(x+iy)\beta/\sqrt{2} \Rightarrow\\
\Rightarrow \delta z\beta_h + \gamma(x-iy)\alpha_h/\sqrt{2} = \delta|0\rangle\beta_h + \gamma|-1\rangle\alpha_h = |\psi_3^h\rangle\\
q_4^v = \delta|0\rangle\beta + \gamma|-1\rangle\alpha = \delta z\beta + \gamma(x-iy)\alpha/\sqrt{2} \Rightarrow
\end{aligned}$$

$$\Rightarrow -\delta z\alpha_h + \gamma(x+iy)\beta_h/\sqrt{2} = -\delta|0\rangle\alpha_h - \gamma|1\rangle\beta_h = -|\psi_4^h\rangle,$$

followed by the second pair of states (q_3^v, q_5^v) that transform also as Γ_7, the "C"-band:

$$\begin{aligned} q_3^v &= \delta|1\rangle\beta - \gamma|0\rangle\alpha = -\gamma z\alpha - \delta(x+iy)\beta/\sqrt{2} \Rightarrow \\ &\Rightarrow -\gamma z\beta_h + \delta(x-iy)\alpha_h/\sqrt{2} = -\gamma|0\rangle\beta_h + \delta|-1\rangle\alpha_h = -|\psi_5^h\rangle \\ q_5^v &= \gamma|0\rangle\beta - \delta|-1\rangle\alpha = \gamma z\beta - \delta(x-iy)\alpha/\sqrt{2} \Rightarrow \\ &\Rightarrow -\gamma z\alpha_h - \delta(x+iy)\beta_h/\sqrt{2} = -\gamma|0\rangle\alpha_h + \delta|1\rangle\beta_h = |\psi_6^h\rangle. \end{aligned} \tag{7.2.8}$$

When comparing the valence-band wave functions in the "hole" representation with that of the "defect-electron" representation one finds the following identity:

$$\begin{aligned} |\psi_1^h\rangle &= q_6^v \\ |\psi_2^h\rangle &= q_1^v \\ |\psi_3^h\rangle &= q_4^v \\ |\psi_4^h\rangle &= q_2^v \\ |\psi_5^h\rangle &= q_5^v \\ |\psi_6^h\rangle &= q_3^v. \end{aligned} \tag{7.2.9}$$

This equivalence between the two representations is similar to that given in Eq. (4.1.5) for the case of zincblende structure.

Concerning the exciton wave functions we obtain in the electron-hole basis the following expressions:

For the"A"-band states, originating from the valence-band states (q_1^v, q_6^v):

$$\begin{aligned} P_1^{ex} &= q_1^v\beta_e \Rightarrow -|-1\rangle\beta_h\beta_e = -|\psi_1^h\rangle\beta_e \\ P_2^{ex} &= q_1^v\alpha_e \Rightarrow -|-1\rangle\beta_h\alpha_e = -|\psi_1^h\rangle\alpha_e \\ P_3^{ex} &= q_6^v\beta_e \Rightarrow |1\rangle\alpha_h\beta_e = |\psi_2^h\rangle\beta_e \\ P_4^{ex} &= q_6^v\alpha_e \Rightarrow |1\rangle\alpha_h\alpha_e = |\psi_2^h\rangle\alpha_e. \end{aligned}$$

Then one obtains the exciton states:

$$\begin{aligned} |x\rangle_A^{ex} &= (q_6^v\beta_e - q_1^v\alpha_e)/\sqrt{2} \Rightarrow (|1\rangle\alpha_h\beta_e + |-1\rangle\beta_h\alpha_e)/\sqrt{2} = (|\psi_2^h\rangle\beta_e + |\psi_1^h\rangle\alpha_e)/\sqrt{2} \\ |y\rangle_A^{ex} &= i(q_6^v\beta_e + q_1^v\alpha_e)/\sqrt{2} \Rightarrow i(|1\rangle\alpha_h\beta_e - |-1\rangle\beta_h\alpha_e)/\sqrt{2} = i(|\psi_2^h\rangle\beta_e - |\psi_1^h\rangle\alpha_e)/\sqrt{2} \\ |t_1\rangle_A^{ex} &= (q_6^v\alpha_e - q_1^v\beta_e)/\sqrt{2} \Rightarrow (|1\rangle\alpha_h\alpha_e + |-1\rangle\beta_h\beta_e)/\sqrt{2} = (|\psi_2^h\rangle\alpha_e + |\psi_1^h\rangle\beta_e)/\sqrt{2} \\ |t_2\rangle_A^{ex} &= -i(q_6^v\alpha_e + q_1^v\beta_e)/\sqrt{2} \Rightarrow -i(|1\rangle\alpha_h\alpha_e - |-1\rangle\beta_h\beta_e)/\sqrt{2} = -i(|\psi_2^h\rangle\alpha_e - |\psi_1^h\rangle\beta_e)/\sqrt{2}. \end{aligned} \tag{7.2.10}$$

The "B"- and "C"-exciton states are constructed in the same way. One obtains for the "B"-exciton states:

$$
\begin{aligned}
|x\rangle_B^{ex} &= (-q_2^v\beta_e + q_4^v\alpha_e)/\sqrt{2} = (-(\delta|0\rangle\alpha + \gamma|1\rangle\beta)\beta_e + (\delta|0\rangle\beta + \gamma|-1\rangle\alpha)\alpha_e)/\sqrt{2} \Rightarrow \\
&\Rightarrow (-(\delta|0\rangle\beta_h + \gamma|-1\rangle\alpha_h)\beta_e + (-\delta|0\rangle\alpha_h - \gamma|1\rangle\beta_h)\alpha_e)/\sqrt{2} = (-|\psi_3^h\rangle\beta_e - |\psi_4^h\rangle\alpha_e)/\sqrt{2} \\
|y\rangle_B^{ex} &= i(q_2^v\beta_e + q_4^v\alpha_e)/\sqrt{2} = i((\delta|0\rangle\alpha + \gamma|1\rangle\beta)\beta_e + (\delta|0\rangle\beta + \gamma|-1\rangle\alpha)\alpha_e)/\sqrt{2} \Rightarrow \\
&\Rightarrow i((\delta|0\rangle\beta_h + \gamma|-1\rangle\alpha_h)\beta_e + (-\delta|0\rangle\alpha_h - \gamma|1\rangle\beta_h)\alpha_e)/\sqrt{2} = i(|\psi_3^h\rangle\beta_e - |\psi_4^h\rangle\alpha_e)/\sqrt{2} \\
|z\rangle_B^{ex} &= (q_2^v\alpha_e + q_4^v\beta_e)/\sqrt{2} = ((\delta|0\rangle\alpha + \gamma|1\rangle\beta)\alpha_e + (\delta|0\rangle\beta + \gamma|-1\rangle\alpha)\beta_e)/\sqrt{2} \Rightarrow \\
&\Rightarrow ((\delta|0\rangle\beta_h + \gamma|-1\rangle\alpha_h)\alpha_e + (-\delta|0\rangle\alpha_h - \gamma|1\rangle\beta_h)\beta_e)/\sqrt{2} = (|\psi_3^h\rangle\alpha_e - |\psi_4^h\rangle\beta_e)/\sqrt{2} \\
|t\rangle_B^{ex} &= -i(q_2^v\alpha_e - q_4^v\beta_e)/\sqrt{2} = -i((\delta|0\rangle\alpha + \gamma|1\rangle\beta)\alpha_e - (\delta|0\rangle\beta + \gamma|-1\rangle\alpha)\beta_e)/\sqrt{2} \Rightarrow \\
&\Rightarrow -i((\delta|0\rangle\beta_h + \gamma|-1\rangle\alpha_h)\alpha_e - (-\delta|0\rangle\alpha_h - \gamma|1\rangle\beta_h)\beta_e)/\sqrt{2} = -i(|\psi_3^h\rangle\alpha_e + |\psi_4^h\rangle\beta_e)/\sqrt{2}.
\end{aligned}
\tag{7.2.11}
$$

The "C"-exciton states become in the electron-hole representation:

$$
\begin{aligned}
|x\rangle_C^{ex} &= (-q_3^v\beta_e - q_5^v\alpha_e)/\sqrt{2} = (-(-\gamma|0\rangle\alpha + \delta|1\rangle\beta)\beta_e - (\gamma|0\rangle\beta - \delta|-1\rangle\alpha)\alpha_e)/\sqrt{2} \Rightarrow \\
&\Rightarrow (-(-\gamma|0\rangle\beta_h + \delta|-1\rangle\alpha_h)\beta_e - (-\gamma|0\rangle\alpha_h + \delta|1\rangle\beta_h)\alpha_e)/\sqrt{2} = (|\psi_5^h\rangle\beta_e - |\psi_6^h\rangle\alpha_e)/\sqrt{2} \\
|y\rangle_C^{ex} &= i(q_3^v\beta_e - q_5^v\alpha_e)/(\sqrt{2}) = i((-\gamma|0\rangle\alpha + \delta|1\rangle\beta)\beta_e - (\gamma|0\rangle\beta - \delta|-1\rangle\alpha)\alpha_e)/\sqrt{2} \Rightarrow \\
&\Rightarrow i((-\gamma|0\rangle\beta_h + \delta|-1\rangle\alpha_h)\beta_e - (-\gamma|0\rangle\alpha_h + \delta|1\rangle\beta_h)\alpha_e)/\sqrt{2} = i(-|\psi_5^h\rangle\beta_e - |\psi_6^h\rangle\alpha_e)/\sqrt{2} \\
|z\rangle_C^{ex} &= (-q_3^v\alpha_e + q_5^v\beta_e)/\sqrt{2} = (-(-\gamma|0\rangle\alpha + \delta|1\rangle\beta)\alpha_e + (\gamma|0\rangle\beta - \delta|-1\rangle\alpha)\beta_e)/\sqrt{2} \Rightarrow \\
&\Rightarrow (-(-\gamma|0\rangle\beta_h + \delta|-1\rangle\alpha_h)\alpha_e + (-\gamma|0\rangle\alpha_h + \delta|1\rangle\beta_h)\beta_e)/\sqrt{2} = (|\psi_5^h\rangle\alpha_e + |\psi_6^h\rangle\beta_e)/\sqrt{2} \\
|t\rangle_C^{ex} &= -i(q_3^v\alpha_e + q_5^v\beta_e)/\sqrt{2} = -i((-\gamma|0\rangle\alpha + \delta|1\rangle\beta)\alpha_e + (\gamma|0\rangle\beta - \delta|-1\rangle\alpha)\beta_e)/\sqrt{2} \Rightarrow \\
&\Rightarrow -i((-\gamma|0\rangle\beta_h + \delta|-1\rangle\alpha_h)\alpha_e + (-\gamma|0\rangle\alpha_h + \delta|1\rangle\beta_h)\beta_e)/\sqrt{2} = -i(-|\psi_5^h\rangle\alpha_e + |\psi_6^h\rangle\beta_e)/\sqrt{2}.
\end{aligned}
\tag{7.2.12}
$$

The representation in Eqs. (7.2.10)–(7.2.12) is interesting since it shows directly the origin of the notation "spin-singlet" states (the dipole-active exciton-states labeled $|x\rangle^{ex}$, $|y\rangle^{ex}$, $|z\rangle^{ex}$) and the "spin-triplet" states $|t_1\rangle^{ex}$ and $|t_2\rangle^{ex}$, which are forbidden in optical transitions. As it is introduced in atomic physics the spins of two electrons are anti-parallel in the case of singlet states, while they are parallel in triplet states. This convention is also used in exciton physics when speaking of the orientation of the spins of an electron and of the associated hole which together make up the considered exciton.

We have stated above that the spin-orbit and crystal-field splitting of the valence band lead to an energy splitting of the different exciton series. In general, however, the spin-orbit and crystal-field splitting of the exciton series can be slightly modified with respect to the values of the valence band by the electron-hole exchange-interaction. This is due to the fact that exchange-interaction can lead to a mixture of the exciton states of the different series. We are going to discuss some aspects of electron-hole exchange-interaction in the following section.

7.3 Electron-Hole Exchange-Interaction in the Exciton Ground State of Wurtzite-Type Material

The last term in Eq. (7.1.4) H^{ex}_{ech} concerns the exchange interaction between a conduction-band electron and a hole in the valence band of wurtzite-type material. As mentioned before, in order to determine the full description of the exchange interaction in the exciton problem, one has to calculate the Kronecker product of the matrices described by Eqs. (6.3.4) and (7.1.2). This procedure has been discussed in detail in Sect. 4.3 for zincblende-type material. In this context we have stated already that the electron-hole exchange-interaction gives usually only rise to small corrections of the exciton-energy fine-structure that is mainly determined by the crystal-field and spin-orbit interactions. These two interactions dominate also the energy fine structure of the valence band.

In wurtzite-type semiconductors different exchange-interaction terms can be constructed but, on the contrary to zincblende structure, the material is uniaxial and thus the exchange interaction is also anisotropic. According to the transformation properties of the basis matrices given in Eqs. (6.3.4) and (7.1.2) nine different terms can now be established. They are:

$$
\begin{aligned}
H^{ex}_{ech1z} &= a^{1z}_{ech}(T_{21}\text{''} \otimes T_e) \\
H^{ex}_{ech1xy} &= a^{1xy}_{ech}(X_{51}\text{''} \otimes X_e + Y_{51}\text{''} \otimes Y_e) \\
H^{ex}_{ech2z} &= a^{2z}_{ech}(T_{12}\text{''} \otimes T_e) \\
H^{ex}_{ech2xy} &= a^{2xy}_{ech}(X_{15}\text{''} \otimes X_e + Y_{15}\text{''} \otimes Y_e) \\
H^{ex}_{ech3z} &= a^{3z}_{ech}(T_{1a2}\text{''} \otimes T_e) \\
H^{ex}_{ech3xy} &= a^{3xy}_{ech}(X_{1a5}\text{''} \otimes X_e + Y_{1a5}\text{''} \otimes Y_e) \\
H^{ex}_{ech4z} &= a^{4z}_{ech}(T_{5a5}\text{''} \otimes T_e) \\
H^{ex}_{ech4xy} &= a^{4xy}_{ech}(X_{5a2}\text{''} \otimes X_e + Y_{5a2}\text{''} \otimes Y_e) \\
H^{ex}_{ech5xy} &= a^{5xy}_{ech}(X_{65}\text{''} \otimes X_e + Y_{65}\text{''} \otimes Y_e).
\end{aligned}
\tag{7.3.1}
$$

In this representation the first eight terms H^{ex}_{ech1z} to H^{ex}_{ech4xy} have the same origin as the corresponding terms H^{ex}_{ech1} to H^{ex}_{ech4} given in Eq. (4.3.1) for the zincblende structure. In wurtzite-type material they can be different along the **c**-direction and perpendicular to it, leading to a higher number of possible terms. Only the term H^{ex}_{ech5xy} is new in wurtzite structure and vanishes in zincblende structure. This can be easily inferred e.g. according to the subscripts of the relevant valence-band matrices X_{65}" and Y_{65}": The subscript "6" means that the angular-momentum components transform as Γ_6, while in zincblende structure the basis matrices transform only as $\Gamma_1, \Gamma_3, \Gamma_4$, or Γ_5 (see Eq. (3.1.5)). All terms in Eq. (7.3.1) are the signature of the indiscernibility of electrons.

As discussed above H^{ex}_{ech2z} and H^{ex}_{ech2xy} as well as H^{ex}_{ech3z} and H^{ex}_{ech3xy} describe the direct exchange interaction between valence-band and conduction-band electrons, which are in the same spin state. In this exchange interaction the orbital-angular momentum of the states is not involved. On the other hand, H^{ex}_{ech1z} and H^{ex}_{ech1xy} describe the spin-orbit coupling between the angular momentum of the valence band, which interacts directly with the spin of the conduction-band electron.

The other terms are more complicated. They involve in addition to the exchange interaction the spin-orbit coupling between the hole spin and angular momentum. In an approach applying perturbation theory, they would be of higher order than the foregoing terms. When considering electron-hole exchange interaction in the exciton problem, one would therefore restrict to the terms H^{ex}_{ech2z} and H^{ex}_{ech2xy} as well as to H^{ex}_{ech3z} and H^{ex}_{ech3xy} in the case of small spin-orbit coupling and to H^{ex}_{ech1z} and H^{ex}_{ech1xy} if spin-orbit coupling largely exceeds the electron-hole exchange interaction.

7.4 Symmetry-Breaking Effects and Exciton Exchange Interaction in Wurtzite-Type Semiconductors

The effect of magnetic fields on energy levels of both "A" (Γ_9 valence band) and "B" (Γ_7 valence band) excitons in hexagonal CdSe was studied by Komarov et al. [3]. Generally speaking, the interaction of carriers/excitons with external magnetic fields and the crystal field is rather weak and sometimes difficult to be observed experimentally. It was, however, revealed that doping of CdSe with magnetic impurities (here Mn ions) considerably enhanced the effect of symmetry breaking on excitons

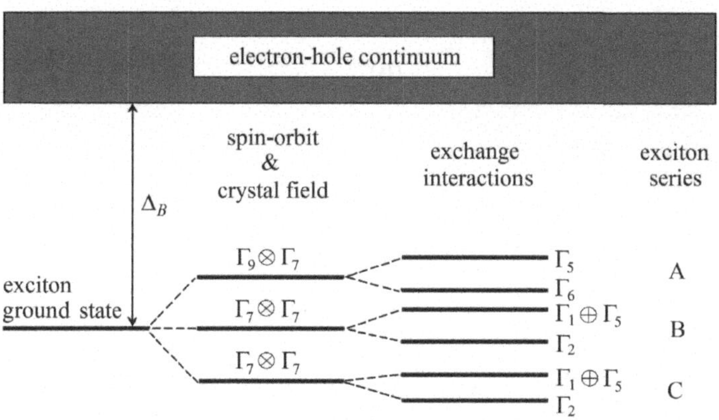

Fig. 7.1 Energy level scheme of excitons resulting from a conduction band with Γ_7-symmetry and a valence bands with Γ_9 and two others with Γ_7-symmetries in simple semiconductors with wurtzite structure (C_{6v} point-group symmetry). Δ_B indicates the exciton-binding energy determined with respect to the energy minimum of the electron-hole continuum of states. The twelve-fold degeneracy of the exciton ground state is partly lifted by crystal-field and spin-orbit interactions into the three exciton ground states labeled "A", "B", and "C". Their energy is further split by the different electron-hole exchange interactions. Exciton states with Γ_1 and Γ_5 symmetry are dipole-active states. See text

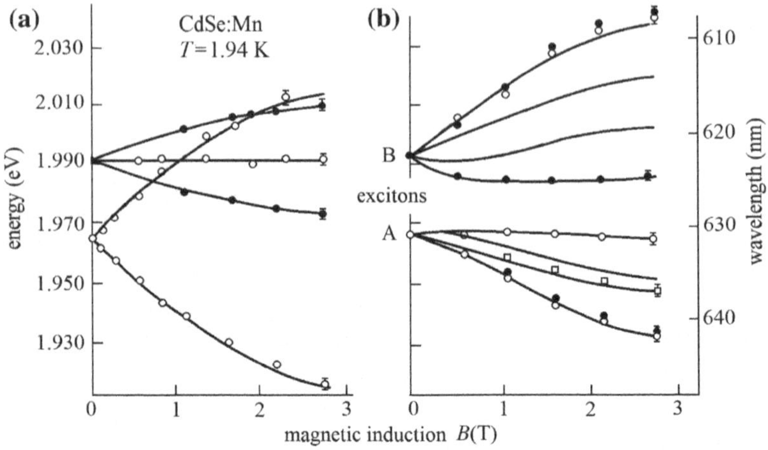

Fig. 7.2 Positions of the maxima of the A and B excitonic reflection singularities in hexagonal CdSe:Mn versus external magnetic induction B for (a) $\boldsymbol{B} \parallel \boldsymbol{c}$ and (b) $\boldsymbol{B} \perp \boldsymbol{c}$. Different experimental points •, ○ and □ correspond to different orientations of the electric light field vector $\boldsymbol{E}$ with respect to $\boldsymbol{c}$-axis and to the direction of $\boldsymbol{B}$. Temperature $T = 1.94$ K, solid curves represent results of calculation. According to Ref. [3]

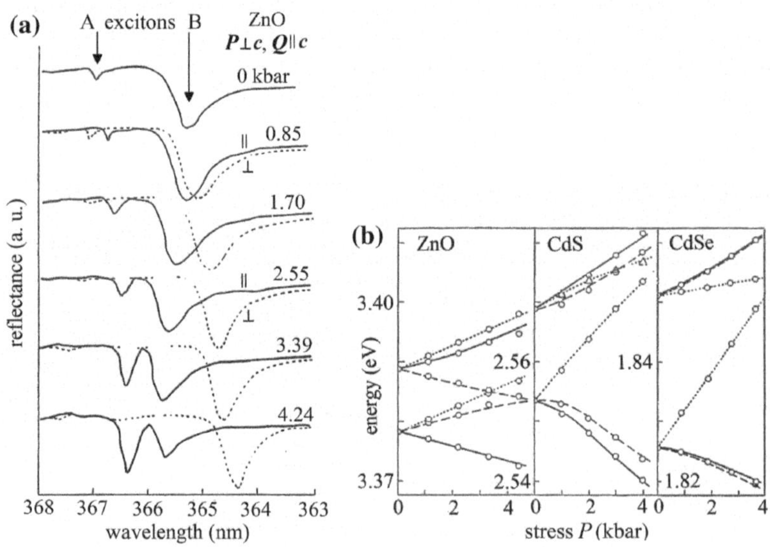

Fig. 7.3 (**a**) Excitonic reflectivity curves for uniaxially stressed hexagonal ZnO at $T = 1.8$ K. Values of applied uniaxial pressure $\boldsymbol{P}$ are indicated at each curve. Geometry of experiment: $\boldsymbol{P} \perp \boldsymbol{c}$ and $\boldsymbol{Q} \parallel \boldsymbol{c}$. Symbols (⊥) and (∥) in the panel denote orientations of electric light field vector $\boldsymbol{E}$ with respect to $\boldsymbol{P}$. (**b**) Curves of averaged measurements on ZnO, CdS and CdSe. Solid and dashed lines mean different geometry of experiment in terms of $\boldsymbol{E}$, $\boldsymbol{P}$, and $\boldsymbol{c}$. Points are results of theoretical calculations. According to Ref. [4]

in a magnetic field. As a result, a giant splitting of excitonic states was observed via measuring optical reflection spectra at $T = 1.94$ K in a CdSe sample containing a Mn concentration of 10 Mol. percent. From typical reflection singularities (c.f. Figs. 1.2 and 1.3) plots were obtained as shown in Fig. 7.2, which represent splitting of the A and B excitonic levels as a function of external magnetic induction $\boldsymbol{B}$. (It is worth mentioning that a difference in wavelength from 617 nm (~2.010 eV) up to 649 nm (~1.910 eV) is really "giant".) The reason for this unusually large spectral splitting of ~32 nm (about 100 meV) consists in the strong exchange interaction of electrons with the 3d orbitals of Mn ions; in fact, in an external magnetic field that polarizes spins of the impurities, this exchange interaction becomes equivalent to the action of an additional effective magnetic field. Full curves in Fig. 7.2 represent results of calculations.

Symmetry-breaking effects due to a strain-dependent action exerted on excitons in wurtzite-type semiconductors is demonstrated by Fig. 7.3. Again, a very important role is played (as we shall see shortly) by the exchange interaction; this time by electron-hole interaction within an exciton. Panel (a) in Fig. 7.3 displays optical reflectivity curves measured at $T = 1.8$ K in the excitonic region of a hexagonal ZnO crystal exposed to different uniaxial pressures. Panel (b) of this figure reviews splittings and shifts of the lines as a function of the applied stress together with similar results obtained on hexagonal CdS and CdSe [4]. The overall situation being quite complicated (splitting of excitonic lines occurred only in a particular experimental geometry, different split components exhibited various polarizations etc.), we do not intend to go into details. We wish to stress one point only: A customary interpretation of these results in terms of a deformation potential concept was not successful. Only when, in addition, also electron-hole exchange interaction was taken into account very good agreement between theory (open circles in Fig. 7.3b) and experiment was achieved. Then, the maximal wavelength difference between split lines in Fig. 7.3b amounts to ~7 nm (about 30 meV) only, which contrasts heavily with the giant splitting in Fig. 7.2.

References

1. Cho, K.: Phys. Rev. B **14**, 4463 (1976)
2. Koster, J.F., Dimmock, J.O., Wheeler, R.G., Statz, H.: Properties of the Thirty-two Point Groups. MIT Press, Cambridge, Mass. (1963)
3. Komarov, A.V., Ryabchenko, M., Semenov, Yu.G., Shanina, B.D., Vitrikhovskil, N.I.: Sov. Phys. JETP **52**, 783 (1980)
4. Koda, T., Langer, D. W., Euwema, R. N.: IX. International Conference on the Physics of Semiconductors, Moscow 1968, vol. 1, p. 242. Nauka, Leningrad (1968)

Chapter 8
Light-Matter Interaction and Exciton-Polaritons in Semiconductors

8.1 Propagating Electromagnetic Light Fields in a Dielectric Medium

Light corresponds to a transverse electromagnetic radiation field. It is characterized by its electric- $\boldsymbol{e}$ and magnetic-polarization vectors $\boldsymbol{h}$. $\boldsymbol{e}$ and $\boldsymbol{h}$ are both oscillating in time with frequency ω and propagating in space, characterized by the wave-vector $\boldsymbol{q}$. The fields obey the wave equation and will be described below by harmonic plane waves. The electric- $\boldsymbol{e}$ and magnetic-polarization vectors $\boldsymbol{h}$ are special forms of electric field $\boldsymbol{E}$ and magnetic induction vectors $\boldsymbol{B}$ whose evolution in space and time is governed by Maxwell's equations.

The wave aspect of an electromagnetic radiation field can be most easily derived from Maxwell's equations for the electric field and magnetic induction vector, written in their differential form. Maxwell's equations then read in terms of $\boldsymbol{E}$ and $\boldsymbol{B}$:

$$\nabla \cdot (\epsilon\epsilon_0)\boldsymbol{E} = \rho \tag{8.1.1}$$

$$\nabla \cdot \boldsymbol{B} = 0 \tag{8.1.2}$$

$$\nabla \times \boldsymbol{E} = -\frac{\partial \boldsymbol{B}}{\partial t} \tag{8.1.3}$$

$$\nabla \times 1/(\mu\mu_0)\boldsymbol{B} = (\boldsymbol{J} + (\epsilon\epsilon_0)\frac{\partial \boldsymbol{E}}{\partial t}), \tag{8.1.4}$$

where "∇" is the nabla operator. Maxwell's equations involve the following quantities:

- $\boldsymbol{E}$: Electric field vector
- $\boldsymbol{B}$: Magnetic induction vector
- $\boldsymbol{J}$: Electric current density vector

© Springer International Publishing AG, part of Springer Nature 2018

B. Hönerlage and I. Pelant, *Symmetry and Symmetry-Breaking in Semiconductors*, Springer Tracts in Modern Physics 279,
https://doi.org/10.1007/978-3-319-94235-3_8

- ρ: charge density
- ϵ_0: vacuum permittivity
- ϵ: relative permittivity or dielectric function
- μ_0: vacuum permeability
- μ: relative permeability

Equations (8.1.1)–(8.1.4) describe the response of a material on an applied electric or magnetic field. Therefore, one introduces usually the following quantities:

- $\boldsymbol{P} = \epsilon_0 \chi \boldsymbol{E}$: Electric polarization vector
- $\boldsymbol{D} = \epsilon \epsilon_0 \boldsymbol{E} = \epsilon_0 \boldsymbol{E} + \boldsymbol{P}$: Electric displacement vector
- $\boldsymbol{H} = 1/(\mu\mu_0)\boldsymbol{B}$: Magnetic field vector
- $\boldsymbol{M} = \boldsymbol{B} - \mu_0 \boldsymbol{H}$: Magnetic polarization vector

In nonmagnetic semiconductors, in which we are interested here, the response to magnetic fields is only slightly different from that of vacuum. In general it is not necessary to go into details but it is sufficient to introduce the magnetic polarization vector $\boldsymbol{M}$ as the magnetic dipole moment per unit volume. As stated above, in homogeneous, linear (non ferromagnetic), and isotropic media the magnetic field $\boldsymbol{H}$ and the magnetic induction vector $\boldsymbol{B}$ are parallel and the relative permeability μ is a scalar, not a tensor.

Since the response to magnetic fields is usually not important in nonmagnetic semiconductors we will not discuss problems connected to the magnetic response function μ in more detail. This is different for electric fields where the material's response is described by the relative permittivity (or dielectric function) ϵ, on which we will concentrate.

The dielectric function $\epsilon = (1 + \chi)$ is determined by the dielectric susceptibility χ. This quantity establishes in a general way the relation between the electric field vector $\boldsymbol{E}$ and the electric polarization vector $\boldsymbol{P}$ that the field has induced in the material. It should be carried in mind that χ (and thus ϵ) are nonlinear functions of $\boldsymbol{E}$ and have tensor character in anisotropic materials. For the sake of simplicity this aspect is neglected throughout this section.

The electric current density vector $\boldsymbol{J}$ in Eq. (8.1.4) is connected to the electric field vector $\boldsymbol{E}$ through Ohm's law

$$\boldsymbol{J} = \sigma \boldsymbol{E}, \tag{8.1.5}$$

where σ denotes the conductivity of the material. Since $\boldsymbol{J}$ is characterized by the presence of free carriers, it is also called the "displacement current".

The optical properties of intrinsic or only weakly doped semiconductors can be described by a vanishing electric current density $\boldsymbol{J}$ or a low conductivity. The electric current density is not very important in the context of this chapter concerning exciton-polaritons in intrinsic semiconductors. The second term of Eq. (8.1.4) is, however, more subtle since it originates from the presence of bound charges, which can be displaced from their equilibrium position by the electric field $\boldsymbol{E}$. This contribution is often called the "polarization current". It involves the dielectric function ϵ and depends on the time variation of the electric field. It is therefore of crucial importance when discussing the optical properties of materials that are nonmagnetic and nonconducting.

Let us first consider the propagation of electromagnetic fields in vacuum. For vacuum, the following conditions hold:

$$\boldsymbol{J} = 0;\ \rho = 0;\ \boldsymbol{P} = 0;\ \boldsymbol{M} = 0. \tag{8.1.6}$$

They correspond to $\epsilon = \mu = 1$ and lead with Eq. (8.1.1) to $\nabla \cdot \boldsymbol{E} = 0$. In vacuum one obviously has $\boldsymbol{D} \parallel \boldsymbol{E}$ and $\boldsymbol{B} \parallel \boldsymbol{H}$. Introducing the conditions Eq. (8.1.6) into Eqs. (8.1.3) and (8.1.4) we obtain

$$\nabla \times \boldsymbol{E} = -\mu_0 \frac{\partial \boldsymbol{H}}{\partial t} \text{ and } \nabla \times \boldsymbol{H} = \epsilon_0 \frac{\partial \boldsymbol{E}}{\partial t}. \tag{8.1.7}$$

Applying the differentiations $(\nabla \times)$ and $(\frac{\partial}{\partial t})$ to Eq. (8.1.7) and interchanging temporal and spatial differentiations lead to

$$\nabla \times \nabla \times \boldsymbol{E} = -\mu_0 \nabla \times \frac{\partial \boldsymbol{H}}{\partial t} \text{ and } \nabla \times \frac{\partial \boldsymbol{H}}{\partial t} = \epsilon_0 \frac{\partial^2 \boldsymbol{E}}{\partial t^2}. \tag{8.1.8}$$

We now use the properties of the nabla operator:

$$\nabla \times \nabla \times \boldsymbol{E} = \nabla(\nabla \cdot \boldsymbol{E}) - \nabla^2 \boldsymbol{E}. \tag{8.1.9}$$

And since $\nabla \cdot \boldsymbol{E} = 0$ we obtain from Eq. (8.1.8)

$$\nabla^2 \boldsymbol{E} = \mu_0 \epsilon_0 \frac{\partial^2 \boldsymbol{E}}{\partial t^2}, \tag{8.1.10}$$

where "$\nabla^2 = \Delta$" is the Laplace operator. The same differential equation, which is the well-known wave equation in the absence of any damping process, is also obtained for the magnetic field vector $\boldsymbol{H}$.

In conclusion, we find that when solving Maxwell's equations in vacuum, all components Ψ_i of the electromagnetic radiation field vectors $\boldsymbol{h}$ and $\boldsymbol{e}$ obey the wave equation, which reads in Cartesian coordinates (x, y, z) in its general form

$$\frac{\partial^2 \Psi_i}{\partial x^2} + \frac{\partial^2 \Psi_i}{\partial y^2} + \frac{\partial^2 \Psi_i}{\partial z^2} = (1/v^2) \frac{\partial^2 \Psi_i}{\partial t^2}. \tag{8.1.11}$$

The electromagnetic radiation field can be described by harmonic plane waves of the form

$$\begin{gathered} \boldsymbol{e}(\boldsymbol{r}, t) = \boldsymbol{e_0} \exp[i(\boldsymbol{qr} - \omega t)] \\ \text{and} \\ \boldsymbol{h}(\boldsymbol{r}, t) = \boldsymbol{h_0} \exp[i(\boldsymbol{qr} - \omega t)], \end{gathered} \tag{8.1.12}$$

which are solutions of the wave equation (8.1.10). The fields are characterized by a wavenumber vector $\boldsymbol{q}$, the frequency ω, and the electric and magnetic field vectors $\boldsymbol{e}_0$ and $\boldsymbol{h}_0$, respectively.

The field propagates in space and time, "v" being the phase velocity of the oscillation. Comparing Eqs. (8.1.10) and (8.1.11) one finds for the phase velocity $v = \sqrt{1/(\mu_0\epsilon_0)}$. This is the light velocity "c" in vacuum, which is one and the same constant for all frequencies.

We will now consider in more detail the orientation of the field vectors of the propagating electromagnetic field. Inserting the plane-wave function Eq. (8.1.12) into the condition $\nabla \cdot \boldsymbol{E} = 0$ one obtains

$$\nabla \cdot \boldsymbol{e}(\boldsymbol{r}, t) = i\boldsymbol{q} \cdot \boldsymbol{e} = i\boldsymbol{q} \cdot \boldsymbol{e}_0 \exp[i(\boldsymbol{q}\boldsymbol{r} - \omega t)] = 0, \tag{8.1.13}$$

which results in $\boldsymbol{e}_0 \perp \boldsymbol{q}$. Similarly, we obtain from Eq. (8.1.7)

$$\nabla \times \boldsymbol{e}(\boldsymbol{r}, t) = -\mu_0 \frac{\partial \boldsymbol{h}(\boldsymbol{r}, t)}{\partial t}. \tag{8.1.14}$$

The differentiation leads to the relation

$$\boldsymbol{q} \times \boldsymbol{e}(\boldsymbol{r}, t) = \boldsymbol{q} \times \boldsymbol{e}_0 \exp[i(\boldsymbol{q}\boldsymbol{r} - \omega t)] = \mu_0 \omega \boldsymbol{h}(\boldsymbol{r}, t) = \mu_0 \omega \boldsymbol{h}_0 \exp[i(\boldsymbol{q}\boldsymbol{r} - \omega t)] \tag{8.1.15}$$

or

$$\boldsymbol{q} \times \boldsymbol{e}_0 = \mu_0 \omega \boldsymbol{h}_0. \tag{8.1.16}$$

This indicates that for the magnetic polarization vector conditions $\boldsymbol{h}_0 \perp \boldsymbol{e}_0$ and $\boldsymbol{h}_0 \perp \boldsymbol{q}$ hold.

Consequently, in vacuum the orientation of the three vectors $\boldsymbol{h}, \boldsymbol{e}$, and $\boldsymbol{q}$ is mutually perpendicular to each other:

$$\boldsymbol{e} \perp \boldsymbol{h} \perp \boldsymbol{q} \perp \boldsymbol{e}. \tag{8.1.17}$$

Equation (8.1.17) remains valid throughout the propagation of the fields.

The term "Photons" is used to describe propagating light quanta (the quanta of the electromagnetic radiation field), which transport energy in a well defined direction. Photons are treated as quasi-particles: The photon energy is quantized and given by $\hbar\omega$ and the propagation of photons is ruled by their quasi-momentum labeled $\hbar\boldsymbol{q}$. Photons have in addition an angular momentum $J = 1$. However, it is important to remember Eq. (8.1.17) here: It indicates that the electric and magnetic fields of the propagating light quanta are perpendicular to the wave-vector. Therefore, when taking the direction of propagation given by $\boldsymbol{q}$ as direction of quantization, photons are characterized by magnetic quantum numbers ($M_J = \pm 1$); a photon state characterized by $M_J = 0$ does not exist since it would correspond to a longitudinal field.

Applying Eq. (8.1.12) (describing the propagating electromagnetic radiation field) to the wave equation Eq. (8.1.10) results into

$$c^2 = 1/(\mu_0 \epsilon_0) = (\omega/q)^2, \tag{8.1.18}$$

indicating that the light velocity is determined by the wavenumber vector and the frequency of oscillation. Both quantities are not independent, free variables but are linked with each other through the phase velocity. This fact is known as a "dispersion relation", which connects the wavenumber vector to the frequency of oscillation or the wave-vector of the radiation to the photon energy.

In problems concerning wave propagation one considers not only the phase velocity but one introduces also a velocity, called "group velocity" $\boldsymbol{v}_g$, with which a "wave packet" (an ensemble of partial waves or "wavelets" around a central frequency) propagates. Such a superposition of partial waves can e.g. describe a pulsed field. The group velocity $\boldsymbol{v}_g$ is defined through

$$\boldsymbol{v}_g = \frac{\partial \omega}{\partial \boldsymbol{q}} = grad_{\boldsymbol{q}} \omega. \tag{8.1.19}$$

If one prepares at a certain time a wave packet, propagating into a well defined direction at a given place, it will keep its spatial shape if the group velocity is independent of frequency. If $\boldsymbol{v}_g$ is frequency dependent, the partial waves, which build the wave packet, will propagate with different phase velocities and the packet will smear out, i.e. the superposition of the partial waves will change its form as a function of time and place. In vacuum, the group velocity $\boldsymbol{v}_g$ is equal to the phase velocity and $v_g = c$ is also a constant for all frequencies ω. Thus, a wave packet will keep its form during propagation in vacuum.

Let us now consider a homogeneous, isotropic, and non-absorbing dielectric medium in which the electromagnetic radiation field is propagating. The field is still characterized by the oscillation frequency ω and its propagation by a wavenumber vector, which is denoted by $\boldsymbol{q}_p$, where the index reminds that the propagation is not in vacuum but in a medium that can be polarized. The orientations of the three vectors $\boldsymbol{e}_p \perp \boldsymbol{h}_p \perp \boldsymbol{q}_p$ remain mutually perpendicular to each other but the value of $\boldsymbol{q}_p$ and therefore the phase velocity of the propagating light field are different from those in vacuum. This will be discussed in detail in the following.

Similar to the case of vacuum discussed above, the wave propagation can be determined from Maxwell's equations Eqs. (8.1.1)–(8.1.4) but with changed conditions describing the presence of a medium, which is electrically neutral but contains bound charges. The problem is treated in more details in Ref. [1] and literature cited therein. As in vacuum, the medium is considered as being linear, isotropic, nonmagnetic, and without free charges, but it can be polarized by an electric field $\boldsymbol{E}$. Then, the following conditions shall hold:

$$\boldsymbol{J} = 0; \rho = 0; \boldsymbol{P} = \chi \epsilon_0 \boldsymbol{E}; \boldsymbol{M} = 0, \tag{8.1.20}$$

where $\chi \neq 0$. The important assumption is at the moment that the polarization $\boldsymbol{P} \propto \boldsymbol{E}$, i.e. that χ is a scalar function that does not depend on the field strength.

The conditions given above correspond to $\epsilon \neq 1$, $\mu = 1$, but lead again with Eq. (8.1.1) to

$$\nabla \cdot (\epsilon\epsilon_0)\boldsymbol{E} = 0 \text{ or } \nabla \cdot \boldsymbol{D} = 0. \tag{8.1.21}$$

If $\epsilon \neq 0$ this is equivalent to

$$\nabla \cdot \boldsymbol{E} = 0. \tag{8.1.22}$$

Introducing the conditions Eq. (8.1.20) into Eqs. (8.1.3) and (8.1.4) we obtain

$$\nabla \times \boldsymbol{E} = -\mu_0 \frac{\partial \boldsymbol{H}}{\partial t} \text{ and } \nabla \times \boldsymbol{H} = \epsilon\epsilon_0 \frac{\partial \boldsymbol{E}}{\partial t}. \tag{8.1.23}$$

Proceeding as above, one obtains

$$\nabla \times \nabla \times \boldsymbol{E} = -\mu_0 \nabla \times \frac{\partial \boldsymbol{H}}{\partial t} \text{ and } \nabla \times \frac{\partial \boldsymbol{H}}{\partial t} = \epsilon\epsilon_0 \frac{\partial^2 \boldsymbol{E}}{\partial t^2} \tag{8.1.24}$$

and together with Eq. (8.1.9) one obtains

$$\nabla^2 \boldsymbol{E} = \epsilon\mu_0\epsilon_0 \frac{\partial^2 \boldsymbol{E}}{\partial t^2}. \tag{8.1.25}$$

Since in our model only the phase velocity has changed, the same discussion as in vacuum applies for the dielectric material. The most important point is that solutions of Eq. (8.1.24) are still plane waves of the form

$$\begin{aligned} \boldsymbol{e_p}(\boldsymbol{r}, t) &= \boldsymbol{e_0}\exp[i(\boldsymbol{q_p r} - \omega t)] \\ &\text{and} \\ \boldsymbol{h_p}(\boldsymbol{r}, t) &= \boldsymbol{h_0}\exp[i(\boldsymbol{q_p r} - \omega t)], \end{aligned} \tag{8.1.26}$$

characterized by a wavenumber vector $\boldsymbol{q_p}$, the frequency ω, and the polarization vectors $\boldsymbol{e_0}$ and $\boldsymbol{h_0}$, respectively.

Applying Eq. (8.1.26) to the wave equation Eq. (8.1.25) results now into

$$v^2 = 1/(\epsilon\mu_0\epsilon_0) = c^2/\epsilon = (\omega/q_p)^2, \tag{8.1.27}$$

indicating again that the phase velocity v is determined by the wavenumber vector q_p inside the medium and the frequency ω of oscillation. According to Eq. (8.1.27) it is changed when compared to vacuum by the dielectric function ϵ that is given by:

$$\epsilon = c^2 q_p^2/\omega^2 = q_p^2/q^2 \tag{8.1.28}$$

since $c = \omega/q$ (Eq. (8.1.18)).

We have defined $\epsilon = 1$ in vacuum. Normally, in a transparent dielectric medium $\epsilon \geq 1$ holds, implying that $c \geq v$ or with Eq. (8.1.28) $q_p \geq q$. Thus the photon propagation is slower in the medium than in vacuum. This gives rise to light refraction at an interface between two dielectric media with different dielectric constants since the refractive index n of the dielectric medium is related to the dielectric function by

$$n^2 = \epsilon. \tag{8.1.29}$$

At this point one should notice the fact that only transverse electromagnetic waves are solutions of Maxwell's equations in vacuum. In a dielectric medium we find, however, that Eq. (8.1.21) can also be fulfilled if $\epsilon = 0$ at a certain frequency. Then we find:

$$\begin{gathered} \boldsymbol{D} = 0 \text{ and } \nabla \cdot \boldsymbol{D} = 0 \\ \text{or} \\ \nabla \cdot \boldsymbol{D} = (\epsilon\epsilon_0)\nabla \cdot \boldsymbol{E} = (\epsilon\epsilon_0)(i\boldsymbol{q}_p \cdot \boldsymbol{E}). \end{gathered} \tag{8.1.30}$$

Equation (8.1.30) shows that a longitudinal field $\boldsymbol{E} \parallel \boldsymbol{q}_p$ is a possible solution of Maxwell's equations in the dielectric medium if $\epsilon = 0$. Using the definitions of the electric displacement vector $\boldsymbol{D}$ and of the electric polarization $\boldsymbol{P}$ one finds in the case of $\boldsymbol{D} = 0$

$$\boldsymbol{E} = -1/\epsilon_0 \boldsymbol{P}, \tag{8.1.31}$$

i.e. $\boldsymbol{E}$ and $\boldsymbol{P}$ are anti-parallel. On the other hand Eq. (8.1.23) remains still valid. This leads when using the plane-wave solutions for $\boldsymbol{E}$ and $\boldsymbol{H}$ to

$$\boldsymbol{q}_p \times \boldsymbol{E} = \mu_0 \omega \boldsymbol{H} \text{ and } \boldsymbol{q}_p \times \boldsymbol{E} = 0 \tag{8.1.32}$$

since $\boldsymbol{E} \parallel \boldsymbol{q}_p$. We thus find for longitudinal electric fields that $\boldsymbol{h}_0 = 0$ and $\boldsymbol{B} = 0$, indicating that the above solution does not describe an electromagnetic wave but a longitudinal electric polarization wave in which $\boldsymbol{E}$ and $\boldsymbol{P}$ are anti-parallel.

Let us now come back to transverse electromagnetic waves in a dielectric medium. When introducing the definition of the electric displacement vector $\boldsymbol{D} = \epsilon\epsilon_0 \boldsymbol{E} = \epsilon_0 \boldsymbol{E} + \boldsymbol{P}$ into Eq. (8.1.25) we obtain

$$\nabla^2 \boldsymbol{E} - \mu_0 \epsilon_0 \frac{\partial^2 \boldsymbol{E}}{\partial t^2} = \mu_0 \frac{\partial^2 \boldsymbol{P}}{\partial t^2}, \tag{8.1.33}$$

which relates the electric polarization vector (i.e. the response of the dielectric medium) to the time dependent electric field. Equation (8.1.33) is very general and follows immediately from Maxwell's equation.

In order to determine the response of the coupled system of dielectric medium and photons one has to define the properties of the medium. Let us first consider a simple situation where oscillators (e.g. atoms) with eigenfrequency ω_0 are coupled linearly to the propagating electromagnetic field. The oscillators are not interacting with each other, and the medium is assumed to be homogeneous, nonmagnetic, and isotropic. As discussed above, the electromagnetic light field is described by plane waves with wavenumber vector $\boldsymbol{q}_p$ and frequency ω. The atoms have a mass m and an electric charge e. They can be polarized by the electric field and each atom has an electric dipole moment

$$\boldsymbol{p} = e\boldsymbol{s}, \tag{8.1.34}$$

where $\boldsymbol{s}$ is the elongation of the charge with respect to its equilibrium position. The elongation is driven by the electric field and, in the absence of damping effects, the equation of the harmonic oscillator is given by

$$m\frac{\partial^2 \boldsymbol{s}}{\partial t^2} + \beta \boldsymbol{s} = e\boldsymbol{E} \tag{8.1.35}$$

where β is the spring constant of the oscillator. Introducing the eigenfrequency of the oscillator $\omega_0 = \sqrt{\beta/m}$ and using the dipole moment instead of the elongation as variable one obtains

$$\frac{\partial^2 \boldsymbol{p}}{\partial t^2} + \omega_0^2 \boldsymbol{p} = (e^2/m)\boldsymbol{E} \tag{8.1.36}$$

For a macroscopic system one rather introduces the polarization $\boldsymbol{P}$ and the density of atoms N. Then one obtains in addition to Eq. (8.1.33) a second equation, which relates the electric polarization vector to the time dependent electric field:

$$\frac{\partial^2 \boldsymbol{P}}{\partial t^2} + \omega_0^2 \boldsymbol{P} = (Ne^2/m)\boldsymbol{E}. \tag{8.1.37}$$

Eliminating from Eq. (8.1.33) and (8.1.37) the electric polarization $\boldsymbol{P}$ and using the relations

$$\frac{\partial^2 \boldsymbol{E}}{\partial t^2} = -\omega^2 \boldsymbol{E} \text{ and } \frac{\partial^2 \boldsymbol{P}}{\partial t^2} = -\omega^2 \boldsymbol{P},$$

one obtains a differential equation for the electric field:

$$\left(\frac{\partial^2}{\partial t^2} + \omega_0^2\right)\left(\nabla^2 \boldsymbol{E} - \mu_0\epsilon_0 \frac{\partial^2 \boldsymbol{E}}{\partial t^2}\right) = (\mu_0 N e^2/m)\frac{\partial^2 \boldsymbol{E}}{\partial t^2}. \tag{8.1.38}$$

Using the plane waves given in Eq. (8.1.26) as solution ansatz for $\boldsymbol{E}$, one finds that the plane waves are solutions of Eq. (8.1.38) if the constants fulfill the following condition:

$$(-\omega^2 + \omega_0^2)(q_p^2 - \mu_0\epsilon_0\omega^2) = \mu_0 N e^2 \omega^2/m, \tag{8.1.39}$$

where $\mu_0\epsilon_0 = 1/c^2$. When introducing the atomic polarizability as

$$\alpha = \mu_0 N e^2 c^2/m = Ne^2/(m\epsilon_0)$$

one obtains an equation relating the parameters of the dielectric material

$$c^2q_p^2/\omega^2 = 1 + \alpha/(\omega_0^2 - \omega^2). \tag{8.1.40}$$

In the framework of this "one-oscillator model" Eq. (8.1.40) indicates that q_p and ω are not independent free variables but have to fit the material parameters in order to obtain a propagating wave. It establishes the dispersion relation $\omega(q_p)$ of our model material.

The dispersion relation, which we have already introduced for vacuum in connection with Eq. (8.1.18), is an important relation that characterizes the optical properties and the field propagation in a dielectric material. We will discuss it in detail in the next sections.

The dielectric function $\epsilon(\omega)$ governs the light-matter interaction in a general way in a homogeneous system under stationary (non-transient) excitation conditions. The interaction scenario has to be revised under pulsed excitation and for inhomogeneous systems. In the following discussions we assume that a homogeneous material is interacting with a stationary electromagnetic field.

Before considering more complicated material properties let us first discuss the simple atomic model described above. According to Eqs. (8.1.28), (8.1.40) gives also the dielectric function $\epsilon(\omega)$ of the model material:

$$\epsilon(\omega) = c^2q_p^2/\omega^2 = 1 + \alpha/(\omega_0^2 - \omega^2). \tag{8.1.41}$$

Figure 8.1 shows the typical form of the dielectric function $\epsilon(\omega)$ in the case of the one-oscillator model. First, it is important to notice that in a dielectric material $\epsilon(\omega)$ is not a constant as in vacuum but depends on the photon energy $\hbar\omega$ through the material parameters α and ω_0. For a dielectric material α is usually a positive constant and $\epsilon(\omega) > 1$ for small values of ω. Then, $\epsilon(\omega)$ is first increasing with increasing ω and has a singularity at the resonance frequency ω_0. Above the resonance $\epsilon(\omega)$ is negative in the frequency range $\omega_0 < \omega < \sqrt{\omega_0^2 + \alpha}$. As discussed with Eq. (8.1.30) $\epsilon(\omega) = 0$ holds for longitudinal electric waves, which oscillate at the frequency ω_L. We will come back to this point later. Here we find that $\epsilon(\omega_L) = 0$ for $\omega_L = \sqrt{\omega_0^2 + \alpha}$. When increasing ω further, ϵ is positive and monotonously increasing again. For $\omega \longrightarrow \infty$ one obtains $\epsilon(\omega) \longrightarrow 1$, indicating that the field is no longer influenced by the resonance but completely decoupled from the oscillator.

Equation (8.1.41) gives the most simple model for simulating a dielectric material. Let us now discuss several possible extensions of this model, which may be relevant in the context of our work. In the presence of several spectrally separated resonances "j" with polarizabilities α_j and eigenfrequencies ω_{j0} Eq. (8.1.41) can be generalized to

$$\epsilon(\omega) = 1 + \sum_j \left(\alpha_j/(\omega_{j0}^2 - \omega^2)\right). \tag{8.1.42}$$

Then the scenario for $\epsilon(\omega)$ is very similar to that of the one-oscillator model: for small frequencies $\epsilon(\omega) = 1 + \sum_j \left(\alpha_j/(\omega_{j0}^2)\right) > 1$. Then $\epsilon(\omega)$ has singularities at each

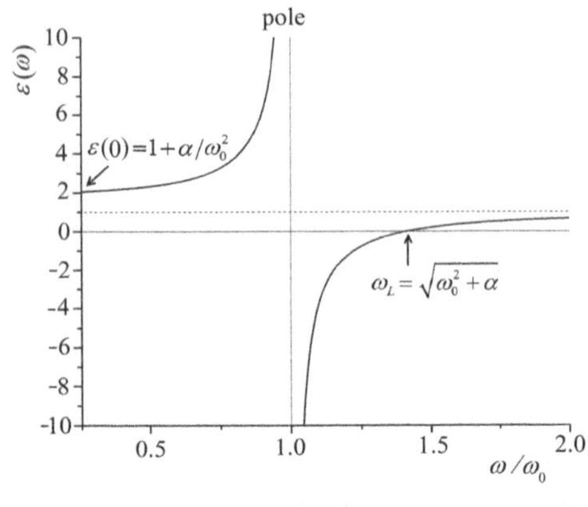

Fig. 8.1 Typical form of the dielectric function $\epsilon(\omega)$ in the case of the one-oscillator model without damping and spatial dispersion. With α: polarizability; ω_0: resonance frequency (frequency of transverse mode); ω_L: frequency of longitudinal mode. (In a quantum mechanical formulation one uses: oscillator strength f_s; energy of transverse mode E_T; energy of longitudinal mode E_L. Then: $\alpha \rightarrow f_s$; $\hbar\omega_0 \rightarrow E_T$; $\hbar\omega_L = \hbar\sqrt{\omega_0^2 + \alpha} \rightarrow E_L$)

resonance frequency ω_{j0}, followed by a frequency region where $\epsilon(\omega)$ is negative, before becoming positive again. As for the one-oscillator model we obtain $\epsilon(\omega) \longrightarrow 1$ when $\omega \longrightarrow \infty$. This scenario indicates that the electromagnetic field is mainly influenced by resonances that have a higher frequency than that of the field, i.e. a classical oscillator decouples from a driving field that oscillates too rapidly such that the oscillator cannot follow.

We have considered the two simplest cases in which independent (not mutually coupled) oscillators with the same or different eigenfrequencies are coupled to the electromagnetic radiation field. This approach can be easily generalized to several interesting cases, including damping or the mutual coupling between oscillators: If a linear damping of the form $(\gamma m \frac{ds}{dt})$ is added to Eqs. (8.1.35), (8.1.41) takes the form

$$\epsilon(\omega) = 1 + \alpha/(\omega_0^2 - \omega^2 - i\omega\gamma). \tag{8.1.43}$$

In this case, $\epsilon(\omega)$ is a complex function, the real part giving the index of refraction and the imaginary part the absorption of the system. The case of several, spectrally separated resonances "j" including damping constants γ_j, may be treated in the same way.

If the frequency ω is close to a single, isolated resonance with frequency ω_s this resonance dominates all other contributions and determines the dielectric function. When including damping constants γ_j for the different resonances, Eq. (8.1.43) then reads

$$\epsilon(\omega) = 1 + \sum_j \left(\alpha_j/(\omega_{j0}^2 - \omega^2 - i\omega\gamma_j)\right). \tag{8.1.44}$$

The dielectric function can then be approximated by replacing in Eq. (8.1.44) ω by ω_s in all terms, which are not resonant, i.e. whose contributions to $\epsilon(\omega)$ do not change considerably if ω is slightly changed in the vicinity of the resonance frequency ω_s. One then obtains:

$$\epsilon(\omega) = 1 + \sum_{j \neq s} \left(\alpha_j/(\omega_j^2 - \omega_s^2 - i\omega_s\gamma_j)\right) + \alpha_s/(\omega_s^2 - \omega^2 - i\omega\gamma_s) = \\ = \epsilon_b + \alpha_s/(\omega_s^2 - \omega^2 - i\omega\gamma_s), \tag{8.1.45}$$

where

$$\epsilon_b = 1 + \sum_{j \neq s} \left(\alpha_j/(\omega_j^2 - \omega_s^2 - i\omega_s\gamma_j)\right) \tag{8.1.46}$$

denotes the "background dielectric constant" close to the resonance "s". This approximation has the advantage that the dielectric function is now described as in the one-oscillator model in which only the constant term is modified by introducing the dielectric background, which is independent of frequency ω.

Let us now go back to the most simple one-oscillator model for a dielectric medium in the absence of damping. As discussed above, Eq. (8.1.40) determines not only the dielectric function $\epsilon(\omega)$, but its solution gives also the "dispersion relation $\omega(\boldsymbol{q_p})$" of the dielectric medium. The value of $\boldsymbol{q_p}$ is either real, corresponding to the wavenumber vector of propagating waves, or purely imaginary. An imaginary wavenumber vector $\boldsymbol{q_p}$ describes evanescent waves, which appear in the frequency range where $\epsilon(\omega)$ is negative, i.e. close to the resonance frequency ω_0. (See Ref. [1].) Electromagnetic waves cannot propagate in this spectral region and, when impinging onto a surface of a dielectric material, the wave undergoes total reflection. Inside the material is, however, an evanescent wave generated and $\mid 1/q_p \mid$ determines the penetration depth of the light field.

Figure 8.2 shows the dispersion relation $\omega(q_p)$ as given by Eq. (8.1.40). We see that if $\omega \approx 0$, i.e. if $\omega << \omega_0$ the dispersion relation simplifies to $\omega = (c/\sqrt{1 + \alpha/\omega_0^2})q_p$. This means that $\omega(q_p)$ is given by a straight line with slope $c/\sqrt{1 + \alpha/\omega_0^2} = c/\sqrt{\epsilon(0)}$. The dispersion bends down close to the resonance at $\omega = \omega_0$, where the group velocity $v_g = \dfrac{\partial \omega}{\partial q} = 0$. At higher frequencies ω is purely imaginary but becomes real again if $\alpha/(\omega_0^2 - \omega^2) > -1$. We see that this frequency ω is equal to $\omega_L = \sqrt{\omega_0^2 + \alpha}$ defined above. It characterizes the eigenfrequencies of longitudinal electric waves, which are solutions of Maxwell's equations. When $\omega = \omega_L$ the wavenumber vector $q_p = 0$ and the group velocity $v_g = \dfrac{\partial \omega}{\partial q} = 0$ again. At very high frequencies, i.e. if $\omega_0 << \omega$ the dispersion relation simplifies again to a straight line with slope c. We see that the asymptotic values of the group velocity v_g is smaller below an electronic resonance than above the resonance.

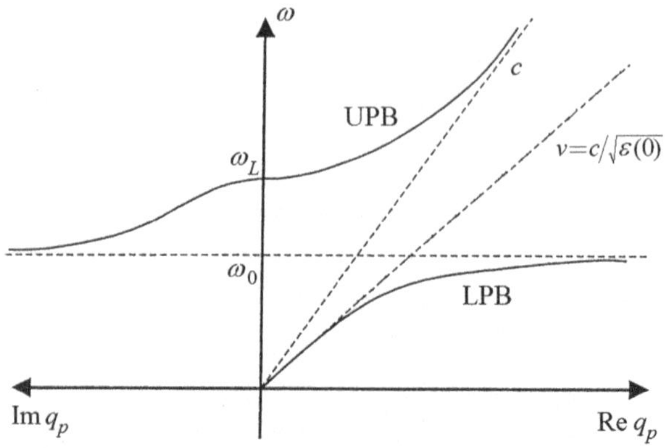

Fig. 8.2 Typical form of the dispersion relation $\omega(q_p)$ in the case of the one-oscillator model without damping and spatial dispersion. ω_0: resonance frequency; α: polarizability; $\omega_L = \sqrt{\omega_0^2 + \alpha}$: frequency of longitudinal mode; LPB (UPB): lower (upper) polariton branch; slope of the dotted line: light velocity c; slope of the dashed-dotted line: polariton phase velocity $v = c/\sqrt{\epsilon(0)}$

If different resonances are considered we obtain the same general features as for a single resonance: For a given ω the solutions for $\boldsymbol{q_p}$ are either real or imaginary, indicating either propagating or evanescent waves. If a finite damping is considered, the wavenumber vector $\boldsymbol{q_p}$ is a complex number, indicating that the field is attenuated during its propagation. The equation to be solved is then

$$c^2 q_p^2/\omega^2 = 1 + \sum_j \left(\alpha_j/(\omega_j^2 - \omega^2 - i\omega\gamma_j)\right). \tag{8.1.47}$$

Let us now go back to Eq. (8.1.40) but consider the dispersion relation in the following interesting case: If the atoms, which give rise to the electric polarization $\boldsymbol{P}$ are not independent but are interacting directly with each other, their eigenfrequency ω_0 has no longer a constant values, but ω_0 becomes a function of the wavenumber vector $\boldsymbol{q_p}$. Or, to put it differently, $\omega(\boldsymbol{q_p})$ is no longer a single valued function but has for a given frequency several solutions for the wavenumber vector $\boldsymbol{q_p}$. Due to their mutual interaction the oscillator modes noted $\omega_0(\boldsymbol{q_p})$ are said to show "spatial dispersion". Their dispersion is first determined and the interaction with the electromagnetic radiation field is then considered in a second step. This concept is used throughout this article when modes with spatial dispersion are coupled to the electromagnetic radiation field.

The dispersion relation $\omega(\boldsymbol{q_p})$ of the interacting oscillators coupled to the light field is now given by the solution of Eq. (8.1.40), which takes the form:

$$c^2 q_p^2/\omega(\boldsymbol{q_p})^2 = 1 + \alpha/(\omega_0(\boldsymbol{q_p})^2 - \omega(\boldsymbol{q_p})^2). \tag{8.1.48}$$

A typical form of the resulting dispersion relation using the one-oscillator model is shown in Fig. 8.9 and more complex cases are discussed in Sect. 8.4.

Up to now, we have treated atoms in the framework of classical mechanics. Our discussion remains valid, however, if atoms and optical transitions are treated by quantum mechanics. Fields are then still described by classical electrodynamics, which is sufficient in this work. This is the approach in Hopfield's theory [2]. In order to obtain a quantum mechanical description of the problem, one has to perform, however, several modifications (for more details see Ref. [1]):

First, one has to replace the oscillators by (in our case electronic) transitions and to introduce an oscillator strength f_s instead of the atomic polarizability α_s. In Eq. (8.1.37) the term (Ne^2/m) describes the coupling strength of the oscillators to the electromagnetic radiation field. It is replaced by the oscillator strength, which measures the coupling of the electromagnetic field to the electron that undergoes a dipole-active transition between two states $|i\rangle$ and $|j\rangle$. The transition probability is proportional to the corresponding dipole-matrix element squared

$$\left|H_{ij}^D\right|^2 = \left|\langle i|H^D|j\rangle\right|^2,$$

where H^D denotes the dipole operator, which induces the transition. The dipole-matrix element depends on the symmetry of the states $|i\rangle$ and $|j\rangle$ and on the field polarization. Following Ref. [1] the oscillator strength f_s then takes the form

$$f_s(\omega_0) = 2N\omega_0 \left|H_{ij}^D\right|^2 / (\hbar\epsilon_0),$$

which replaces the atomic polarizability α_s. (We note here that some authors define the oscillator strength as a dimensionless quantity $f' = 2m\omega_0 \left|H_{ij}^D\right|^2 / (\hbar e^2)$, which multiplies the polarizability $\alpha = Ne^2/(m\epsilon_0)$ in order to obtain f_s given above. For sake of simplicity, we follow here Ref. [2].).

Second, one considers energies rather than frequencies. We therefore note in the following by E_s the energy $\hbar\omega_s$ of the considered resonant transition "s", i.e. the energy difference between the two states, which are involved in the electronic transition. Since the energy-quantum E_s is given to the material during the excitation by the electromagnetic field (the system has a higher energy after the excitation than before, and this energy remains in the system during a certain time) this process is considered as an "elementary excitation" of the system. As discussed in preceding chapters this elementary excitation is considered as a quasi-particle, E_s being its energy (Here we are mainly considering excitons in the following but our approach is also valid for other elementary excitations as e.g. dipole-active phonons.).

In order to relate the oscillator strength $f_s(\omega_0)$ to measured quantities let us return to the one-oscillator model given in Eq. (8.1.45) for an isolated resonance ω_s, but in the absence of damping. The dielectric material is assumed to be homogeneous and isotropic. Starting from Eqs. (8.1.45) and (8.1.46) we introduce instead of

$$\epsilon(\omega) = \epsilon_b + \alpha_s/(\omega_s^2 - \omega^2)$$

the oscillator strength f_s and write:

$$\epsilon(\omega) = \epsilon_b + f_s/(\omega_s^2 - \omega^2). \tag{8.1.49}$$

One replaces now the resonance frequency ω_s by the energy of the considered transverse exciton state E_T that couples to the light field. Then, the polarization and the electric field lead to the following equation

$$\epsilon(\omega) = \epsilon_b + \hbar^2 f_s/(E_T^2 - (\hbar\omega)^2). \tag{8.1.50}$$

Figure 8.1 shows ϵ as a function of ω. At $\hbar\omega = 0$ the dielectric function turns to $\epsilon(0) = \epsilon_b + \hbar^2 f_s/(E_T^2)$. $\epsilon(\omega)$ has a singularity at $\hbar\omega_T = E_T$ and is then negative, but changes its sign again at

$$\hbar\omega = \sqrt{E_T^2 + \hbar^2 f_s/\epsilon_b}. \tag{8.1.51}$$

As discussed with Eq. (8.1.30) $\epsilon(\omega) = 0$ holds also for longitudinal electric waves. Therefore the energy of the transverse coupled modes given above is identical with the energy of the longitudinal exciton E_L (which corresponds to a longitudinal electric wave). We thus find

$$E_L = \sqrt{E_T^2 + \hbar^2 f_s/\epsilon_b}, \tag{8.1.52}$$

which implies immediately with Eq. (8.1.50)

$$\hbar^2 f_s = \epsilon_b(E_L^2 - E_T^2). \tag{8.1.53}$$

Equation (8.1.53) relates the oscillator strength directly to the longitudinal and transverse eigenfrequencies of the oscillation. This situation follows from a general rule (see Ref. [3]) according to which the frequencies of longitudinal and transverse excitations of a system are the signature of a zero and a singularity (pole) of the dielectric function, respectively. Figure 8.1 illustrates this rule. For $\hbar\omega > E_L$ the dielectric function $\epsilon(\omega)$ increases monotonously to ϵ_b when $\hbar\omega >> E_L$.

Let us now come back to the dispersion relation and consider the quasi-momentum $\hbar\boldsymbol{Q}$ of the coupled elementary excitations in the absence of spatial dispersion (i.e. E_L and E_T have constant values). (In order to stress that we consider elementary excitations as quasi-particles we denote their wave-vectors by $\boldsymbol{Q}$ and their quasi-momenta by $\hbar\boldsymbol{Q}$. The wave-vector $\boldsymbol{Q}$ is directly given by the wave-vector of the photons $\boldsymbol{q}_p$ that excite the polarizable medium in which they propagate.) Then the polarization and the electric field lead to the following equation, which determines the dispersion relation of the elementary excitations:

$$c^2\hbar^2 Q^2/E_i(Q)^2 = \epsilon_b + \hbar^2 f_s/(E_T^2 - E_i(Q)^2) = \epsilon_b \left(1 + (E_L^2 - E_T^2)/(E_T^2 - E_i(Q)^2)\right), \tag{8.1.54}$$

where $E_i(\boldsymbol{Q}) = \hbar\omega_i(\boldsymbol{Q})$ denotes the energy of this mode "i". (We have introduced an index "i" since the transverse modes can be degenerate and may depend on the polarization of the field as will be discussed later.) Since the polarizable state coupled to the light field is considered as a quasi-particle, it is called a "polariton". As in the classical case discussed above, $E_i(\boldsymbol{Q})$ is obtained from the solution of Eq. (8.1.54) and gives the polariton dispersion.

Again Fig. 8.2 shows its general form after considering the modifications mentioned above. (In addition, when compared to Eq. (8.1.40) we have introduced in Eq. (8.1.54) the background dielectric constant ϵ_b.) The dispersion $E_i(\boldsymbol{Q})$ has two branches where $\boldsymbol{Q}$ is real, which are separated in energy by a solution where $\boldsymbol{Q}$ is imaginary. The two real branches of $\boldsymbol{Q}$ are called the "lower" and "upper" polariton branch and are labeled "LPB" and "UPB", respectively, in Fig. 8.2. The slopes of these branches give the polariton group velocity. The lower polariton branch (LPB) starts with a linear slope at $\boldsymbol{Q} = 0$, $E_i(\boldsymbol{Q}) = 0$ and approaches the energy value $E_T = \hbar\omega_0$ from below for large wave-vectors $\boldsymbol{Q}$, where the polariton group velocity is equal to 0.

The upper polariton branch (UPB) starts at $\boldsymbol{Q} = 0$ and energy $E_L = \hbar\omega_L$ with a group velocity $v_g = 0$. It then varies linearly in energy for larger wave-vectors $\boldsymbol{Q}$. There, its slope is $c/\sqrt{\epsilon_b}$, i.e. it increases when compared to the lower polariton branch.

In order to better understand the identification of $E_L = \sqrt{E_T^2 + \hbar^2 f_s/\epsilon_b}$ given above, consider the following situation: An oscillating dipole is embedded in a homogeneous, isotropic material. ω_E is its eigenfreqency. At rest, the eigenfrequency is, by symmetry, independent of the dipole orientation in space, spanned by the Cartesian coordinates $(\boldsymbol{x},\boldsymbol{y},\boldsymbol{z})$. If, however, the oscillator is propagating with wave-vector $\boldsymbol{Q}$ along the z-direction, the wave-vector breaks the spherical symmetry of the system that has now cylindrical symmetry. Then the eigenfreqency of the oscillator may be different when oscillating along the z-direction (parallel to $\boldsymbol{Q}$) or perpendicular to $\boldsymbol{Q}$. Under cylindrical symmetry $\boldsymbol{x}$ and $\boldsymbol{y}$ directions remain equivalent, i.e. transverse oscillations along $\boldsymbol{x}$ or $\boldsymbol{y}$ are different modes, but they obey the same dispersion relation.

In the exciton problem described above, "longitudinal" and "transverse" modes can be only distinguished if $Q \neq 0$ since then the spherical symmetry is broken. For $Q = 0$, however, we have spherical symmetry and "longitudinal" and "transverse" modes are indistinguishable and their energy is therefore degenerate. This implies for our example that longitudinal excitons and transverse upper-branch exciton-polaritons have the same energy E_L at the Γ-point.

Polaritons with a small group velocity are said to be "exciton-like" and when the slope of the dispersion relation is large they are called "photon-like" since excitons and photons behave in that way. It is interesting to notice that the dispersions of transverse excitons and photons (the latter is given by the dotted line with slope

"c" in Fig. 8.2) cross, but for polariton branches this crossing is forbidden. This "anti-crossing" behavior (hybridization) of dispersions is well known in quantum mechanics for interacting, almost degenerate states that have the same symmetry.

Concerning the energetic region where $\boldsymbol{Q}$ is imaginary, it makes no sense to interpret it in terms of "polaritons". Polaritons are propagating quasi-particles consisting of the electromagnetic radiation field, which excites electronic transitions. Within the polariton concept, this range corresponds rather to an energy gap where wave propagation is forbidden. In this region, this energy gap allows the presence of evanescent waves.

8.2 Exciton-Polaritons in Direct-Gap Semiconductors

Let us now leave the general problem of propagating electromagnetic fields in a dielectric medium and focus on the physical situation we are interested in: excitonic polaritons in direct semiconductors close to the fundamental band gap. Then our discussion will concern photon energies of several eV (about 1 to 3 eV). Using typical values of $\epsilon = (2 \text{ to } 10)$ for semiconductors, Eq. (8.1.40) leads to photon-wave numbers q_p of several 10^5 to 10^6 cm^{-1}. When comparing to the extension of reduced Brillouin zones in simple crystals (i.e. wave numbers at the zone boundary q_B of the order of some 10^8 cm^{-1}) we see that the photon-wave numbers are very small, i.e. $q_p \approx 0$ when discussing them in comparison to q_B. This statement is equivalent to the fact that the wavelength of the plane waves considered here is long when compared to the inter-atomic distances (which gives also the order of magnitude of wave numbers at the zone boundary q_B in simple crystals). Therefore, semiconductor crystals can be considered as being homogeneous materials with respect to propagating photons. Then light diffraction due to the translational invariant atomic structure does not take place for the wavelengths considered here and impurity scattering can be neglected in a good approximation. Of course, this discussion applies only to systems whose dimensions are infinite and where surface effects and dislocations can be neglected.

Since we will discuss electronic transitions from one state to other states, we will mainly be interested in bound electron-hole states, the excitons. As discussed in the foregoing chapters excitons, having a dipole moment, are characterized by a total angular momentum $J = 1$. Because of this symmetry, transverse excitons couple to photons, which have equally an angular momentum $J = 1$ and the dipole-matrix element between a valence-band state and a conduction-band state may be $\neq 0$. This means that transverse excitons can be excited optically and the resulting elementary excitations have mixed exciton and photon character.

Direct optical transitions discussed here take place with conservation of energy and momentum and do not involve other quasi-particles as e.g. phonons or impurity scattering. The electron spin is not changed by the transition since the interaction between electrons and photons involves only the orbital part of the electron-wave functions. In a one-electron band-structure schema $E(\boldsymbol{k})$ transitions of an electron from an occupied valence-band state (characterized by its energy $E_v(\boldsymbol{k}_v)$, wave-vector $\boldsymbol{k_v}$, and spin σ_v) to an unoccupied conduction-band state (characterized by $E_c(\boldsymbol{k_e})$, $\boldsymbol{k_e}$, and σ_e) by absorption of a photon with energy $\hbar\omega$ and wave-vector $\boldsymbol{Q}$ leads to

$$\begin{aligned} E_c(\boldsymbol{k_e}) &= E_v(\boldsymbol{k_v}) + \hbar\omega \\ \boldsymbol{k_e} &= \boldsymbol{k_v} + \boldsymbol{Q} \\ &\text{and} \\ \sigma_e &= \sigma_v. \end{aligned} \tag{8.2.1}$$

According to the above discussion $Q \ll k_v$ and optical transitions are almost vertical when represented in the one-electron $E(\boldsymbol{k})$ band-structure diagram. Then Eq. (8.2.1) leads to

$$\boldsymbol{k_e} = \boldsymbol{k_v}. \tag{8.2.2}$$

Above we have introduced a model, which can be used to determine the polariton dispersion relation $E_i(\boldsymbol{Q})$ and the dielectric function $\epsilon(\omega)$ of a material. Let us now discuss some general aspects of coupled polarization-photon systems: In optics, when shining light of frequency ω onto a material, $\epsilon(\omega)$ or the dispersion relation $E_i(\boldsymbol{Q})$ impose the wave-vector (i.e. both its absolute value and the direction) that governs the propagation of the plane waves. At a given frequency and for a given polarization the dielectric function $\epsilon(\omega)$ or $E_i(\boldsymbol{Q})$ are determined by the electronic states of a material. The material becomes polarized by the light field and the electric charges can change from one state to another. But electronic transitions between states correspond to frequency resonances in the dielectric function, which depends on the photon energy $\hbar\omega$, the propagation wave-vector $\boldsymbol{Q}$ and the electric polarization vector $\boldsymbol{e_p}$ with respect to the crystal axis, the intensity of the light field, etc. Furthermore, $\epsilon(\omega)$ is a complex function since materials are usually absorbing.

In general, $\epsilon(\omega)$ and the dispersion relation $E_i(\boldsymbol{Q})$ are therefore complicated functions of the material parameters. Several measurement techniques have been developed for semiconductors and insulators in order to determine exciton-polariton structures in the presence of spatial dispersion and symmetry breaking effects.

To determine a dispersion relation needs the simultaneous knowledge of the energy E_i and of the wave-vector $\boldsymbol{Q}$ of the elementary excitation. Since propagating light fields are used in these measurement techniques, they have all to overcome an inherent difficulty: The fields are propagating inside a sample, which one wants to study, but the measuring procedure is performed outside of it. Thus, the studied signal has to be transmitted through a surface. At the surface, the propagating polaritons are partly reflected and partly transmitted and propagate as photons outside the sample. While the frequency (or photon energy) of the elementary excitation is conserved throughout the transmission process, the wave-vector is not a conserved quantity since space (at the surface) is not homogeneous. Therefore, methods have to be developed, which allow to reconstruct the wave-vector $\boldsymbol{Q}$ inside the sample from quantities that can be determined outside the sample. This is the reason why simple straightforward optical methods as one-photon absorption or conventional emission (luminescence) measurements are not suitable for determination of the polariton dispersion.

Experimental methods developed for momentum-space or "$\boldsymbol{Q}$-space spectroscopy" are among others:

- Analysis of Fabry-Perot modes
- Time-of-flight method
- Two- or three-photon spectroscopy
- Two-photon Raman scattering (denoted by some authors also as hyper-Raman scattering)
- Resonant Brillouin scattering
- Two-phonon Raman scattering
- Thin-prism method

Many details and results of the experimental techniques are easily found in literature, and we give here only some few examples in order to explain the measuring procedure.

When shining white light on a thin sample with flat, parallel surfaces Fabry-Perot modes appear in the transmission and reflection spectra of the sample. They are due to the interference of partial waves, transmitted and/or reflected from the different surfaces. Successive maxima or minima of the Fabry-Perot fringes are separated by $\Delta Q = \pi/d$ where $\boldsymbol{Q}$ is the propagation wave-number of the polariton and d the geometrical thickness of the sample. Far from a resonance, $\epsilon(\omega)$ of a material can be easily measured and for a given frequency ω the polariton wave-number $\boldsymbol{Q}$ can be determined from Eq. (8.1.28). Starting from such a point in the $(\omega, \boldsymbol{Q})$ diagram, the polariton dispersion can be reconstructed from the spectral positions of the Fabry-Perot fringes (see Refs. [4–8]). Let us consider CuCl as an example. As we will discuss in more detail in Sect. 8.3, CuCl is a semiconductor, in which the lowest lying exciton series can be well described by the one-oscillator model. As shown in Fig. 8.3 one clearly observes Fabry-Perot interference fringes in the light intensity, transmitted (T_{FP}) and reflected (R_{FP}) by the sample. From these fringes the dispersion of the upper and lower polariton branch (UPB and LPB, respectively) has been reconstructed and visualized on the right panel of Fig. 8.3 by the bold line.

As stated in Eq. (8.1.19) the group velocity $\boldsymbol{v}_g = \dfrac{\partial \omega}{\partial \boldsymbol{Q}}$ determines the propagation of a polariton through a sample. As shown in Fig. 8.4 for the example of CuCl, the variation of v_g/c as a function of photon energy is quite important in the vicinity of the exciton resonance, resulting in a "slowing down" of the propagating quasi-particles. This region is therefore also called the "bottleneck region" of the dispersion. As shown in Fig. 8.4, very small values of $v_g/c \approx 10^{-5}$ to 10^{-4} are observed when the photon energy of the exciting laser pulses is tuned through the bottleneck region. The thickness of the sample being known, the polariton group velocity can be determined by measuring the time-of-flight of a light pulse through a sample as a function of frequency. Then the dispersion relation is obtained through integration of Eq. (8.1.19) (see Refs. [9–11]).

In two-photon spectroscopy (see Fig. 8.5 as an example) one uses a spectrally broad test and a narrow pump beam of frequency ω_p to excite a sample at spatial and temporal coincidence. In Fig. 8.5 Θ is the angle between the exciting beams inside

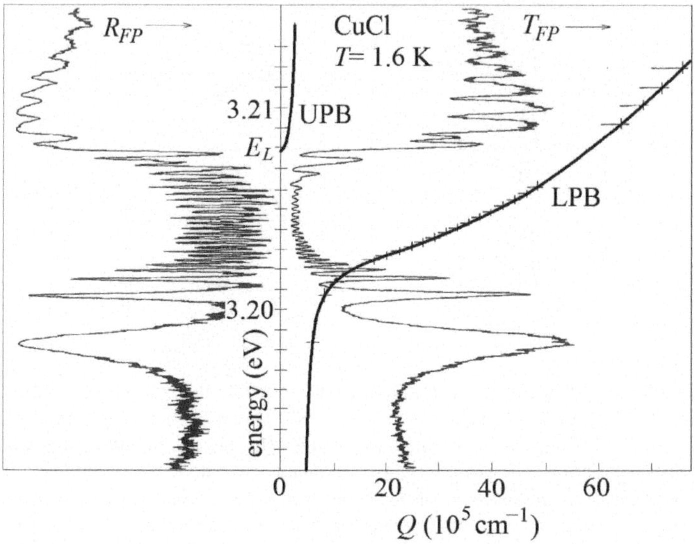

Fig. 8.3 Analysis of Fabry-Perot modes in reflection (R_{FP}, left panel) and transmission (T_{FP}, right panel) of a 0.15 μm thick CuCl platelet at a temperature $T = 1.6$ K in the Z_3-exciton series. Using the one-oscillator model the polariton dispersion is reconstructed from the interference fringes. The polariton dispersion is superimposed on the right panel and visualized by the bold line. E_L denotes the energy of longitudinal exciton. LPB (UPB) denote the lower (upper) polariton branch. (Adapted after Refs. [4, 7])

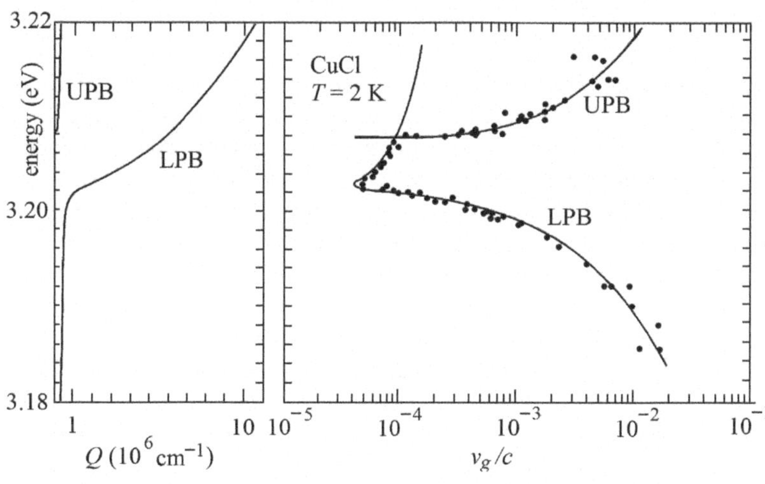

Fig. 8.4 Time-of-flight method through a semiconductor sample. The polariton dispersion (left panel) $E(Q)$ (energy E as function of wave-number Q) and the resulting relative polariton-group velocity v_g/c (right panel) against photon energy are shown together with experimental results of time-of-flight measurements on a CuCl platelet at $T = 2$ K. v_g denotes the polariton group velocity; c: light velocity; LPB (UPB) lower (upper) polariton branch. (Adapted after Refs. [1, 9, 10])

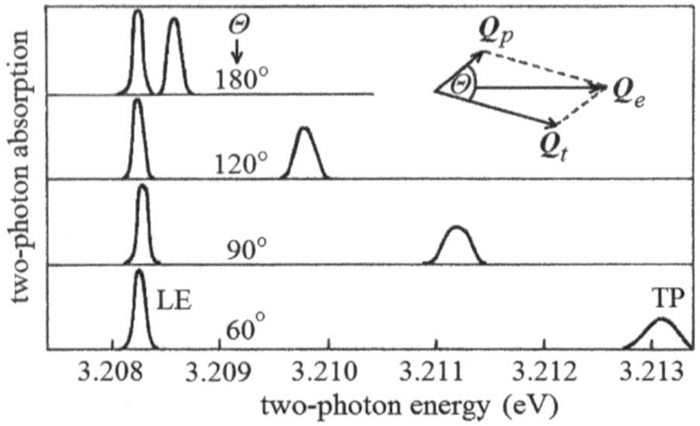

Fig. 8.5 Two-photon absorption measurements for different angular configurations as a function of the two-photon energy in CuCl (temperature $T = 1.5$ K). The inset shows the scattering configuration. The polariton wave-vectors of test and pump beams $\boldsymbol{Q}_t$, $\boldsymbol{Q}_p$ are calculated from the angles of incidence and the refraction index of the sample, Θ denotes the angle between the exciting beams. Their sum determines $\boldsymbol{Q}_e$, the final-state polariton wave-vector, created by the two-photon absorption process. In the test-beam spectrum, transmitted through a sample, two-photon absorption gives rise to absorption peaks that are due to the creation of a longitudinal exciton (LE) and to a polariton (TP) on the upper polariton branch. While the peak labeled LE stays at a fixed spectral position, the position of peak TP changes when Θ is varied, i.e. if the module of $\boldsymbol{Q}_e$ is changed. This allows to reconstruct the polariton dispersion by adjusting the parameters of Eq. (8.1.54). Since the two-photon absorption coefficients are small (of the order of 10^{-9} cmW^{-1}), high intensity laser pulses (of about 20 ns duration and several MW/cm^2 maximum peak intensity) and samples, having a thickness of several mm, have to be used. Experiments with these characteristics are possible since the photon energies of the beams are outside the exciton resonance. (Adapted after Refs. [12, 13])

the crystal. The pump beam changes the transmission of the test beam through the sample and leads to spectral hole burning at frequency ω_t. By changing the photon energy of the pump beam one verifies that an induced absorption change is due to a two-photon absorption process and not to population of exciton or impurity states, which may also be induced by the pump beam. (In a two-photon absorption process the sum $\omega_e = \omega_t + \omega_p$ remains almost constant if ω_p is changed. If an exciton population is created by the pump beam the spectral hole burning at frequency ω_t remains at a fixed frequency, independent of the frequency ω_p.) Thus, the frequency ω_e of the final state in the two-photon absorption process is measured.

If the frequency of both beams is outside of resonances, the dielectric function is constant and as mentioned above, the wave-numbers Q_t and Q_p of test and pump pulses can be calculated from ω_t and ω_p, respectively. One measures the angles of incidence of test and pump pulses, determines from the dielectric function $\epsilon(\omega) = n^2$ the corresponding refractive indexes n_t and n_p, and calculates the wavenumber vectors (direction and absolute value) $\boldsymbol{Q}_t$ and $\boldsymbol{Q}_p$ of test and pump polaritons. The sum of $\boldsymbol{Q}_t$ and $\boldsymbol{Q}_p$ determines the wavenumber vector $\boldsymbol{Q}_e$ of the final-state polariton, excited in this "two-photon absorption" process (see inset of Fig. 8.5). When changing the angle Θ between the exciting beams $\boldsymbol{Q}_e$ is varied and different points on the polariton dispersion can be reached, which can thus be partly reconstructed. One should notice that this technique does not allow to attain the lower polariton branch (see Refs. [12, 13]) but only higher polariton branches as well as longitudinal excitons. Although not dipole active, the latter can be optically excited by two-photon absorption in a non-collinear configuration.

The sample can be excited similarly in three-photon spectroscopy by three spectrally narrow laser beams. Usually, one of them has a fixed photon energy while the second one is tunable. The latter is split into two beams and the angular configuration of the three beams can be chosen. As discussed above it determines the wavenumber vector of the final state $\boldsymbol{Q}_e$ that is reached by the three-photon absorption process. In this way elementary excitation populations can be excited resonantly. They relax to other states, which give rise to photoluminescence (PL). The intensity of PL lines that are sensible to such three-photon absorption processes is measured and their maxima determined, i.e. the three-photon absorption maximum is determined for a fixed angular configuration by photoluminescence excitation (PLE) spectroscopy. As described above the energy and the wavenumber vector $\boldsymbol{Q}_e$ of the final state reached by the three-photon absorption process can be determined. Varying the angular configuration allows then to reconstruct the polariton dispersion. For details see e.g. Refs. [14, 15].

Two-photon Raman (or hyper-Raman) scattering is a quite efficient method in which one excites an almost resonant intermediate state (mostly an excitonic molecule or biexciton) virtually by two-photon absorption. One may either use one or two different, spectrally narrow, tunable laser beams to excite the sample. (The one-beam configuration is discussed in the following.) The intermediate state recombines into two other polaritons or one longitudinal exciton and one polariton, obeying energy and wave-vector conservation of the two incoming and outgoing quasi particles. In Fig. 8.6 Θ is the scattering angle between exciting and the observed polariton. The sample emission is measured spectrally resolved in a well defined direction for a fixed angle of incidence of the exciting beam. The observed emission lines shift spectrally with the photon energy of excitation. These measurements are repeated for different angels of incidence and observation in forward (dashed line in Fig. 8.6) or backward (full line in Fig. 8.6) scattering configurations. The parametrized polariton dispersion is then used in a self-consistent way to calculate from the measured angles and photon energies the wavenumber vectors $\boldsymbol{Q}_i$ ($i = 1$ to 4) of the four quasi particles involved in the scattering process. The parameters defining the dispersion relation are then adjusted such that the experimental results are reproduced. (See Refs. [16–20].)

In resonant Brillouin scattering Ref. [21] a spectrally very narrow, tunable laser excites close to the exciton resonance a polariton state in the sample. As indicated in Fig. 8.7 it scatters by emission of an acoustical phonon to another polariton state, obeying to energy and momentum conservation. The scattered polariton transforms into a photon outside the sample, which is then detected. (One often works in a backscattering configuration with phonon emission (Stokes scattering); scattering involving phonon absorption (Anti-Stokes scattering) is, however, also possible.) If the frequency of the excitation source is changed, different polariton states are excited that scatter to new final states and involve other phonons with different frequencies.

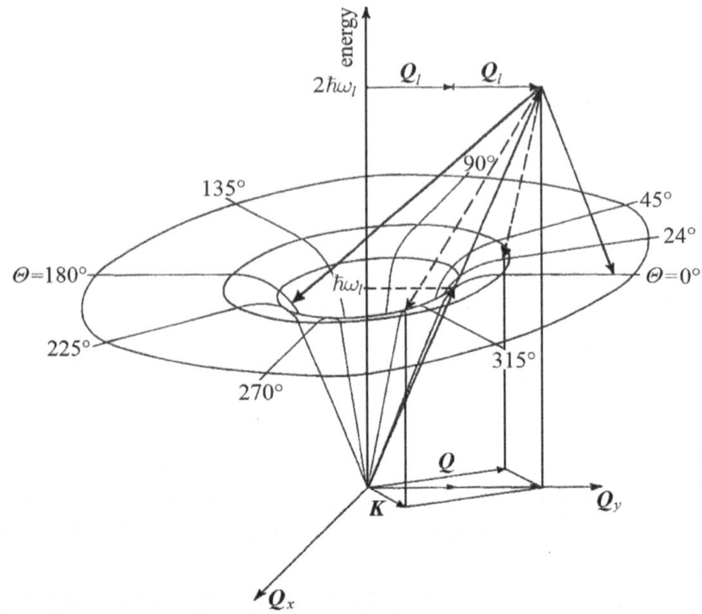

Fig. 8.6 Scheme of two-photon Raman scattering via biexcitons (see Ref. [16]) in a forward (dashed lines) and backward (full lines) scattering configuration. Two polaritons with photon energy $\hbar\omega_l$ and wave-vector $\boldsymbol{Q}_l$ excite biexcitons virtually. They recombine under energy and wave-vector conservation (here into two LBP polaritons) with wave-vector $\boldsymbol{Q}$ and $\boldsymbol{K}$, respectively. Θ is the scattering angle inside the crystal between the exciting polaritons and the final-state polariton with wave-vector $\boldsymbol{Q}$. This polariton is observed outside the crystal as a photon, emitted by the sample. The second quasi-particle with wave-vector $\boldsymbol{K}$, which is not observed in the experiment, can be either a polariton on the lower polariton branch or a longitudinal exciton

The dispersion of acoustical phonons being known, the excitonic polariton dispersion can be reconstructed close to the exciton resonance (see e.g. Refs. [22–27]).

A slightly modified scattering mechanism is used in two-phonon Raman scattering (c.f. Refs. [28, 29]). In this process, in a first step, the exciting polariton is first scattered by a longitudinal or transverse acoustical phonon to an intermediate polariton state. This scattering process can take place with absorption or emission of the acoustical phonon. The intermediate-state polariton decays under emission of a longitudinal optical phonon to a final-state polariton, which is situated on the lower polariton branch and transforms into a photon, which is then detected. Since the energy of longitudinal optical phonons is constant in the wave-vector interval relevant in two-phonon Raman processes, the excitonic polariton dispersion can be reconstructed from the known dispersion of the acoustical phonons.

The thin-prism method is a well known method to determine the refractive index n of transparent materials. A parallel light beam is transmitted trough a sample with a prismatic shape. The prism angle being known, the refractive index of the material can be determined from the deviation of the light beam and the wave-number calculated from Eq. (8.1.28) and the frequency of the light (c.f. Refs. [30, 31]). As the analysis

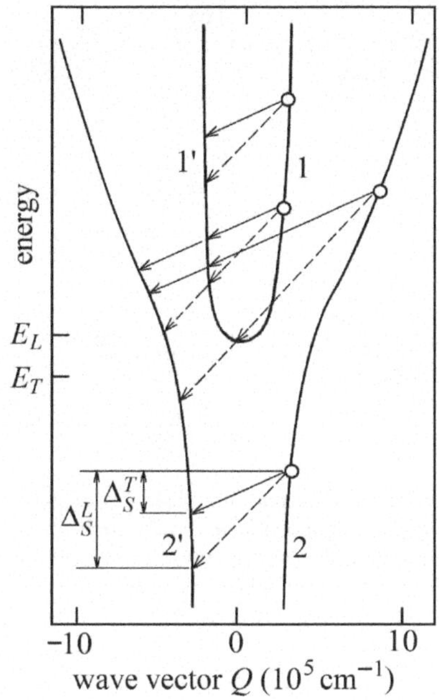

Fig. 8.7 Scheme of resonant Brillouin scattering (Stokes scattering) in a backward configuration involving the emission of longitudinal (L, dashed arrows) and transverse (T, full arrows) acoustical phonons. The full lines give the polariton dispersion $E(Q)$ of a one-oscillator model. (1, 2) and (1', 2') indicate the branches of the exciting and final-state polaritons and (1, 1') or (2, 2') denote the upper or lower polariton branches, respectively. E_L and E_T indicate the energies of the longitudinal and transverse excitons. Δ_S^L and Δ_S^T visualize the Stokes shift of the Brillouin emission lines due to a L or T phonon-emission process, respectively. (Adapted after Refs. [4, 22, 40])

of Fabry-Perot modes and the time-of-flight method, the thin-prism method is quite difficult to apply to materials with strong excitonic resonances since such materials have also a strong absorption. Then the transmitted beams are highly attenuated and the imaginary part of the dielectric function can no longer be neglected compared to its real part.

All we have said so far about polaritons has been concerned with direct band-gap semiconductors (after all, as in all the preceding chapters). Now it might be important to ask the following question: What about polaritons in indirect band-gap semiconductors, where the bottom of the conduction-band and the top of the valence-band are located at different wave-vectors $\boldsymbol{Q}$ in the first Brillouin zone?

We have sketched both cases (direct and indirect band-gap materials) in Fig. 8.8. In this figure, exciton energy levels are displayed as an extension to $\boldsymbol{Q} \neq 0$ of the relevant exciton features from Fig. 4.2 (in other words, these are the exciton dispersion curves). Along with these features, also photon dispersion curves, represented by straight lines with slopes driven by the speed of light v in the given material (the phase velocity v is defined in Eq. (8.1.27)) are shown. Here we obtain

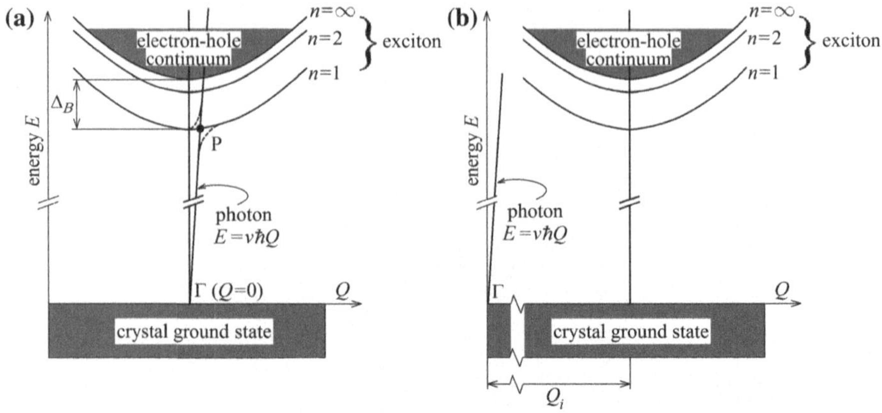

Fig. 8.8 Schematic representation of the exciton- and photon-dispersion curves in (**a**) direct band-gap and (**b**) indirect band-gap semiconductors. (**a**) While polaritons (dashed lines) are formed in the cross-over region of exciton and photon dispersion curves (around point "P") in direct semiconductors, such a cross-over region (see panel (**b**)) does not exist in indirect semiconductors. (See text)

$$v = c/\sqrt{\epsilon}$$
$$\text{and} \tag{8.2.3}$$
$$E = \hbar\omega = v\hbar Q.$$

In the left panel (a) of Fig. 8.8, i.e. in a direct band-gap semiconductor, the exciton and photon dispersion curves cross each other. If both quasi particles have the same symmetry (i.e. in the present case: if the excitons are dipole-active) they give rise in this way to degenerate states that belong to both systems, excitons and photons. Then, as is well known from quantum mechanics, the dispersion curves must split at the point of intersection P in order to remove the degeneracy (leading to the "anti-crossing" behavior mentioned already above). This is illustrated in panel (**a**) by the dashed curves around P. The region in the vicinity of P then schematically represents polariton-dispersion curves, which have been introduced in this section and will be treated in detail in the rest of this chapter (as discussed above "polaritons" are understood as mixed states of excitons and photons).

On the other hand, in an indirect semiconductor the dispersion curves of excitons and photons do not intersect, as shown in panel (**b**) of Fig. 8.8. There are two reasons for this: First, since the speed of light is extremely high, the slope of the photon dispersion is also very large and the "photon-like region" is limited to a narrow interval of wave-vector Q values (with $0 < Q < 10^5\ \text{cm}^{-1}$) around the Γ-point. Second, in virtually all known semiconductors with indirect band gap the Q_i-point (this is the point of minimum energy of the indirect exciton) is located very far from the Γ-point, namely, at (or close to) the first Brillouin-zone boundary ($Q_i \approx 10^8\ \text{cm}^{-1}$). For these reasons the exciton and photon states do not enter into an immediate interaction (interaction may be mediated e.g. by phonon fields) and thus there is no reason to consider exciton-polariton states in indirect band-gap materials.

Perhaps a last note is worth mentioning at this point. The optical properties of semiconductors are usually treated in the so-called dipole approximation $\boldsymbol{Q} = 0$ (see Refs. [1, 32]). This approximation covers most of the excitonic properties and experimental manifestations, but not all of them. First, because $\boldsymbol{Q}$ is in fact a little bit larger than zero, the photon dispersion curve in direct-gap semiconductors must cross that of the excitons (see Fig. 8.8a), leading quite naturally to the polariton concept. Second, the diffusion of electronic excitations (free excitons) in real space over a distance of $1/Q \approx 10^{-5}$ cm (= 100 nm) causes non-local effects known as "polariton spatial dispersion" (or wave-vector dependence of the dielectric constant). From the point of view of symmetry these phenomena can be regarded as "symmetry breaking" due to the final magnitude (even if very small) of the wave-vector $\boldsymbol{Q}$. Thus, the polariton concept goes beyond the common dipole approximation but, at the same time, it is important to realize that this concept is limited to a very narrow range of wave-vectors $\boldsymbol{Q}$ close to the Γ point.

8.3 Exciton-Polaritons in the $\Gamma_6 \otimes \Gamma_7$ Subspace in Zincblende-Type Semiconductors

As already discussed in Sect. 4.1, excitons are bound states of an electron and a hole in a semiconductor. Besides the band indexes of the states, excitons are usually characterized according to their center-of-mass wave-vector $\boldsymbol{Q}$ and their spin σ. In the electron-hole representation wave-vector and spin of the excitons are given by

$$\boldsymbol{Q} = \boldsymbol{k}_e + \boldsymbol{k}_h \quad \text{and} \quad \sigma = \sigma_e + \sigma_h, \tag{8.3.1}$$

$\boldsymbol{k}_e$ and σ_e being the wave-vector and spin of the electron. Since it will be important in the following discussion we recall that the hole wave-vector $\boldsymbol{k}_h$ and spin σ_h are obtained from the wave-vector $\boldsymbol{k}_v$ and spin σ_v of the missing electron in the valence band by Kramers' conjugation, i.e.

$$\boldsymbol{k}_h = -\boldsymbol{k}_v \quad \text{and} \quad \sigma_h = -\sigma_v. \tag{8.3.2}$$

Using Eq. (8.2.2) together with Eq. (8.3.1) we see that the center-of-mass wave-vector $\boldsymbol{Q}$ of excitons interacting with the electromagnetic radiation field in direct semiconductors is very small when compared with the wave-vector characterizing the boundary of the Brillouin zone q_B, i.e. $Q/q_B \approx 10^{-3}$ for dipole-active excitons in which we are interested here.

Electrons and holes being charged quasi-particles, they interact with each other via Coulomb interaction and they are subject to the periodic crystal potential. The electron-hole interactions, which possess the full crystal symmetry have been discussed in Chap. 4 and especially in Sect. 5.1. Using the $|J, M_J\rangle$ pseudo-spin notation for the exciton states in the electron-hole representation, one obtains the four states given in Eqs. (5.1.7) and (5.1.8).

While the attractive direct Coulomb interaction leads to the exciton-binding energy, which is the same for the $J = 0$ and $J = 1$ exciton states, the repulsive electron-hole exchange-interaction splits the singlet-exciton states (the $J = 1$ states) with Γ_5 symmetry from the triplet-exciton state with Γ_2 symmetry (see Fig. 4.3). The spin-triplet exciton state corresponds to the eigenstate of pseudo-spin ($J = 0$, $M_J = 0$), which does not carry a dipole moment. It is not affected by the exchange interaction.

$J = 1$ exciton states posses a dipole moment. The eigenstates consist of two transverse exciton states and a longitudinal one, in which the dipole moment is orientated along the center-of-mass wave-vector $\boldsymbol{Q}$. The dipole moment of the two transverse exciton states is orientated perpendicular to $\boldsymbol{Q}$. For completeness, we mention at this point that one can define transverse and longitudinal exciton states only if the center-of-mass wave-vector $\boldsymbol{Q} \neq 0$ and if it is a well defined quantity. This is the case in bulk crystals but it is no longer the case in very small nanostructures. This point is important since the electron-hole exchange interaction has two contributions as discussed in Sect. 4.3 (see also Ref. [33]): an analytical one, which is independent of the orientation of the dipole moment with respect to the wave-vector $\boldsymbol{Q}$ and a non-analytical exchange interaction, which acts only on longitudinal excitons. Thus the energies of transverse and longitudinal exciton states increase (with respect to the spin-triplet exciton state) with increasing analytical exchange interaction. In addition the energy of longitudinal excitons is further increased by the non-analytical exchange interaction. This results in the fact that longitudinal and transverse excitons are no longer degenerate in bulk material. Thus, at $\boldsymbol{Q} = 0$, the electron-hole exchange-interaction splits in the $\Gamma_6 \otimes \Gamma_7$ subspace of zincblende-type semiconductors the exciton states into a triplet-exciton state with Γ_2 symmetry, two transverse exciton states with Γ_{5T} symmetry and a longitudinal exciton states with Γ_{5L} symmetry (see Fig. 4.3 and also Fig. 4.1).

As discussed in connection with Eq. (8.1.53), dipole active transverse excitons have a so-called "oscillator strength" f_s, which is proportional to the splitting between longitudinal and transverse-exciton states. In fact if one neglects the wave-vector dependence of the longitudinal (E_L) and transverse exciton energies (E_T) one finds for the "one-oscillator model" (a situation where only one isolated oscillator interacts with the light field and all other resonances are accounted for by a background dielectric constant ϵ_b): $\hbar^2 f_s = \epsilon_b(E_L^2 - E_T^2)$. We see that the non-analytical part of the exchange interaction accounts for the oscillator strength of excitons.

Transverse $J = 1$ excitons and photons (the quanta of the electromagnetic radiation field) have in zincblende-type semiconductors the same symmetry properties: both types of quasi-particles possess a dipole moment, which is orientated perpendicular to their propagation direction and the states transform as Γ_5. Quasi-particles having the same symmetry are said to be in the "strong-coupling regime". This means that these elementary excitations, "exciton-polaritons", have mixed exciton and photon character. As discussed above and displayed e.g. in Figs. 8.2 and 8.8, because of the strong coupling between excitons and photons, the dispersion of the coupled system is strongly modified in the cross-over region, where exciton and photon states have similar energies. Hybridization or "anti crossing" of the dispersions is observed in such situations. In general, only dispersion relations of elementary excitations having different symmetries can cross each other.

In most cases the longitudinal and transverse exciton energies depend on the wave-vector, i.e. they show spatial dispersion: Excitons are a collective excitation of a crystal and correspond classically to an excitation of coupled oscillators. Therefore, one has to consider in the formation of polaritons that E_T is a function of $\boldsymbol{Q}$. As discussed in Sect. 5.2 the finite wave-vector $\boldsymbol{Q}$ breaks the point-group symmetry of the crystal and the exciton dispersion is described by a power-law:

$$E_T(\boldsymbol{Q}) = E_T + \sum_n a_n \boldsymbol{Q}^n. \tag{8.3.3}$$

In the $\Gamma_6 \otimes \Gamma_7$ subspace the leading power is $n = 2$. As given in Eq. (5.2.3), different wave-vector dependent direct and exchange electron-hole interactions terms may exist at this order. In the following we will neglect the (even anisotropic) exchange interactions, which break the point-group symmetry. As discussed in connection with Eq. (5.2.4) the factor $a_2 = \hbar^2/(2M_{ex})$ is determined by the effective exciton center of mass value M_{ex}. This term is isotropic and does not mix the different exciton states. Thus, no oscillator strength is transferred from the transverse exciton state with a given polarization to the other states. In the framework of this approximation (i.e. when neglecting all different symmetry breaking interactions and considering only the term proportional to $a_2 = \hbar^2/(2M_{ex})$) the polariton problem reduces in the $\Gamma_6 \otimes \Gamma_7$ subspace to the one-oscillator problem, since the other excitons do not couple to the light field with the considered polarization. One then obtains for $E_T(\boldsymbol{Q})$:

$$E_T(\boldsymbol{Q}) = E_T + (\hbar^2/2M_{ex})\,\boldsymbol{Q}^2. \tag{8.3.4}$$

The polariton dispersions $E_i(\boldsymbol{Q})$ are given for the different branches $i = (1, 2)$ by the solution of the dispersion equation Eq. (8.1.54):

$$\begin{aligned} \hbar^2 c^2 \boldsymbol{Q}^2/E_i(\boldsymbol{Q})^2 &= \epsilon_b + \hbar^2 f_s/(E_T(\boldsymbol{Q})^2 - E_i(\boldsymbol{Q})^2) \\ &= \epsilon_b(1 + (E_L^2 - E_T^2)/(E_T(\boldsymbol{Q})^2 - E_i(\boldsymbol{Q})^2)), \end{aligned} \tag{8.3.5}$$

c being the light velocity in vacuum and ϵ_b the background dielectric constant due to the other oscillators that are not explicitly considered. Usually Eq. (8.3.5) can be simplified by neglecting the wave-vector dependence of the exciton energies in the numerator. Then the solution of Eq. (8.3.5) is given by:

$$E_i(Q) = (1/\sqrt{2})\sqrt{(A \pm \sqrt{A^2 - 4B^2})}$$
where
$$A = \hbar^2 c^2 Q^2/\epsilon_b + E_T(Q)^2 (E_L^2/E_T^2) \tag{8.3.6}$$
and
$$B^2 = \hbar^2 c^2 Q^2 E_T(Q)^2/\epsilon_b.$$

Spatial dispersion modifies considerably the polariton dispersion as shown in Fig. 8.9: For small energies Eq. (8.3.5) has only one solution for real wave-vectors. This branch corresponds to the lower polariton branch, which is sketched in Fig. 8.2 in the absence of spatial dispersion. Above E_T the energy gap (leading to total reflection as discussed with Eq. (8.1.54)) disappears and the system has always at least one propagating solution. Above E_L the system shows two real (propagating) modes,

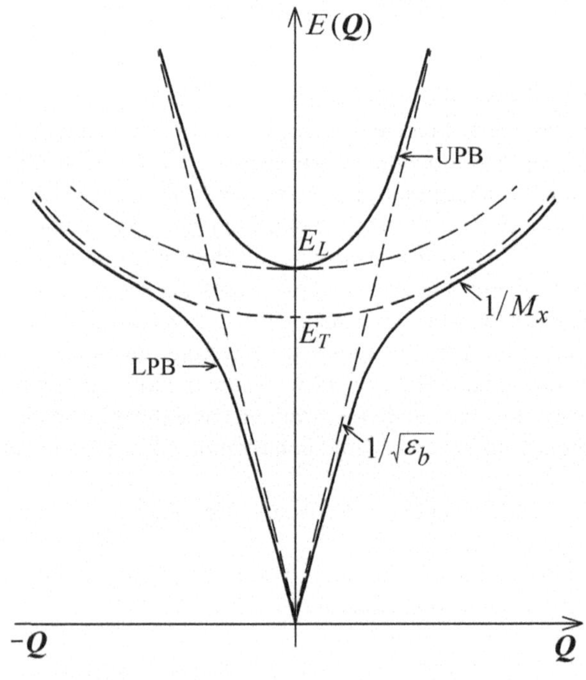

Fig. 8.9 Polariton dispersion $E(Q)$ (full lines) in the one-oscillator model with spatial dispersion that applies to the $\Gamma_6 \otimes \Gamma_7$ subspace in zincblende-type semiconductors. LPB (UPB): lower (upper) polariton branch. E_L and E_T indicate the energies of the longitudinal and transverse excitons, respectively. Dashed curves (straight line) indicate the exciton (photon) dispersions. M_x denotes the effective exciton center-of-mass value. $1/\sqrt{\epsilon_b}$ is proportional to the high-energy polariton group velocity of the UPB. At high energies, upper branch polaritons are photon like, lower branch polaritons have excitonic character. LPB states are photon like for $E(Q) \ll E_T$. It is important to notice that there is no energy gap between E_T and E_L if spatial dispersion is taken into account, i.e. if $0 < M_x < \infty$. (According to Ref. [4], see text)

one corresponding to the upper polariton branch, but the second one (the lower polariton branch) still exists. (This leads in semiconductor optics to the well known problem that transmission and reflection coefficients of light at interfaces cannot be determined from the electromagnetic theory alone but the so-called "additional boundary conditions (abc)" (Refs. [34–36]) are needed in this spectral region.) The upper polariton branch finds its analytical continuation to energies below E_L in a purely imaginary wave-vector branch, but its discussion is beyond the scope of this book.

Experimental results have often been analyzed in the framework of the one-oscillator model since it is the most simple model describing the exciton-photon system in the strong-coupling regime. CuCl is a semiconductor, in which the lowest lying exciton series can be well described by this model since it is well-separated in energy from the higher-lying exciton states. It is acting in the $\Gamma_6 \otimes \Gamma_7$ subspace, all other exciton states being neglected. As discussed already in the beginning of Sect. 8.2, this simple polariton dispersion has been successfully studied by the method of Fabry-Perot modes (see Ref. [7] and Fig. 8.3) and the time-of-flight method (see Refs. [9, 10] and Fig. 8.4) .

Nonlinear optical spectroscopic methods have also been applied to determine the exciton-polariton dispersion relation. Results of measurements obtained by two-photon absorption spectroscopy (c.f. Refs. [12, 13]) and two-photon Raman scattering (Refs. [16, 17]) via biexcitons have been used to construct parts of the polariton dispersion and that of longitudinal excitons (see Figs. 8.10 and 8.11).

Depending on the presence of symmetry breaking interactions between electrons and/or holes and intrinsic or extrinsic fields, the different exciton states can be mixed. Thus, oscillator strength is transferred from the transverse state under consideration to the other states. Therefore, the electromagnetic radiation field may also couple to these states and the polariton dispersion becomes more complex. Such a situation will be considered in the next section where exciton-polaritons in the $\Gamma_6 \otimes \Gamma_8$ subspace of zincblende-type semiconductors are discussed in detail.

8.4 Exciton-Polaritons in the $\Gamma_6 \otimes \Gamma_8$ Subspace in Zincblende-Type Semiconductors

Polaritons are compound quasi-particles with mixed exciton and photon character, which propagate as polarization waves through the dielectric medium. We have discussed in the preceding section the formation of exciton-polaritons in the simple one-oscillator model where only one dipole-active exciton state (called "oscillator state" here) couples to the electromagnetic radiation field. All other exciton states, which may be almost degenerate in energy with this "oscillator state", have a different symmetry or polarization. Thus they do not couple to the light field and are called "dark states" in the following.

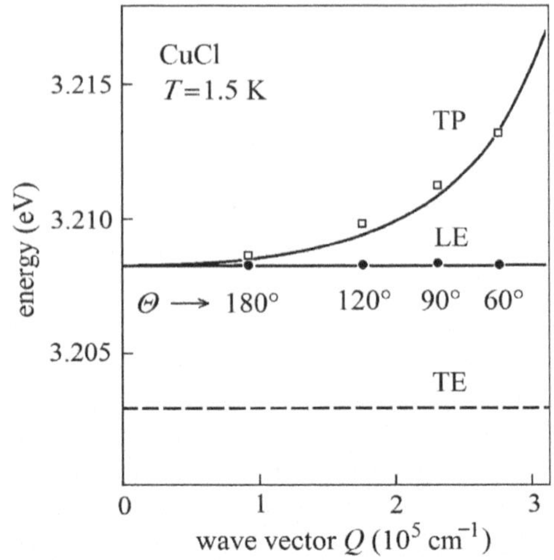

Fig. 8.10 Polariton and longitudinal-exciton dispersion $E(Q)$ of CuCl, reconstructed from the two-photon absorption measurements of Fig. 8.5 at a temperature $T = 1.5$ K. Θ denotes the angle between the exciting beams outside the crystal. TP indicates the spectral position of the upper branch polaritons and LE that of the longitudinal excitons, created in the two-photon absorption process. The dashed line indicates the spectral position of the transverse exciton (TE). The resulting lower polariton branch is not accessible in two-photon absorption measurements. (Adapted after Refs. [12, 13])

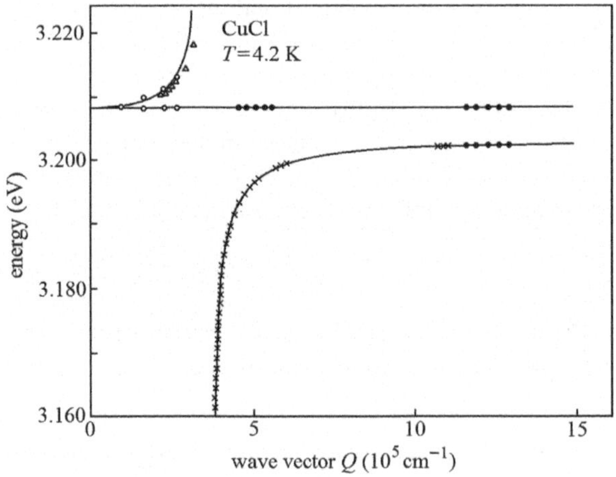

Fig. 8.11 Polariton and longitudinal-exciton dispersion $E(Q)$ of CuCl. The reconstruction makes use of data obtained by two-photon Raman (hyper-Raman) scattering via biexcitons (full circles and crosses) at temperature $T = 4.2$ K (see Ref. [16]) and of the data from Fig. 8.10, obtained by two-photon absorption (see Ref. [12, 13], open circles). Additional points determined by two-photon Raman scattering (see Ref. [17], triangles) are added. (Adapted after Refs. [4, 12, 16, 17], see text)

As we have seen, if the point-group symmetry is broken due to interactions between electrons and/or holes with intrinsic or extrinsic fields the different exciton states can be mixed. Thus, oscillator strength is transferred from the transverse "oscillator state" to the other states that become also dipole-active, i.e. the electromagnetic radiation field couples to the previously "dark states". The oscillator strength of the new "mixed states" depends on the amount of admixture of the "oscillator state" to the "dark states". As a consequence the polariton dispersion becomes more complex and is a function of the strength of the symmetry-breaking interactions.

We will discuss in the following exciton-polaritons in the $\Gamma_6 \otimes \Gamma_8$ subspace, which is spanned by eight exciton states. As discussed in Sect. 5.3 we will start with $J = 1$ and $J = 2$ exciton states and adapt them to the T_d point-group symmetry. Then the states given in Eqs. (5.3.10), (5.3.15) and (5.3.16) are these symmetry-adapted exciton states in which the direct Coulomb and the electron-hole exchange-interactions are diagonal. Using these basis functions, the polariton dispersion can be studied in detail as a function of symmetry-breaking interactions.

Before we go to a discussion of polaritons we are about to discuss dispersion relations of "bare" excitons for wave-vectors $\boldsymbol{Q} \neq 0$, i.e. out of the Γ-point of the first Brillouin zone. As discussed in Sect. 5.3 the effective Hamiltonian H_8^{ex} of the exciton ground states reads in the $\Gamma_6 \otimes \Gamma_8$ pseudo-spin subspace:

$$\begin{aligned} H_8^{ex} = {} & \Delta_{08} \mathbf{1}_{v8} \otimes \mathbf{1}_e + \Delta_{18} (j_v^x \otimes \sigma_e^x + j_v^y \otimes \sigma_e^y + j_v^z \otimes \sigma_e^z) + \\ & + \Delta_{28} \left[(j_v^x)^3 \otimes \sigma_e^x + (j_v^y)^3 \otimes \sigma_e^y + (j_v^z)^3 \otimes \sigma_e^z \right]. \end{aligned} \tag{8.4.1}$$

The electron-hole basis functions, which diagonalize the exchange interaction $H_{ech18}^{ex} = \Delta_{18} (j_v^x \otimes \sigma_e^x + j_v^y \otimes \sigma_e^y + j_v^z \otimes \sigma_e^z)$, are the $J = 1$ states (having Γ_5 symmetry)

$$\begin{aligned} v_{118}^{exd} &= (|\Psi_3\rangle + \sqrt{3}|\Psi_8\rangle)/2 = |1, 1\rangle \\ v_{218}^{exd} &= -(|\Psi_2\rangle + |\Psi_7\rangle)/\sqrt{2} = |1, 0\rangle \\ v_{318}^{exd} &= (|\Psi_6\rangle + \sqrt{3}|\Psi_1\rangle)/2 = |1, -1\rangle \end{aligned} \tag{8.4.2}$$

and the $J = 2$ states

$$\begin{aligned} v_{418}^{exd} &= (|\Psi_8\rangle - \sqrt{3}|\Psi_3\rangle)/2 = |2, 1\rangle \\ v_{518}^{exd} &= |\Psi_4\rangle = |2, 2\rangle \\ v_{618}^{exd} &= (|\Psi_2\rangle - |\Psi_7\rangle)/\sqrt{2} = |2, 0\rangle \\ v_{718}^{exd} &= (\sqrt{3}|\Psi_6\rangle - |\Psi_1\rangle)/2 = |2, -1\rangle \\ v_{818}^{exd} &= -|\Psi_5\rangle = |2, -2\rangle. \end{aligned} \tag{8.4.3}$$

Since we will consider in the following mainly light fields with linear polarization, we rather introduce, instead of the states given in Eq. (8.4.2), the exciton states $(|x\rangle, |y\rangle, |z\rangle)$ that transform as the Cartesian coordinates (x, y, z):

$$
\begin{aligned}
|x\rangle &= (-|1,1\rangle + |1,-1\rangle)/\sqrt{2} = (-(|\Psi_3\rangle + \sqrt{3}|\Psi_8\rangle)/2 + (|\Psi_6\rangle + \sqrt{3}|\Psi_1\rangle)/2)/\sqrt{2} \\
|y\rangle &= i(|1,1\rangle + |1,-1\rangle)/\sqrt{2} = i((|\Psi_3\rangle + \sqrt{3}|\Psi_8\rangle)/2 + (|\Psi_6\rangle + \sqrt{3}|\Psi_1\rangle)/2)/\sqrt{2} \\
|z\rangle &= |1,0\rangle = -(|\Psi_2\rangle + |\Psi_7\rangle)/\sqrt{2}.
\end{aligned} \tag{8.4.4}
$$

Usually the anisotropic exchange-interaction term Δ_{28} in Eq. (8.4.1) is small when compared to the isotropic interactions Δ_{08} and Δ_{18}. Therefore in most direct semiconductors with T_d point-group symmetry the $J = 2$ states remain degenerate at the Γ point. It is nevertheless interesting to consider the anisotropic exchange-interaction Δ_{28} since it allows to identify easily the symmetry adapted exciton wave-functions with Γ_3 and Γ_4 symmetry. When taking into account the cubic exchange-interaction term Δ_{28} and diagonalizing

$$
H^{ex}_{ech28} = \Delta_{28}\left[(j^x_v)^3 \otimes \sigma^x_e + (j^y_v)^3 \otimes \sigma^y_e + (j^z_v)^3 \otimes \sigma^z_e\right]
$$

the $J = 1$ states given in Eq. (8.4.4) are not modified and diagonalize also the cubic exchange-interaction.

The exciton states with Γ_3 and Γ_4 symmetry are split in energy by H^{ex}_{ech28} and linear combinations of $J = 2$ states have to be used in order to diagonalize the direct Coulomb and the electron-hole exchange interactions of Eq. (8.4.1), which are compatible with the T_d point-group symmetry. The $J = 2$ states, which are adapted to the crystal symmetry have been introduced in Eqs. (5.3.15) and (5.3.16). We obtained for the states with Γ_3 symmetry

$$
\begin{aligned}
|2,+\rangle &= (|2,2\rangle + |2,-2\rangle)/\sqrt{2} = (|\Psi_4\rangle - |\Psi_5\rangle)/\sqrt{2} \\
|2,0\rangle &= (|\Psi_2\rangle - |\Psi_7\rangle)/\sqrt{2}
\end{aligned} \tag{8.4.5}
$$

and for those with Γ_4 symmetry

$$
\begin{aligned}
|1,+\rangle &= -(|2,1\rangle + |2,-1\rangle)/\sqrt{2} = -((|\Psi_8\rangle - \sqrt{3}|\Psi_3\rangle)/2 + (\sqrt{3}|\Psi_6\rangle - |\Psi_1\rangle)/2)/\sqrt{2} \\
|1,-\rangle &= i(|2,-1\rangle - |2,1\rangle)/\sqrt{2} = i((\sqrt{3}|\Psi_6\rangle - |\Psi_1\rangle)/2 - (|\Psi_8\rangle - \sqrt{3}|\Psi_3\rangle)/2)/\sqrt{2} \\
|2,-\rangle &= (|2,2\rangle - |2,-2\rangle)/\sqrt{2} = (|\Psi_4\rangle + |\Psi_5\rangle)/\sqrt{2}.
\end{aligned} \tag{8.4.6}
$$

Keeping in mind the symmetry adapted exciton wave functions of Eq. (8.4.5) and Eq. (8.4.6) let us neglect in the following the anisotropic exchange interaction, i.e. take $\Delta_{28} = 0$. We then obtain with Eq. (5.3.5) for the energies of the exciton states $E_{ex8}(J)$ with $(J = 1, 2)$ at wave-vector $\boldsymbol{Q} = 0$, i.e. at the center of the Brillouin zone:

$$
\begin{aligned}
E_{ex8}(J=2) &= E_G + \Delta_{08} + (3/2)\Delta_{18} \text{ for the } J = 2 \text{ exciton states} \\
E_{ex8}(J=1) &= E_G + \Delta_{08} - (5/2)\Delta_{18} \text{ for the } J = 1 \text{ exciton states}\,.
\end{aligned} \tag{8.4.7}
$$

One should recall here that in the considered direct energy-gap semiconductors our "zero" of energy has been fixed at the "crystal ground state", i.e. in the absence of any electronic excitation, which may be thought to be situated at the Γ-point of the Brillouin zone. The minimum of the conduction band (also at the Γ-point) has the energy E_G. Excitons are bound states of an electron and a hole and the exciton binding energy is thus negative, i.e. $E_{ex8}(J) < E_G$. The exciton binding energy can be introduced by considering all direct interactions between the electron and the hole and the surrounding semiconductor. Thus, Δ_{08} can be interpreted as representing the exciton binding energy in this approximation (i.e. relevant to the holes generated in the Γ_8 valence band).

Considering the electron spin leads to additional interactions: the exchange interactions between electrons and defect electrons (which are the Kramer's conjugated states of holes). These interactions are due to the quantum statistics of electrons: Since electrons are Fermions, two electrons cannot be in quantum states, characterized by the same set of quantum numbers. Therefore, a repulsive interaction between these electrons appears: the exchange-correlation energy, which depends on the states under consideration. Since one cannot calculate this energy in a general way (independent of the states) we will discuss it in the following in an approximative form and consider mainly the localized charge density of electron states.

In exciton problems the exchange part of the Coulomb interaction J_{ex} plays a central role: J_{ex} is determined by the spin-singlet part of the localized charge density of the exciton state (corresponding to its dipole moment). As we have seen, the $J = 1$ exciton states transforming as the Cartesian coordinates (x, y, z) posses a dipole moment. Spin-triplet excitons, to which the $J = 2$ exciton states also belong, have a vanishing singlet charge density and are not affected by this exchange interaction (see Ref. [33]). Therefore,

$$E^{g}_{ex8} = E_G + \Delta_{08} + (3/2)\Delta^{(J=2)}_{18} \text{ for the } J = 2 \text{ exciton states} \tag{8.4.8}$$

will be considered in the following as the ground-state energy E^{g}_{ex8} (including exchange interaction) of excitons, which have no dipole moment.

The exchange part of the Coulomb interaction J_{ex} for $J = 1$ excitons (having a dipole moment as discussed above) has in the present approximation two different contributions (c.f. Ref. [29]): It consists of an analytic part that acts on all $J = 1$ exciton states, i.e. it acts equally on the longitudinal and transverse dipole-active excitons states. But it has also a non-analytic part that acts only on the longitudinal excitons and not on the other states. Therefore the exchange part of the Coulomb interaction can only be evaluated if the direction of the wave-vector $\boldsymbol{Q}$ is specified since it defines the notions of a "longitudinal" and "transverse" dipole moment of the excitons states.

If the wave-vector $\boldsymbol{Q}$ is finite the interaction energy $\Delta^{(J=1)}_{18}$ in Eq. (8.4.8) can be decomposed into an analytic Δ^{a}_{18} and a non-analytic Δ^{na}_{18} contribution, namely

$$\Delta^{(J=1)}_{18} = \Delta^{a}_{18} + \Delta^{na}_{18}. \tag{8.4.9}$$

The analytic part of J_{ex} shifts all $J = 1$ exciton states to higher energies compared to the exciton ground-state energy E^g_{ex8}. Using this notation it follows from Eq. (8.4.7) that the singlet-triplet splitting of the $\Gamma_6 \otimes \Gamma_8$ subspace excitons Δst is given by

$$\Delta st = 4\Delta^a_{18}. \tag{8.4.10}$$

The non-analytic part of J_{ex} leads then to an additional splitting of longitudinal and transverse exciton states (Ref. [33]), which are separated by Δ_{LT} with

$$\Delta_{LT} = 4\Delta^{na}_{18}. \tag{8.4.11}$$

We thus obtain instead of Eq. (8.4.7) the following exciton energy-level scheme (see Fig. 8.12):

$$\begin{aligned} E_{ex8}(J = 2) = E^g_{ex8} \text{ for the } J = 2 \text{ exciton states} \\ \text{and} \\ E^T_{ex8}(J = 1) = E^g_{ex8} + \Delta_{st} \text{ for the transverse } J = 1 \text{ exciton states} \\ E^L_{ex8}(J = 1) = E^g_{ex8} + \Delta_{st} + \Delta_{LT} \text{ for the longitudinal } J = 1 \text{ exciton state}\,. \end{aligned} \tag{8.4.12}$$

In comparison with Fig. 4.4, as mentioned above, a small splitting between the Γ_3 and Γ_4 levels due to the cubic exchange interaction is neglected here.

We will now use the symmetry adapted exciton wave functions given in Eq. (8.4.4) to Eq. (8.4.6) as a basis to develop the exciton-polariton dispersion in the presence of symmetry-breaking exciton-interactions. E^g_{ex8} is taken as the origin of energy scale.

We start with the exciton dispersion $E_i(\boldsymbol{Q})$ in the $\Gamma_6 \otimes \Gamma_8$ exciton subspace, where $\boldsymbol{Q}$ denotes the exciton center-of-mass wave-vector. The $\boldsymbol{Q}$-dependent interaction terms up to second order in $\boldsymbol{Q}$ are discussed in Sect. 5.4. They include direct terms as well as the wave-vector dependence of the electron-hole exchange-interaction.

Terms that one may consider in the exciton-polariton problem are given in Eqs. (5.4.3)–(5.4.5): The most important ones are obtained directly from the dispersion relation of the conduction and valence bands. These direct terms are diagonal within the electron states. We find a $\boldsymbol{Q}$-linear term (resulting from the valence-band hole Eq. (3.4.6)), which is given by Eq. (5.4.3):

$$H^{ex8}_{Q1} = C^{ex8}_{Q1}[\{j^x_v, ((j^y_v)^2 - (j^z_v)^2)\}Q_x + \{j^y_v, ((j^z_v)^2 - (j^x_v)^2)\}Q_y + \{j^z_v, ((j^x_v)^2 - (j^y_v)^2)\}Q_z] \otimes \mathbf{1_e} \tag{8.4.13}$$

and terms proportional to $\boldsymbol{Q}^2$, which read (see Eq. (5.4.4))

$$\begin{aligned} H^{ex8}_{Q2} = G^{ex8}_{1Q2}Q^2\mathbf{1_{v8}} \otimes \mathbf{1_e} + \\ + G^{ex8}_{2Q2}[3(Q^2_x - Q^2_y)((j^x_v)^2 - (j^y_v)^2) + (2Q^2_z - Q^2_x - Q^2_y)(2(j^z_v)^2 - (j^x_v)^2 - (j^y_v)^2)] \otimes \mathbf{1_e} + \\ + 2G^{ex8}_{3Q2}[Q_yQ_z\{j^y_v, j^z_v\} + Q_zQ_x\{j^z_v, j^x_v\} + Q_xQ_y\{j^x_v, j^y_v\}] \otimes \mathbf{1_e}. \end{aligned} \tag{8.4.14}$$

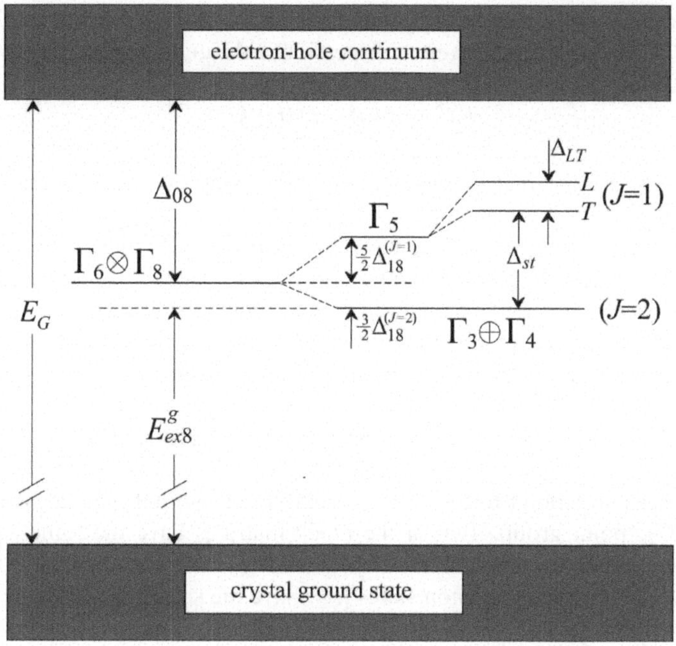

Fig. 8.12 Energy level scheme of excitons arising from holes in the Γ_8 valence band and electrons in the Γ_6 conduction band in zincblende semiconductors. In comparison with Fig. 4.4, here a small splitting between the Γ_3 and Γ_4 levels due to the cubic exchange interaction is neglected (see text)

In addition, one may consider a second group of terms that describes the wave-vector dependence of the exchange interaction. As discussed in Sect. 5.4 they are usually quite small and the most important ones are bi-linear in $\boldsymbol{j}_{v8}$ and $\boldsymbol{\sigma}_e$. These terms are given by (see Eq. (5.4.5)):

$$\begin{aligned} H^{ex8}_{Q2e} &= \delta^{ex8}_{1Q2} Q^2 (j^x_v \otimes \sigma^x_e + j^y_v \otimes \sigma^y_e + j^z_v \otimes \sigma^z_e) + \\ &+ \delta^{ex8}_{2Q2} [3(Q_x^2 - Q_y^2)((j^x_v \otimes \sigma^x_e - j^y_v \otimes \sigma^y_e) + (2Q_z^2 - Q_x^2 - Q_y^2)(2j^z_v \otimes \sigma^z_e - j^x_v \otimes \sigma^x_e - j^y_v \otimes \sigma^y_e)] + \\ &+ 2\delta^{ex8}_{3Q2} [Q_y Q_z [j^y_v, \sigma^z_e] + Q_z Q_x [j^z_v, \sigma^x_e] + Q_x Q_y [j^x_v, \sigma^y_e]]. \end{aligned} \tag{8.4.15}$$

We will call in the following "H^{ex8}_{Q}" the exciton Hamiltonian defined in the $\Gamma_6 \otimes \Gamma_8$ subspace, which includes the symmetry-breaking interactions enumerated above:

$$H^{ex8}_{Q} = H^{ex8}_{Q1} + H^{ex8}_{Q2} + H^{ex8}_{Q2e}. \tag{8.4.16}$$

Usually only some of the terms of H^{ex8}_{Q} are used in literature (see Refs. [4, 33]) to calculate the exciton dispersion. The procedure that we adopt is described in detail in Ref. [37] for CuBr where the $\Gamma_6 \otimes \Gamma_8$ exciton subspace is considered, but this procedure is valid also in the general case. One first establishes the interaction matrix "I_Q^{ex8}" defined for the exciton dispersion in the eight-fold degenerate $\Gamma_6 \otimes \Gamma_8$ subspace through

$$I_Q^{ex8} = Matrix\left(\langle ij||kl\rangle\right), \tag{8.4.17}$$

where $\langle ij||kl\rangle$ are the matrix elements of H_Q^{ex8}, using the basis functions $|ij\rangle$ and $|kl\rangle$ given in Eqs. (8.4.4)–(8.4.6).

For an arbitrary direction of the wave-vector Eq. (8.4.17) has to be diagonalized numerically. In a high symmetry direction it can be given, however, in a block-diagonal form (for detailed calculations concerning CuBr see Refs. [37, 38]). In the [001] direction (i.e. for $\boldsymbol{Q} \parallel z$) the matrix Eq. (8.4.17) falls e.g. into four blocks if the basis functions are transformed in the following way:

$$|\xi_\pm\rangle = (|x\rangle \pm |y\rangle)/\sqrt{2} \text{ and } |\eta_\pm\rangle = (|1+\rangle \pm |1-\rangle)/\sqrt{2}. \tag{8.4.18}$$

These blocks correspond to the four irreducible representations Δ_i (with $i = 1$ to 4) of the C_{2v} point group. Two of the block matrices have the same eigenvalues. The dispersions of the exciton states $|\eta_\pm\rangle$ and $|\xi_\pm\rangle$ remain therefore degenerate. Especially, the transverse exciton states $|\xi_\pm\rangle$ have the same dispersion.

Similar block matrices can be obtained if the exciton wave-vector is aligned to the high symmetry directions [111] or [110]. The exciton-wave functions that have to be used to obtain these block matrices are given in Table 8.1. If $\boldsymbol{Q} \parallel [111]$ direction, which is the new quantization axis ($z' \parallel [111]$), the blocks correspond to the irreducible representations Λ_1, Λ_2 and Λ_3 of the point group C_{3v}. Λ_1 contains as element the wave function $|z'\rangle$ of Table 8.1 and Λ_2 the one noted $|2-'\rangle$. While the representations Λ_1 and Λ_2 are not degenerate, the representation Λ_3 gives rise to three exciton branches, which are each two times degenerate. To these two rows of the representation Λ_3 belong the wave-function elements [$|2+'\rangle$, $|1+'\rangle$, and $|y'\rangle$ (first row)] and [$|20'\rangle$, $|1-'\rangle$, and $|x'\rangle$ (second row)], respectively. Their dispersion

Table 8.1 Exciton wave-functions in crystals with T_d point-group symmetry for $\boldsymbol{Q}$ along the high symmetry directions [110] and [111], which are taken as quantization axes. From Ref. [4]

Linear combination of basis functions		
New basis	$\boldsymbol{Q} \parallel [110]$	$\boldsymbol{Q} \parallel [111]$
$\lvert 2+'\rangle$:	$(-\sqrt{3}\lvert 20\rangle + i\lvert 2-\rangle)/2$	$\lvert 2+\rangle$
$\lvert 20'\rangle$:	$-(\lvert 20\rangle + i\sqrt{3}\lvert 2-\rangle)/2$	$\lvert 20\rangle$
$\lvert 1+'\rangle$:	$(\lvert 1+\rangle + \lvert 1-\rangle)/\sqrt{2}$	$(\lvert 1+\rangle - \lvert 1-\rangle)/\sqrt{2}$
$\lvert 1-'\rangle$:	$-i\lvert 2+\rangle$	$(2\lvert 2-\rangle - \lvert 1+\rangle - \lvert 1-\rangle)/\sqrt{6}$
$\lvert 2-'\rangle$:	$(\lvert 1+\rangle - \lvert 1-\rangle)/\sqrt{2}$	$(\lvert 2-\rangle + \lvert 1+\rangle + \lvert 1-\rangle)/\sqrt{3}$
$\lvert x'\rangle$:	$(-\lvert x\rangle + \lvert y\rangle)/\sqrt{2}$	$(-\lvert x\rangle + \lvert y\rangle)/\sqrt{2}$
$\lvert y'\rangle$:	$\lvert z\rangle$	$(2\lvert z\rangle - \lvert x\rangle - \lvert y\rangle)/\sqrt{6}$
$\lvert z'\rangle$:	$(\lvert x\rangle + \lvert y\rangle)/\sqrt{2}$	$(\lvert x\rangle + \lvert y\rangle + \lvert z\rangle)/\sqrt{3}$

has to be determined numerically, but again the two transverse exciton states have the same dispersion.

It is interesting to notice that in the high symmetry directions $\boldsymbol{Q} \parallel [111]$ or $\boldsymbol{Q} \parallel [100]$ the longitudinal and the transverse $J = 1$ excitons are in different irreducible representation. Therefore, as we will discuss in connection with the polariton problem, longitudinal excitons are not mixed with transverse excitons and are not dipole active, i.e. they do not couple to the light field. This is different if $\boldsymbol{Q} \parallel [110]$. Taking the [110] direction as new quantization axis, the interaction matrix falls only into two blocks corresponding to the two irreducible representations Σ_1 and Σ_2 of the point group C_s. Both irreducible representations contain 4 elements. The wave-functions are given in Table 8.1. The representation Σ_1 is spanned by the wave-function elements $[|1-'\rangle, |2-'\rangle, |y'\rangle$, and $|z'\rangle]$ and Σ_2 by $[|2+'\rangle, |20'\rangle, |1+'\rangle$, and $|x'\rangle]$. Due to the low symmetry of the [110]-direction, the longitudinal and a transverse exciton become mixed at finite wave-vectors in the Σ_1 representation. Since the two transverse exciton states belong to two different irreducible representations they are no longer degenerate in the [110]-direction, i.e. they have different dispersion relations and a cubic crystal is birefringent in this (or an arbitrarily low) crystal direction. This effect is in general small but may become important close to exciton resonances.

It is worth while to mention here that two orthogonal, linearly polarized transverse charge vibrations, which propagate in the [001]-direction of a cubic system see the same charge distribution of the dielectric medium. Therefore, their interaction with the environment is the same and the dispersion of these transverse modes is the same. If one considers, however, propagation of transverse modes in the [110]-direction and their elongations in the $[1\bar{1}0]$-direction or the [001]-direction, it is evident that the charge distribution for both modes is different. One therefore expects that the dispersion of such charge waves depends on the direction of propagation and the direction of the linear polarization, leading to birefringence and dichroism.

Once the bare exciton wave functions and their dispersions are known one can introduce the polariton concept, i.e. consider the problem of excitons that are coupled to the light field. We first solve the Schrödinger equation for the exciton problem:

$$H_{\boldsymbol{Q}}^{ex8}|\Phi_i(\boldsymbol{Q})\rangle = E_i(\boldsymbol{Q})|\Phi_i(\boldsymbol{Q})\rangle, \tag{8.4.19}$$

where the Hamiltonian $H_{\boldsymbol{Q}}^{ex8}$ is given by Eq. (8.4.17) in its matrix form. The exciton wave-function $|\Phi_i(\boldsymbol{Q})\rangle$ is given as a linear combination of the basis functions $|\Psi_j\rangle$ (see Eqs. (8.4.4)–(8.4.6)) by

$$|\Phi_i(\boldsymbol{Q})\rangle = \sum_j c_{ij}(\boldsymbol{Q})|\Psi_j\rangle. \tag{8.4.20}$$

In this case Hopfield's theory for excitonic polaritons (see Ref. [2]) has to be extended to the multiple oscillator case (as e.g. in Refs. [4, 33, 37, 39]). We thus find the dispersion relation $E_i(\boldsymbol{Q})$ of the eight exciton branches labeled "i". Since only $m \leq 8$ dipole active exciton branches can couple to the light field, we further consider only

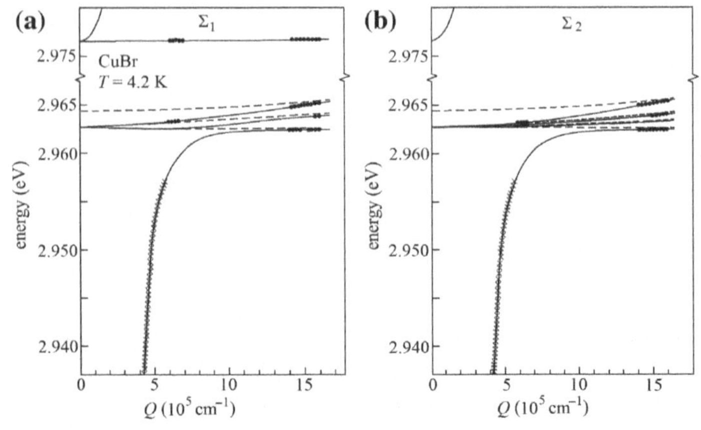

Fig. 8.13 Polariton dispersion relation $E(\boldsymbol{Q})$ in the case of the $\Gamma_6 \otimes \Gamma_8$ subspace in CuBr for (**a**) the Σ_1 irreducible representations, i.e. $\boldsymbol{Q} \parallel [110]$, which is the new quantization axis ($z' \parallel [110]$), and polarization vector $\parallel$ to the [001]-direction and (**b**) the Σ_2 irreducible representations, i.e. as in (**a**) $\boldsymbol{Q} \parallel [110]$ but polarization vector $\parallel$ to the $[1\bar{1}0]$-direction. Measurements were performed in backward scattering (full circles) and in different forward scattering angular configurations (crosses) on CuBr platelets and massive crystals at a temperature $T = 4.2$ K. (Adapted after Refs. [4, 37])

transverse $J = 1$ exciton states labeled "T" where at least one of the $c_{iT}(\boldsymbol{Q}) \neq 0$. The other branches are exciton branches and do not give rise to polaritons as propagating quasi-particles. As discussed above the exciton polarization and the electric field lead then to the following equation

$$\hbar^2 c^2 \boldsymbol{Q}^2 / E_n^{pol}(\boldsymbol{Q})^2 = \epsilon_b \left(1 + \sum_j c_{jT}(\boldsymbol{Q})^2 (E_L(0)^2 - E_T(0)^2)/(E_j(\boldsymbol{Q})^2 - E_n^{pol}(\boldsymbol{Q})^2) \right), \tag{8.4.21}$$

where $E_n^{pol}(\boldsymbol{Q})$ denotes the energy of the n-th polariton branch (with $1 \leq n \leq (m + 1)$) of the irreducible representation considered. As in the classical case $E_n^{pol}(\boldsymbol{Q})$ is obtained from the solution of Eq. (8.4.21) and gives the polariton dispersion (c.f. Refs. [37, 38]).

Similarly to the simple case where the one-oscillator model can be applied, the polariton dispersion can be determined experimentally by hyper-Raman scattering. Using these measurements the polariton dispersion has been determined in a self-consistent procedure (see Ref. [37]). Figure 8.13 shows as an example the dispersion relations $E(\boldsymbol{Q})$ of CuBr for the high symmetry $\boldsymbol{Q} \parallel [110]$-direction, which is the new quantization axis (i.e. $z' \parallel [110]$). Figure 8.13a gives the result for the Σ_1 irreducible representations, i.e. for $\boldsymbol{Q} \parallel [110]$ and polarization vector $\parallel$ to the [001]-direction. In the Σ_2 irreducible representation, i.e. for a polarization vector $\parallel$ to the $[1\bar{1}0]$-direction shown in Fig. 8.13b, the dispersion is qualitatively different.

Such complex excitonic-polariton structures have been analyzed using resonant-Brillouin scattering in several zincblende-type semiconductors as GaAs [22, 40],

CdTe [22], ZnSe [23], or CuBr [24] at low temperatures. As discussed in Sect. 8.2 the phonon velocities being known, the polariton dispersion relation was reconstructed from these measurements.

We have often used the fact that a Hamiltonian is an operator that is invariant under time reversal. The wave-vector $\boldsymbol{Q}$ being an odd function under time reversal, one obtains for dispersion relations $E(\boldsymbol{Q})$ the equality

$$E(\boldsymbol{Q}) = E(-\boldsymbol{Q}). \tag{8.4.22}$$

If the system is interacting in addition with a magnetic field $\boldsymbol{B}$ (which is also an odd function under time reversal), Eq. (8.4.22) has to be completed to (see Refs. [2, 41–43]):

$$E(\boldsymbol{Q}, \boldsymbol{B}) = E(-\boldsymbol{Q}, -\boldsymbol{B}) \neq E(\boldsymbol{Q}, -\boldsymbol{B}) = E(-\boldsymbol{Q}, \boldsymbol{B}). \tag{8.4.23}$$

Thus, $E(-\boldsymbol{Q}, \boldsymbol{B})$ can be reconstructed if $E(\boldsymbol{Q}, \boldsymbol{B})$ is measured as function of $\boldsymbol{B}$ for positive and negative values of $\boldsymbol{B}$.

The inequality of Eq. (8.4.23) can be nicely demonstrated in optical measurements by excitonic polaritons in CuBr, which show important $\boldsymbol{Q}$-linear and $\boldsymbol{B}$-linear interaction terms (Refs. [37, 38, 44–49]).

Since the number of symmetry-breaking interaction terms, which can be constructed in the pseudo-spin formalism is very important, one usually restricts oneself to terms linear in $\boldsymbol{B}$ and linear or quadratic in $\boldsymbol{Q}$ and one neglects the dependence of the exchange interaction on $\boldsymbol{B}$ and $\boldsymbol{Q}$. The most important contributions to the Hamiltonian then read:

$$\begin{aligned}
H^{ex8}_{QB} &= \Delta_{08}\mathbf{1}_{v8} \otimes \mathbf{1}_e + \Delta_{18}(j^x_v \otimes \sigma^x_e + j^y_v \otimes \sigma^y_e + j^z_v \otimes \sigma^z_e)+\\
&+C^{ex8}_{Q1}[\{j^x_v, ((j^y_v)^2 - (j^z_v)^2)\}Q_x + \{j^y_v, ((j^z_v)^2 - (j^x_v)^2)\}Q_y + \{j^z_v, ((j^x_v)^2 - (j^y_v)^2)\}Q_z] \otimes \mathbf{1}_e+\\
&+G^{ex8}_{1Q2}Q^2\mathbf{1}_{v8} \otimes \mathbf{1}_e+\\
&+G^{ex8}_{2Q2}[3(Q_x^2 - Q_y^2)((j^x_v)^2 - (j^y_v)^2) + (2Q_z^2 - Q_x^2 - Q_y^2)(2(j^z_v)^2 - (j^x_v)^2 - (j^y_v)^2)] \otimes \mathbf{1}_e+\\
&+2G^{ex8}_{3Q2}[Q_yQ_z\{j^y_v, j^z_v\} + Q_zQ_x\{j^z_v, j^x_v\} + Q_xQ_y\{j^x_v, j^y_v\}] \otimes \mathbf{1}_e+\\
&+1/2g^c\mu_B\mathbf{1}_{v8} \otimes (B_x\sigma^x_e + B_y\sigma^y_e + B_z\sigma^z_e)+\\
&+\left(a^v_{B1v8a}(B_xj^x_v + B_yj^y_v + B_zj^z_v) + a^v_{B1v8b}(B_x(j^x_v)^3 + B_y(j^y_v)^3 + B_z(j^z_v)^3)\right) \otimes \mathbf{1}_e,
\end{aligned} \tag{8.4.24}$$

where g^c is the Landé factor of the electron in the conduction band and a^v_{B1v8a} and a^v_{B1v8b} are the Landé factors of the holes in the valence band with Γ_8 symmetry. As mentioned previously (see Sect. 3.4), these terms correspond to the parameters κ and q introduced by Luttinger:

$$\begin{aligned}
a^v_{B1v8a} &= -2\mu_B\kappa\\
a^v_{B1v8b} &= -2\mu_B q.
\end{aligned} \tag{8.4.25}$$

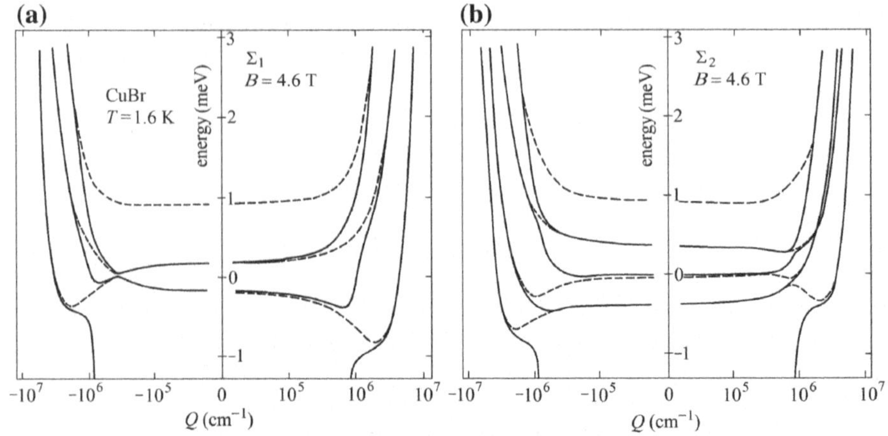

Fig. 8.14 Typical form of the dispersion relation $E(\boldsymbol{Q})$ in the case of the $\Gamma_6 \otimes \Gamma_8$ subspace in zincblende-type semiconductors (CuBr) in a magnetic field $\boldsymbol{B}$. (Dashed lines: exciton dispersions, full lines: polariton dispersions.) $\boldsymbol{Q} \parallel [110]$, which is the new quantization axis ($z' \parallel [110]$), $\boldsymbol{B} \parallel [1\bar{1}0]$. (**a**) Polarization $\boldsymbol{e} \parallel [001]$, corresponding to the irreducible representations Σ_1 of the point group C_s; (**b**) Polarization $\boldsymbol{e} \parallel [1\bar{1}0]$, corresponding to the irreducible representations Σ_2 of the point group C_s. Individual curves represent various exciton/polariton branches. (Adapted after Ref. [49])

Let us consider as an example the case $\boldsymbol{Q} \parallel [110]$-direction. When using the symmetry-adapted exciton-wave functions given in Table 8.1, the interaction matrix corresponding to Eq. (8.4.24) has two diagonal blocks $I^{ex8}_{\Sigma_1}$ and $I^{ex8}_{\Sigma_2}$, giving the energies of the irreducible representations Σ_1 and Σ_2. As described above, after having diagonalized these matrices one obtains the exciton-dispersion relations and from this the polariton dispersions in the presence of a magnetic field. Figure 8.14a, b shows calculated exciton (dashed lines) and polariton (full lines) dispersion relations for CuBr at a fixed magnetic field ($B = 4.6$ T) with $\boldsymbol{B} \parallel [1\bar{1}0]$-direction. In Fig. 8.14a the electric-field polarization $\boldsymbol{e}$ is taken $\boldsymbol{e} \parallel [001]$-direction, Fig. 8.14b shows results for $\boldsymbol{e} \parallel [1\bar{1}0]$. The exciton and polariton dispersions are significantly asymmetric. In Figs. 8.14a, b the upper polariton branches and the longitudinal exciton are not shown. Measurements have been performed on cleaved CuBr platelets by hyper-Raman scattering in a backward configuration at temperature $T = 4.2$ K (see Ref. [49]).

8.5 Exciton-Polaritons in Wurtzite-Type Semiconductors

As indicated in Table 6.1, in wurtzite-type semiconductors, states having Γ_5 or Γ_1 symmetry transform as the components (x, y, z) of the position vector $\boldsymbol{r}$. As indicated in Sect. 7.2, for the different exciton series "i"($i \in (A, B, C)$) the corresponding exciton states are noted $|x\rangle^{ex}_i$, $|y\rangle^{ex}_i$, and $|z\rangle^{ex}_i$, respectively. These states posses a

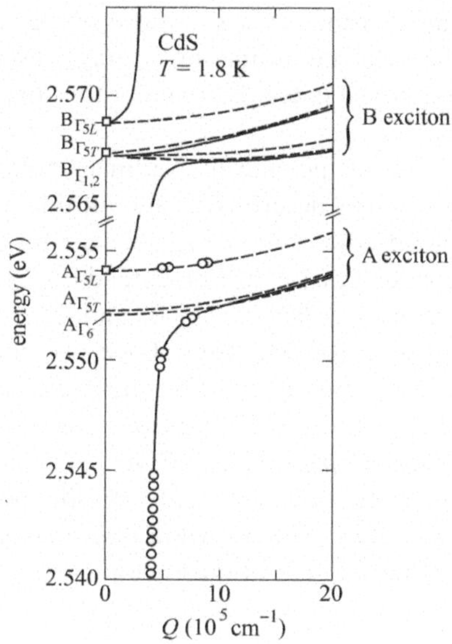

Fig. 8.15 Dispersion relation for ordinary polaritons in wurtzite-type semiconductors (here CdS) where only the A and B exciton resonances were considered. The two-oscillator model with spatial dispersion is calculated from Eq. (8.5.1). Experimental results (circles, squares) are obtained at a temperature of $T = 1.8$ K by hyper-Raman scattering and reflection spectroscopy, respectively. (Adapted from Refs. [4, 50, 51])

dipole-moment, while other states having Γ_6 or Γ_2 symmetry are spin-triplet states (labeled $|t\rangle_i^{ex}$), which are not dipole-active. In the context of exciton-polaritons we are especially interested in the states ($|x\rangle_i^{ex}$, $|y\rangle_i^{ex}$, $|z\rangle_i^{ex}$) since they may couple to the electromagnetic radiation field.

Wurtzite-type crystals are uniaxial crystals, i.e. they have a crystallographic axis, the so called "$\boldsymbol{c}$-axis". In crystal optics this direction defines also the "optical axis" of the crystal. In the following, when using Cartesian coordinates ($\boldsymbol{x}$, $\boldsymbol{y}$, $\boldsymbol{z}$), we choose $\boldsymbol{z} \parallel \boldsymbol{c}$ and $(\boldsymbol{x}, \boldsymbol{y}) \perp \boldsymbol{c}$. Then, the electric transition dipole moment $\boldsymbol{p}$ of an elementary excitation, which is discussed in the beginning of this chapter and defined in Eq. (8.1.34), is decomposed into two components labeled "$\boldsymbol{p}_\parallel$" for the component of $\boldsymbol{p} \parallel \boldsymbol{c}$ and "$\boldsymbol{p}_\perp$" for that of $\boldsymbol{p} \perp \boldsymbol{c}$.

Let us now discuss the properties of an electromagnetic radiation field $\boldsymbol{e}$, which propagates with wave-vector $\boldsymbol{Q}$ inside the crystal. One defines a "principal plane" or "main section" of a crystal as a plane that contains $\boldsymbol{Q}$ and the crystal-optical axis $\boldsymbol{c}$. (For instance, as sketched in the inset of Fig. 8.16 the vectors $\boldsymbol{x}$ and $\boldsymbol{z} \parallel \boldsymbol{c}$ span the principle plane, and the wave-vector $\boldsymbol{Q}$ is inclined by an angle β with respect to the crystal-optical axis $\boldsymbol{c}$.) We will consider first a light field $\boldsymbol{e}$ that is – contrary to the situation just discussed in connection with Fig. 8.16 – linearly polarized perpendicularly to the principal plane. If the wave-vector of the field changes now from $\boldsymbol{Q}$ to a wave-vector $\boldsymbol{Q}_1$ of the same module but with an arbitrary direction within the principal plane, its polarization remains perpendicular to the principal plane and thus to $\boldsymbol{c}$. Then the coupling of the field to the dipole moment, proportional to $\boldsymbol{p}_\perp$, is constant, independent of the angle β between the direction of $\boldsymbol{Q}$ and the optical axis $\boldsymbol{c}$. Consequently, the dielectric function $\epsilon(\omega)$ of the field (as introduced in Eq. (8.1.41)) is independent of the direction of propagation of $\boldsymbol{Q}$. In optics we call these rays "ordinary rays" whose propagation and refraction is governed by the "ordinary dielectric function $\epsilon_{ord}(\omega)$". (We choose the index "*ord*" instead of the usually employed index "*o*" in order to avoid confusion between ϵ_o and vacuum permittivity ϵ_0.)

The dispersion of "ordinary polaritons" is easily described by generalizing the procedure described in the introductory part of the present chapter and that of the one-oscillator model in Sect. 8.3 for the $\Gamma_6 \otimes \Gamma_7$ subspace in zincblende-type semiconductors. Let us first extend Eq. (8.3.5) to the case of three well-separated exciton resonances labeled A, B, and C as it may be the case in wurtzite-type semiconductors. As discussed above, we consider an electromagnetic radiation field with wave-vector $\boldsymbol{Q}$ lying in the principal plane spanned by the Cartesian coordinates $(\boldsymbol{x}, \boldsymbol{c})$. The polarization $\boldsymbol{p}$ is, as the electric field $\boldsymbol{e}$, perpendicular to the principal plane $(\boldsymbol{e}, \boldsymbol{p}) \perp \boldsymbol{c}$, i.e. $(\boldsymbol{e}, \boldsymbol{p}) \parallel \boldsymbol{y}$. All three exciton resonances couple to an electromagnetic radiation field with this polarization, giving rise to propagating ordinary polaritons. The polariton dispersions $E_i^{ord}(\boldsymbol{Q})$ is given for the different branches i by the solution of the equation

$$\hbar^2 c^2 Q^2 / E_i^{ord}(Q)^2 = \epsilon_b \left(1 + \sum_{j=(A,B,C)} (E_L^j(0)^2 - E_T^j(0)^2)/(E_T^j(Q)^2 - E_i^{ord}(Q)^2) \right), \quad (8.5.1)$$

c being the light velocity in vacuum and ϵ_b the background dielectric constant due to the oscillators that are not explicitly considered. $E_T^j(\boldsymbol{Q})$ denotes the dispersion of the exciton "j" when spatial dispersion is considered. The wave-vector dependence of the exciton energies has been neglected in the numerator of Eq. (8.5.1) since $E_L^j(\boldsymbol{Q})$ and $E_T^j(\boldsymbol{Q})$ vary almost in the same manner with $\boldsymbol{Q}$.

Figure 8.15 shows the typical dispersion relation for ordinary polaritons $E_i^{ord}(\boldsymbol{Q})$ of CdS, if only the A- and B-exciton resonances were considered. This dispersion relation is calculated from the two-oscillator model with spatial dispersion given in Eq. (8.5.1).

While the A-exciton of CdS (spanned by the exciton $\Gamma_7 \otimes \Gamma_9$-subspace) contains only second order dispersive terms, the B-exciton that is defined in the $\Gamma_7 \otimes \Gamma_7$-subspace contains also terms, varying linearly with wave-vector $\boldsymbol{Q}$ (see Eq. (6.3.38) and Eq. (6.3.39)). Experimental results are obtained by hyper-Raman scattering (circles) and reflection measurements (squares) (see Refs. [4, 50, 51]).

If the light field $\boldsymbol{e}$ is still linearly polarized but the polarization vector is lying inside the principal plane, i.e. if $\boldsymbol{e} \in (\boldsymbol{x}, \boldsymbol{c})$-plane, the situation is more complicated. Then the coupling strength of the light field with a transition dipole-moment depends on the direction of propagation of the field with respect to the $\boldsymbol{c}$-axis, i.e. on the angle β between the wave-vector $\boldsymbol{Q}$ and the optical axis $\boldsymbol{c}$. This angular-dependent coupling is taken into account by the "extraordinary dielectric function (ϵ_{eo})": In the case $\boldsymbol{Q} \parallel \boldsymbol{c}$ the polarization $\boldsymbol{p}$ is perpendicular to $\boldsymbol{c}$ (i.e. parallel to $\boldsymbol{x}$) and the coupling strength is proportional to $\boldsymbol{p}_\perp$. Then, the value of the dielectric function is given by $\epsilon_{ord}(\omega)$. If, however, the field propagation direction is perpendicular to $\boldsymbol{c}$, the electric

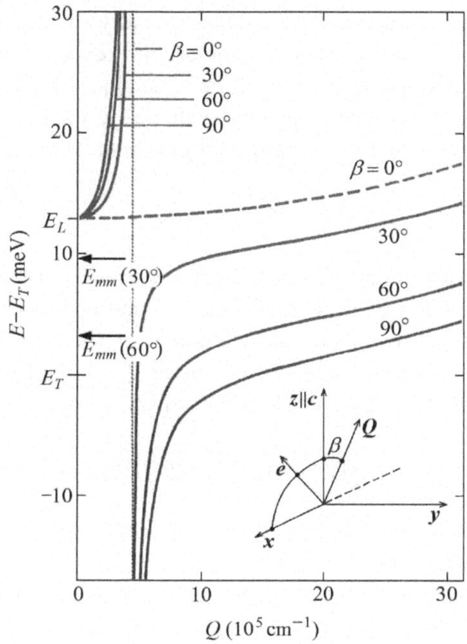

Fig. 8.16 Mixed-mode polariton dispersion in the one-oscillator model with spatial dispersion that applies for example to the B or C subspace exciton polaritons in wurtzite-type semiconductors. The inset shows the principal plane in which the propagation wave-vector $\boldsymbol{Q}$ and the electric field vector $\boldsymbol{e}$ lie. The principal plane is spanned by the vectors $\boldsymbol{x}$ and $z \parallel \boldsymbol{c}$. The wave-vector $\boldsymbol{Q}$ is inclined by the angle β with respect to the crystal-optical axis $\boldsymbol{c}$. A variation of β leads to a changing of the coupling between the exciton with dipole moment $\boldsymbol{p}_{\parallel}$ and the electric field $\boldsymbol{e}$ and thus to the dispersion of mixed-mode polaritons. It can be clearly seen how the longitudinal-transverse splitting diminishes with decreasing angle β. (Adapted from Ref. [1]. See text)

field vector $\boldsymbol{e}$ is parallel to $\boldsymbol{c}$ and therefore the coupling strength is proportional to $\boldsymbol{p}_{\parallel}$. In this situation, i.e. if $\boldsymbol{e} \parallel \boldsymbol{c}$ and $\boldsymbol{Q} \perp \boldsymbol{c}$, the value of the dielectric function is labeled $\epsilon_{eo}(\omega)$. For arbitrary angles β we denote the extra-ordinary dielectric function by $\epsilon_{eo}(\beta, \omega)$, which varies continuously as a function of β between $\epsilon_{ord}(\omega)$ and $\epsilon_{eo}(\omega)$ according to (Ref. [52]):

$$1/\epsilon_{eo}(\beta, \omega) = (\cos\beta)^2/\epsilon_{ord}(\omega) + (\sin\beta)^2/\epsilon_{eo}(\omega). \tag{8.5.2}$$

Let us consider in more detail the situation of the B or C exciton states, which transform as Γ_1 (i.e. like the position vector "z") in wurtzite-type semiconductors, as an example. These excitons have their dipole moment $\boldsymbol{p} \parallel \boldsymbol{c}$, which couples only to the component e_z of the propagating electromagnetic field $\boldsymbol{e}$. Its dipole-matrix element is given by $\boldsymbol{p}_{\parallel}$ if $\beta = 90°$. Otherwise, the dipole-matrix element depends on the projection of $\boldsymbol{e}$ onto the crystallographic $\boldsymbol{c}$-axis, which varies with β. Since polaritons excited in this configuration have a mixed character they are called "mixed-mode polaritons": the wave functions of the optically excited excitons $|z\rangle_i^{ex}$ have neither a longitudinal nor a transverse character with respect to the wave-vector $\boldsymbol{Q}$.

Let us consider only the influence of $\boldsymbol{p}_{\|}$. The dipole-matrix element squared is proportional to the difference in energy realized by the excitons when they are longitudinal or transverse states, i.e. to the longitudinal-transverse splitting Δ_{LT}:

$$\Delta_{LT} = E_L - E_T. \tag{8.5.3}$$

As we have discussed above, longitudinal and transverse exciton states have different energies because of the non-analytic exchange interaction between electron and hole. The exciton energy depends thus on the orientation of the transition dipole moment $\boldsymbol{p}$ with respect to its center-of-mass wave-vector $\boldsymbol{Q}$. Therefore, the longitudinal-transverse splitting of the mixed-mode polariton Δ_{LT}^{mm} as a function of β is given by

$$\Delta_{LT}^{mm}(\beta) = \Delta_{LT}\sin^2(\beta). \tag{8.5.4}$$

The longitudinal-transverse splitting of the mixed mode excitons $\Delta_{LT}^{mm}(\beta)$ is maximal if $(\boldsymbol{Q} \perp \boldsymbol{c})$ and vanishes if $(\boldsymbol{Q} \parallel \boldsymbol{c})$.

The energy of a longitudinal exciton E_L is independent of the light field since longitudinal excitons and the electromagnetic field do not couple, i.e. E_L has a constant value, independent of β. Then, Eqs. (8.5.3) and (8.5.4) determine the energy of the mixed-mode exciton state $E_{mm}(\beta)$ to

$$E_{mm}(\beta) = E_L - \Delta_{LT}^{mm}(\beta) = E_L - \Delta_{LT}\sin^2(\beta). \tag{8.5.5}$$

Following Eq. (8.1.53) the oscillator strength f_s of a dipole-active exciton is given by

$$\hbar^2 f_s = \epsilon_b(E_L^2 - E_T^2). \tag{8.5.6}$$

We thus obtain from Eq. (8.5.5) for the oscillator strength $f_{mm}(\beta)$ of the mixed-mode state

$$\hbar^2 f_{mm}(\beta) = \epsilon_b(E_L^2 - E_{mm}(\beta)^2). \tag{8.5.7}$$

We can now determine the dispersion relation of the mixed-mode polaritons $E_i^{mp}(\beta, \boldsymbol{Q})$, i.e. the energy of the coupled elementary excitations as a function of the wave-vector $\boldsymbol{Q}$, which depends on the angle β between $\boldsymbol{Q}$ and the crystal axis $\boldsymbol{c}$. Let us denote by $E_{mm}(\beta, \boldsymbol{Q})$ the energy of the mixed-mode excitons if we include similarly to Eq. (8.5.1) spatial dispersion. We then obtain the following equation:

$$\hbar^2 c^2 Q^2/E_i^{mp}(\beta, \boldsymbol{Q})^2 = \epsilon_b\left(1 + (E_L^2 - E_{mm}(\beta)^2)/(E_{mm}(\beta, \boldsymbol{Q})^2 - E_i^{mp}(\beta, \boldsymbol{Q})^2)\right). \tag{8.5.8}$$

Its solution determines the energy $E_i^{mp}(\beta, \boldsymbol{Q})$ of the mixed-mode polariton branch "i".

Figure 8.16 shows the typical mixed-mode polariton-dispersion (see Ref. [1]) in the one-oscillator model with spatial dispersion calculated from Eq. (8.5.8) for

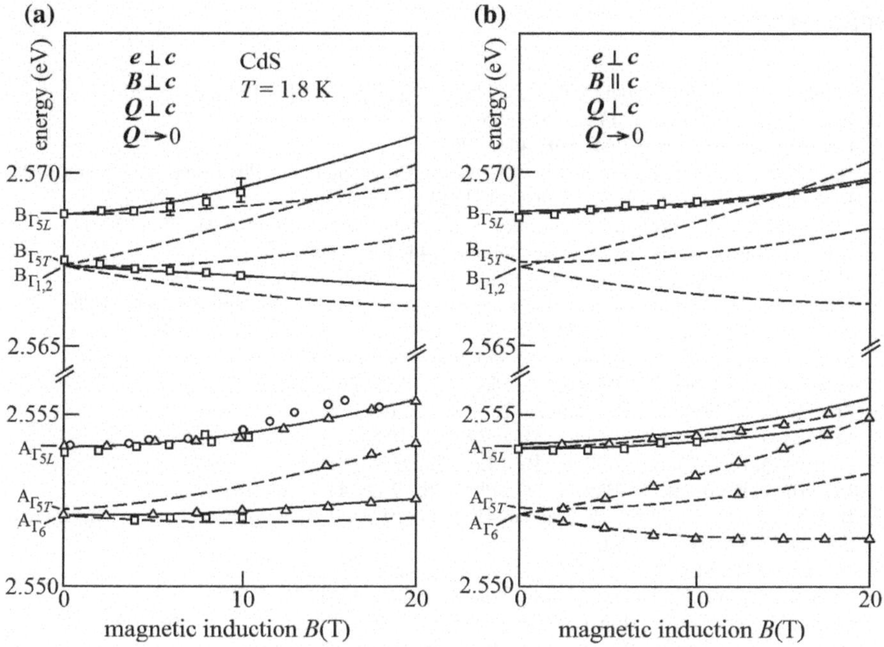

Fig. 8.17 Energetic positions of exciton (dashed lines) and exciton-polariton (full lines) states as a function of magnetic induction B in hexagonal CdS for $\boldsymbol{Q} \rightarrow 0$ in a configuration $\boldsymbol{Q} \perp \boldsymbol{c}$, $\boldsymbol{e} \perp \boldsymbol{c}$. Circles, squares and triangles indicate results obtained at a temperature of $T = 1.8$ K from hyper-Raman scattering, reflection and absorption spectroscopy, respectively. (**a**) $\boldsymbol{B} \perp \boldsymbol{c}$; (**b**) $\boldsymbol{B} \parallel \boldsymbol{c}$. (Adapted from Refs. [4, 50])

various angles β. The model applies to the B- or C-subspace exciton polaritons in wurtzite-type semiconductors, which transform as Γ_1 at wave-vector $\boldsymbol{Q} = 0$.

The above discussion shows that in experiments the samples have to be properly orientated, i.e. the direction of the crystallographic axis has to be well known. Furthermore it is advantageous to choose simple angular configurations: A considered external or internal perturbation $\boldsymbol{\Sigma}$ should be well-orientated with respect to the crystal $\boldsymbol{c}$-axis, mostly $\boldsymbol{\Sigma} \perp \boldsymbol{c}$ or $\boldsymbol{\Sigma} \parallel \boldsymbol{c}$. Then, crystal parameters can be determined when modeling the system within the framework discussed above. Such modeling has been performed for several symmetry-breaking perturbations. Figure 8.17a, b shows the energetic positions of exciton and exciton-polariton states as a function of the magnetic induction B in CdS for $\boldsymbol{Q} \rightarrow 0$ in a configuration $\boldsymbol{Q} \perp \boldsymbol{c}$, $\boldsymbol{e} \perp \boldsymbol{c}$. The orientation of the magnetic field is chosen in Fig. 8.17a $\boldsymbol{B} \perp \boldsymbol{c}$ and in Fig. 8.17b $\boldsymbol{B} \parallel \boldsymbol{c}$. Circles, squares and triangles indicate results from hyper-Raman scattering, reflection and absorption spectroscopy, respectively, while the dashed and full lines represent theoretical modeling. (According to Refs. [4, 50]).

References

1. Klingshirn, C.: Semiconductor Optics, 3rd edn. Springer, Berlin, Heidelberg (2005)
2. Hopfield, J.J.: Phys. Rev. **112**, 1555 (1958)
3. Hubbard, J.: Proc. Phys. Soc. **A68**, 441 (1955)
4. Hönerlage, B., Lévy, R., Grun, J.B., Klingshirn, C., Bohnert, K.: Phys. Rep. **124**, 161 (1985)
5. Kiselev, V.A., Razbirin, R.S., Uraltsev, I.N.: Phys. Status Solidi (b) **72**, 161 (1975)
6. Makarenko, I.V., Uraltsev, I.N., Kiselev, V.A.: Phys. Status Solidi (b) **98**, 773 (1980)
7. Mita, T., Nagasawa, N.: Solid State Commun. **44**, 1003 (1982)
8. Voigt, J., Senoner, M., Rückmann, I.: Phys. Status Solidi (b) **75**, 213 (1976)
9. Matsumoto, Y., Unuma, Y., Tanaka, Y., Shionoya, S.: J. Phys. Soc. Jpn **47**, 1844 (1979)
10. Matsumoto, Y., Unuma, Y., Shionoya, S.: J. Phys. Soc. Japan **49**, Supp. A, 393 (1980)
11. Segawa, Y., Aoyagi, Y., Namba, S.: Solid State Commun. **32**, 229 (1979)
12. Fröhlich, D., Mohler, E., Wiesner, P.: Phys. Rev. Lett. **26**, 554 (1971)
13. Staginnus, B., Fröhlich, D., Lampes, T.: Rev. Sci. Instr. **39**, 1129 (1968)
14. Fröhlich, D., Nieswand, W.: Phil. Mag. **70**, 321 (1994)
15. Beerwerth, F., Fröhlich, D.: Phys. Rev. Lett. **55**, 2603 (1985)
16. Hönerlage, B., Bivas, A., Phach, V.D.: Phys. Rev. Lett. **41**, 49 (1978)
17. Itoh, T., Suzuki, T.: J. Phys. Soc. Japan **45**, 1939 (1978); Itoh, T., Suzuki, T., Ueta, M.: J. Phys. Soc. Japan **42**, 1069 (1977)
18. Nagasawa, N., Mita, T., Ueta, M.: J. Phys. Soc. Japan **41**, 929 (1976)
19. Mita, T., Satome, K., Ueta, M.: Solid State Commun. **33**, 1135 (1980)
20. Phach, V.D., Bivas, A., Hönerlage, B., Grun, J.B.: Phys. Status Solidi (b) **86**, 159 (1978)
21. Brenig, W., Zeyher, R., Birman, J.L.: Phys. Rev. B **6**, 4617 (1972)
22. Ulbrich, R.G., Weisbuch, C.: Phys. Rev. Lett. **38**, 865 (1977)
23. Sermage, B., Fishman, G.: Phys. Rev. Lett. **43**, 1043 (1979); Sermage, B., Fishman, G.: Phys. Rev. B **23**, 5107 (1981)
24. Phach, V.D., Oka, Y., Cardona, M.: Phys. Rev. B **24**, 765 (1981)
25. Broser, I., Rosenzweig, M.: Solid State Commun. **36**, 1027 (1980)
26. Goto, T., Nishina, Y.: Solid State Commun. **31**, 751 (1979)
27. Merle, J.C., Sooryakuma, R., Cardona, M.: Phys. Rev. B **30**, 3261 (1984)
28. Koteles, E.S., Winterling, G.: Phys. Rev. B **20**, 628 (1979)
29. Koteles, E.S.: In: Excitons Rashba, E.I., Sturge, M.D. (eds.) North-Holland Publishing Company, Amsterdam, New York, Oxford (1982)
30. Broser, I., Rosenzweig, M.: J. Phys. Soc. Japan **49**, Suppl. A, 401 (1980)
31. Broser, I., Rosenzweig, M., Beckmann, E., Birkicht, E.: Solid State Commun. **39**, 1209 (1981)
32. Peyghambarian, N., Koch, S.W., Mysyrowicz, A.: Introduction to Semiconductor Optics. Prentice Hall (1993)
33. Cho, K.: Phys. Rev. B **14**, 4463 (1976)
34. Pekar, S.I.: JETP **33**, 1022 (1957); Sov. Phys. JETP **6**, 785 (1957)
35. Ginzburg, V.L.: JETP **34**, 1593 (1958); Sov. Phys. JETP **7**, 813 (1958)
36. Hopfield, J.J., Thomas, D.G.: Phys. Rev. **132**, 563 (1963)
37. Hönerlage, B., Rössler, U., Phach, V.D., Bivas, A., Grun. J.B.: Phys. Rev. B **22**, 797 (1980)
38. Bivas, A., Phach, V.D., Hönerlage, B., Rössler, U., Grun, J.B.: Phys. Rev. B **20**, 3442 (1979)
39. Suga, S., Cho, K., Bettini, M.: Phys. Rev. B **13**, 943 (1976)
40. Ulbrich, R.G., Weisbuch, C.: In: Festkörperprobleme XVIII—Advances in Solid State Physics. J. Treusch edn, p. 217. Vieweg Verlag, Braunschweig (1978)
41. Bir, G.L., Pikus, G.E.: Symmetry and Strain-Induced Effects in Semiconductors. Wiley, New York, Toronto (1974)
42. Agranovich, V.M., Ginzburg, V.L.: In: Spatial Dispersion in Crystal Optics and the Theory of Excitons. Interscience, London, New York, Sidney (1966)
43. Thomas, D.G., Hopfield, J.J.: Phys. Rev. Lett. **5**, 505 (1960). Phys. Rev. **124**, 657 (1961)
44. Merle, J.C., Bivas, A., Wecker, C., Grun, J.B.: Physica **117B/118B**, 296 (1983)

45. Suga, S., Cho, K., Niji, Y., Merle, J.C., Sauder, T.: Phys. Rev. B **22**, 4931 (1980)
46. Fiorini, P., Merle, J.C., Simon, M.: Phys. Rev. B **22**, 4341 (1980)
47. Nozue, Y., Ueta, M.: Solid State Commun. **36**, 781 (1980)
48. Nozue, Y., Miyahara, N., Takagi, S., Ueta, M.: Solid State Commun. **38**, 1199 (1981)
49. Merle, J.C., Bivas, A., Wecker, C., Hönerlage, B.: Phys. Rev. B **27**, 3709 (1983)
50. Blattner, G., Kurtze, G., Schmieder, G., Klingshirn, C.: Phys. Rev. B **25**, 7413 (1982)
51. Lyssenko, V.G., Kempf, K., Bohnert, K., Schmieder, G., Klingshirn, C., Schmitt-Rink, S.: Solid State Commun. **42**, 401 (1982)
52. Born, M.: Optik. Springer, Berlin, Heidelberg, New York (1981)

Appendix A
Example of Matrix Diagonalization

For those readers, who are not familiar with matrix formulation of quantum mechanics we wish to demonstrate how to solve the Schrödinger equation—a crucial task occurring in many places of the book—by hand, i.e. without making use of PC and any dedicated software as "Maple" or "Mathematica". We shall explain the overall procedure step by step, taking as an illustrative example a particular Schrödinger equation as formulated in Eqs. (3.1.9) and (3.1.10). The solution consists in diagonalization of the non-diagonal Hamiltonian matrix Eq. (3.1.9)

$$A' = \begin{pmatrix} z^2 & 0 & x^2 \\ 0 & y^2 & 0 \\ x^2 & 0 & z^2 \end{pmatrix},$$

where we have introduced notation

$$z^2 = 2Q_z^2 - Q_x^2 - Q_y^2\,,\, x^2 = 3(Q_x^2 - Q_y^2), \text{ and } y^2 = -4Q_z^2 + 2Q_x^2 + 2Q_y^2.$$

By performing the diagonalization process we shall obtain three required energy eigenvalues A_{kk} with $k = 1, 2, 3$, namely, three diagonal elements of the resulting diagonal matrix

$$A = \begin{pmatrix} A_{11} & 0 & 0 \\ 0 & A_{22} & 0 \\ 0 & 0 & A_{33} \end{pmatrix} = \begin{pmatrix} A_1 & 0 & 0 \\ 0 & A_2 & 0 \\ 0 & 0 & A_3 \end{pmatrix},$$

where we denoted for the sake of simplicity $A_{kk} = A_k$. Simultaneously, we can (if we wish) obtain the corresponding eigenfunctions $u_k(\boldsymbol{r})$ with $k = 1, 2, 3$ as a linear combination of basis functions $\nu_n(\boldsymbol{r})$:

$$u_k(\boldsymbol{r}) = \sum_{n=1}^{3} S_{kn}^* \nu_n(\boldsymbol{r}). \tag{A.1}$$

© Springer International Publishing AG, part of Springer Nature 2018

B. Hönerlage and I. Pelant, *Symmetry and Symmetry-Breaking in Semiconductors*, Springer Tracts in Modern Physics 279,
https://doi.org/10.1007/978-3-319-94235-3

The basis functions in this particular example are represented by Eq. (3.1.2): $\nu_1(\boldsymbol{r}) = |1\rangle$, $\nu_2(\boldsymbol{r}) = |0\rangle$ and $\nu_3(\boldsymbol{r}) = |-1\rangle$.

Let us start. There is a technical recipe how to diagonalize a non-diagonal matrix A' (see Ref. [1]): The condition

$$det(A'_{ml} - A_{ml}\delta_{ml}) = 0$$

is equivalent to

$$\sum_m S_{km}(A'_{ml} - A_k\delta_{ml}) = 0, \tag{A.2}$$

which has to be fulfilled. S_{km} have been introduced by Eq. (A.1), A'_{ml} and A_k denote elements of the matrices A' and A, respectively, and δ_{ml} represents the Kronecker delta:

$$\delta_{ml} = \begin{cases} 0 & \text{if } m \neq l \\ 1 & \text{if } m = l \end{cases}.$$

The solution of Eq. (A.2) will yield the requested energy eigenvalues A_k.

Because we work in this example in a 3-dimensional subspace, all subscripts in Eq. (A.1) and (A.2) span values 1, 2, and 3. We now write down Eq. (A.2) explicitly for $l = 1$:

$$\begin{aligned} k = 1 : S_{11}(A'_{11} - A_1) + S_{12}A'_{21} + S_{13}A'_{31} &= 0 \\ k = 2 : S_{21}(A'_{11} - A_2) + S_{22}A'_{21} + S_{23}A'_{31} &= 0 \\ k = 3 : S_{31}(A'_{11} - A_3) + S_{32}A'_{21} + S_{33}A'_{31} &= 0. \end{aligned} \tag{A.3}$$

A similar set of equations for $l = 2$ reads

$$\begin{aligned} k = 1 : S_{11}A'_{12} + S_{12}(A'_{22} - A_1) + S_{13}A'_{32} &= 0 \\ k = 2 : S_{21}A'_{12} + S_{22}(A'_{22} - A_2) + S_{23}A'_{32} &= 0 \\ k = 3 : S_{31}A'_{12} + S_{32}(A'_{22} - A_3) + S_{33}A'_{32} &= 0 \end{aligned} \tag{A.4}$$

and for $l = 3$

$$\begin{aligned} k = 1 : S_{11}A'_{13} + S_{12}A'_{23} + S_{13}(A'_{33} - A_1) &= 0 \\ k = 2 : S_{21}A'_{13} + S_{22}A'_{23} + S_{23}(A'_{33} - A_2) &= 0 \\ k = 3 : S_{31}A'_{13} + S_{32}A'_{23} + S_{33}(A'_{33} - A_3) &= 0. \end{aligned} \tag{A.5}$$

From the above system of Eqs. (A.3)–(A.5) one has to select three equations for three unknown quantities. By inspecting Eqs. (A.3)–(A.5) we can see that selection of the three first rows represents a set of three equations for the unknowns S_{11}, S_{12}, and S_{13}:

$$\begin{aligned} S_{11}(A'_{11} - A_1) + S_{12}A'_{21} + S_{13}A'_{31} &= 0 \\ S_{11}A'_{12} + S_{12}(A'_{22} - A_1) + S_{13}A'_{32} &= 0 \\ S_{11}A'_{13} + S_{12}A'_{23} + S_{13}(A'_{33} - A_1) &= 0. \end{aligned}$$

For our particular example this can be rewritten as

$$\begin{aligned} S_{11}(z^2 - A_1) + S_{13}x^2 &= 0 \\ S_{12}(y^2 - A_1) &= 0 \\ S_{11}x^2 + S_{13}(z^2 - A_1) &= 0. \end{aligned} \tag{A.6}$$

This is a homogeneous set of equations, requiring for a non-zero solution $S_{ij} \neq 0$ that the determinant D be equal zero:

$$D = \begin{vmatrix} (z^2 - A_1) & 0 & x^2 \\ 0 & y^2 - A_1 & 0 \\ x^2 & 0 & (z^2 - A_1) \end{vmatrix} = 0$$

or

$$(z^2 - A_1)(y^2 - A_1)(z^2 - A_1) - x^4(y^2 - A_1) = 0.$$

Let $y^2 \neq A_1$. Then $(z^2 - A_1) = \pm x^2$ is the required solution. We shall consider the (+) sign; in this case one obtains

$$A_1 = z^2 - x^2 = (2Q_z^2 - Q_x^2 - Q_y^2) - 3Q_x^2 + 3Q_y^2 = 2Q_z^2 - 4Q_x^2 + 2Q_y^2,$$

which is expression (3.1.11), indeed. (Considering the (−) sign leads to $A_1 = 2Q_z^2 + 2Q_x^2 - 4Q_y^2$, which is nothing else but the eigenvalue A_3, as we will see below.)

Three other equations for the three unknown quantities S_{21}, S_{22}, and S_{23} are defined by the second rows in Eqs. (A.3)–(A.5):

$$\begin{aligned} S_{21}(A'_{11} - A_2) + S_{22}A'_{21} + S_{23}A'_{31} &= 0 \\ S_{21}A'_{12} + S_{22}(A'_{22} - A_2) + S_{23}A'_{32} &= 0 \\ S_{21}A'_{13} + S_{22}A'_{23} + S_{33}(A'_{33} - A_2) &= 0 \end{aligned}$$

or

$$\begin{aligned} S_{21}(z^2 - A_2) + S_{23}x^2 &= 0 \\ S_{22}(y^2 - A_2) &= 0 \\ S_{21}x^2 + S_{23}(z^2 - A_2) &= 0. \end{aligned} \tag{A.7}$$

We put again

$$D = \begin{vmatrix} (z^2 - A_2) & 0 & x^2 \\ 0 & y^2 - A_2 & 0 \\ x^2 & 0 & (z^2 - A_2) \end{vmatrix} = 0$$

and therefore obtain:

$$(z^2 - A_2)(y^2 - A_2)(z^2 - A_2) - x^4(y^2 - A_2) = 0.$$

Let $y^2 \neq A_2$. Then $(z^2 - A_2)^2 = x^4$ and $A_2 = z^2 \pm x^2$ but this is A_1. It follows that $y^2 = A_2$ must hold, or $A_2 = y^2 = 2Q_x^2 + 2Q_y^2 - 4Q_z^2$ as quoted in expression (3.1.11). (Notice that $y^2 \neq A_1$ as it is required in the preceding step.)

The last three equations (for the unknown terms S_{31}, S_{32}, and S_{33}) follow from the third rows of Eqs. (A.3)–(A.5):

$$\begin{aligned} S_{31}(z^2 - A_3) + S_{33}x^2 &= 0 \\ S_{32}(y^2 - A_3) &= 0 \\ S_{31}x^2 + S_{33}(z^2 - A_3) &= 0. \end{aligned} \tag{A.8}$$

We put again

$$D = \begin{vmatrix} (z^2 - A_3) & 0 & x^2 \\ 0 & y^2 - A_3 & 0 \\ x^2 & 0 & (z^2 - A_3) \end{vmatrix} = 0$$

and thus

$$(z^2 - A_3)^2(y^2 - A_3) - x^4(y^2 - A_3) = 0.$$

Obviously $y^2 \neq A_3$ (because $y^2 = A_2$). Then: $(z^2 - A_3)^2 = x^4$ and $A_3 = z^2 + x^2 = 2Q_z^2 - Q_x^2 - Q_y^2 + 3(Q_x^2 - Q_y^2) = 2Q_x^2 - 4Q_y^2 + 2Q_z^2$ as quoted in expression (3.1.11).

In this way we have determined the eigenvalues A_1, A_2, and A_3. In order to know also the relevant eigenfunctions as given in Eq. (A.1), we have to find the coefficients S_{ij}^*. To do so, we substitute into Eq. (A.6) the solution A_1, and obtain thus three equations for three unknown quantities S_{11}, S_{12}, and S_{13}:

$$\begin{aligned} S_{11}x^2 + S_{13}x^2 &= 0 \\ S_{12}(y^2 - z^2 + x^2) &= 0 \\ S_{11}x^2 + S_{13}x^2 &= 0. \end{aligned}$$

Since x^2, y^2, and z^2 can take arbitrary values, it is obvious that in order to satisfy these equations, one has to put:

$$\begin{aligned} S_{12} &= 0 \\ S_{11} &= -S_{13} \neq 0. \end{aligned}$$

It follows then from Eq. (A.1) that the corresponding eigenfunction is equal to

$$u_1(\boldsymbol{r}) = |\Psi_{a3Q2}(1)\rangle = \text{const.}\,(-|1\rangle + |-1\rangle) \tag{A.9}$$

in compliance with Eq. (3.1.11). It is to be noted that in this way the eigenfunctions can be determined within accuracy of a multiplicative constant, which will be discussed soon. Repeating in a similar manner the same procedure, i.e. substituting the solution of A_2 into Eq. (A.7), we obtain:

$$
\begin{gathered}
S_{21}(z^2 - y^2) + S_{23}x^2 = 0 \\
S_{22}(y^2 - y^2) = 0 \\
S_{21}x^2 + S_{23}(z^2 - y^2) = 0.
\end{gathered}
$$

Taking into consideration the first and the third equation only, we have:

$$
\begin{gathered}
S_{21}(z^2 - y^2) + S_{23}x^2 = 0 \\
S_{21}x^2 + S_{23}(z^2 - y^2) = 0,
\end{gathered}
$$

which represents a homogeneous system of two equations for S_{21}, and S_{23}. To obtain a non-zero solution $S_{21} \neq 0$, and $S_{23} \neq 0$, we put

$$
D = \begin{vmatrix} (z^2 - y^2) & x^2 \\ x^2 & (z^2 - y^2) \end{vmatrix} = 0
$$

or

$$
\begin{gathered}
(z^2 - y^2)^2 = x^4 \\
(z^2 - y^2) = \pm x^2.
\end{gathered}
$$

But this relation is in contradiction with the introductory formulation of the problem, namely, with the very form of the matrix A'. Therefore, the only possible solution reads: $S_{21} = S_{23} = 0$ and $S_{22} \neq 0$. Thus we obtain

$$
u_2(\boldsymbol{r}) = |\Psi_{a3Q2}(2)\rangle = \text{const.}|0\rangle \tag{A.10}
$$

again in agreement with Eq. (3.1.11). We let to the reader to show that by substituting the solution for A_3 into Eq. (A.8) one obtains $S_{31} = S_{33} \neq 0$ and $S_{32} = 0$. Consequently

$$
u_3(\boldsymbol{r}) = |\Psi_{a3Q2}(3)\rangle = \text{const.}\left(|1\rangle + |-1\rangle\right). \tag{A.11}
$$

The values of the constants occurring in expressions (A.9)–(A.11) can be determined by making use of the condition of unitarity of the matrix S_{ij} [1]. This is due to the fact that the eigenfunctions of the operators are orthonormalized functions in the vector-space, which is considered. One thus obtains

$$
\sum_n S_{kn} S^*_{ln} = \delta_{kl}. \tag{A.12}
$$

Itemizing this sum for $k = l$ one obtains:

$$
\begin{aligned}
\text{if } k = l = 1: S_{11}S^*_{11} + S_{12}S^*_{12} + S_{13}S^*_{13} &= \delta_{11} = 1 \\
\text{if } k = l = 2: S_{21}S^*_{21} + S_{22}S^*_{22} + S_{23}S^*_{23} &= \delta_{22} = 1 \\
\text{if } k = l = 3: S_{31}S^*_{31} + S_{32}S^*_{32} + S_{33}S^*_{33} &= \delta_{33} = 1.
\end{aligned} \tag{A.13}
$$

The first Eq. (A.13) implies (since $S_{11} = -S_{13}$ and $S_{12} = 0$):

$$S_{11}S_{11}^* + S_{11}S_{11}^* = 1$$
$$2\,|S_{11}|^2 = 1$$
$$S_{11} = 1/\sqrt{2}.$$

At this point one should mention that wave functions have to be square integrable by definition but there is no restriction on their phase, which can be freely chosen. If several degenerate or nearly degenerate wave functions are considered, one has only to pay attention that they are defined with the same phase factor such that they remain orthogonal to each other if a superposition of these function is constructed, which diagonalizes a Hamiltonian. Consequently, we obtain now

$$u_1(\boldsymbol{r}) = |\Psi_{a3Q2}(1)\rangle = (1/\sqrt{2})\,(-|1\rangle + |-1\rangle)\,.$$

Similarly, the second Eq. (A.13) implies (since $S_{21} = S_{23} = 0$)

$$S_{22}S_{22}^* = 1$$
$$S_{22} = 1.$$

Consequently,

$$u_2(\boldsymbol{r}) = |\Psi_{a3Q2}(2)\rangle = |0\rangle.$$

Finally, by applying the third Eq. (A.13) we get (since $S_{32} = 0$ and $S_{31} = S_{33}$):

$$2S_{33}S_{33}^* = 1$$
$$S_{33} = S_{31} = 1/\sqrt{2}$$

and therefore

$$u_3(\boldsymbol{r}) = |\Psi_{a3Q2}(3)\rangle = (1/\sqrt{2})\,(|1\rangle + |-1\rangle)\,.$$

Appendix B
Basis Transformations of Matrices

In quantum mechanics, and in particular in its matrix formulation, it becomes quite often necessary to express various operators (i.e. the relevant matrices) using different basis "vectors". One example of such an operation has been exposed in the present textbook when transforming the Hamiltonian-related matrix P_{34}" (Eq. (3.2.7)) from one system of basis matrices, given by Eq. (3.2.1), to another one described by Eq. (3.2.12). The result of this operation, being displayed as P_{34so}" in Eq. (3.2.20), can be easily obtained by making use of the "Maple" or "Mathematica" software but, when doing so, the mathematics behind remains hidden. It is the purpose of this Appendix to unveil the calculation procedure step by step and to convince the reader that he/she is able to perform the whole computation himself, "by hand", if he/she will not regret the (effectively) sacrificed time.

First, let us formulate the problem. Let us suppose we have two basis systems ν_n and u_k, and let an operator $\hat{A}$ be expressed in each of them via its matrix elements as

$$A_{kl} = \int u_k^* \hat{A} u_l du \text{ and } A'_{kl} = \int \nu_k^* \hat{A} \nu_l d\nu.$$

At the same time there is an interrelationship between the two bases which reads (see Ref. [1])

$$u_k = \sum_n S_{kn}^* \nu_n, \tag{B.1}$$

where S is a matrix of unitary transformation. (We use here "S" as the usual notation for a matrix of unitary transformation. Care has to be taken not to confound this notation with the matrices S and S' presented in Tables 2.1, 2.2, 2.3 and 6.1–6.3.)

The A matrix (formulated using the u_k-basis) can be expressed through the A' matrix (ν_n-basis) as follows (see Ref. [1]):

$$A = SA'S^{-1} = SA'S^{+}, \tag{B.2}$$

© Springer International Publishing AG, part of Springer Nature 2018

B. Hönerlage and I. Pelant, *Symmetry and Symmetry-Breaking in Semiconductors*, Springer Tracts in Modern Physics 279,
https://doi.org/10.1007/978-3-319-94235-3

where S^{-1} denotes an inverted matrix of S, and S^{+} is the Hermitian conjugate of S. Now, in our particular case, let the ν_n-basis be represented by the six "vectors" defined in Eq. (3.2.1):

$$\nu_1^v = |1\rangle\alpha,\ \nu_2^v = |1\rangle\beta,\ \nu_3^v = |0\rangle\alpha,\ \nu_4^v = |0\rangle\beta,\ \nu_5^v = |-1\rangle\alpha,\ \nu_6^v = |-1\rangle\beta.$$

In a matrix form, they can be written down (in their "own" basis) also like this (we omit the superscript for the sake of simplicity):

$$\nu_n = \begin{pmatrix} \nu_1 & 0 & 0 & 0 & 0 & 0 \\ 0 & \nu_2 & 0 & 0 & 0 & 0 \\ 0 & 0 & \nu_3 & 0 & 0 & 0 \\ 0 & 0 & 0 & \nu_4 & 0 & 0 \\ 0 & 0 & 0 & 0 & \nu_5 & 0 \\ 0 & 0 & 0 & 0 & 0 & \nu_6 \end{pmatrix}.$$

Vectors of the u_k-basis in our example are defined by the columns of the matrix given by Eq. (3.2.12):

$$u_k = \begin{pmatrix} 0 & 0 & 0 & 0 & 0 & 1 \\ 0 & -\sqrt{6}/3 & 0 & 0 & \sqrt{3}/3 & 0 \\ 0 & \sqrt{3}/3 & 0 & 0 & \sqrt{6}/3 & 0 \\ -\sqrt{3}/3 & 0 & 0 & \sqrt{6}/3 & 0 & 0 \\ \sqrt{6}/3 & 0 & 0 & \sqrt{3}/3 & 0 & 0 \\ 0 & 0 & 1 & 0 & 0 & 0 \end{pmatrix}. \tag{B.3}$$

Let us consider the matrix A' expressed in Eq. (3.2.7) to be transformed with the aid of Eq. (B.2) from the ν_n- to the u_k-basis as an example:

$$A' = P_{34}\text{''} = \begin{pmatrix} 0 & -1 & 0 & 0 & 0 & 3 \\ -1 & 0 & 0 & 0 & 3 & 0 \\ 0 & 0 & 0 & 2 & 0 & 0 \\ 0 & 0 & 2 & 0 & 0 & 0 \\ 0 & 3 & 0 & 0 & 0 & -1 \\ 3 & 0 & 0 & 0 & -1 & 0 \end{pmatrix}. \tag{B.4}$$

In order to be able to exploit Eq. (B.2), we are now in a position to find the matrix S of an unitary transformation between ν_n and u_k. The general relation (B.1) can be in our case specified as

$$u_k = \sum_{n=1}^{6} S_{kn}^{*} \nu_n \text{ with } k = 1, \ldots, 6$$

and combining this expression with matrix (B.3), it is easy to itemize this sum (and thus to obtain the following six vectors u_1 to u_6, expressed in the basis of the ν_n) to

$$
\begin{aligned}
u_1 = -(\sqrt{3}/3)|0\rangle\beta + (\sqrt{6}/3)|-1\rangle\alpha &= S^*_{14}\nu_4 + S^*_{15}\nu_5 \\
u_2 = -(\sqrt{6}/3)\nu_2 + (\sqrt{3}/3)\nu_3 &= S^*_{22}\nu_2 + S^*_{23}\nu_3 \\
u_3 = \nu_6 &= S^*_{36}\nu_6 \\
u_4 = (\sqrt{6}/3)\nu_4 + (\sqrt{3}/3)\nu_5 &= S^*_{44}\nu_4 + S^*_{45}\nu_5 \\
u_5 = (\sqrt{3}/3)\nu_2 + (\sqrt{6}/3)\nu_3 &= S^*_{52}\nu_2 + S^*_{53}\nu_3 \\
u_6 = \nu_1 &= S^*_{61}\nu_1,
\end{aligned}
$$

which corresponds exactly to Eq. (3.2.13). (For the sake of simplicity and clearness we have chosen here the notation "u_k" instead of the v_k^v-basis functions given in Chap. 3.) It follows immediately that $S \equiv S^*$ (there are only real coefficients in the above expressions for u_1 to u_6) and this matrix reads

$$
S \equiv S^* = \begin{pmatrix}
0 & 0 & 0 & -\sqrt{3}/3 & \sqrt{6}/3 & 0 \\
0 & -\sqrt{6}/3 & \sqrt{3}/3 & 0 & 0 & 0 \\
0 & 0 & 0 & 0 & 0 & 1 \\
0 & 0 & 0 & \sqrt{6}/3 & \sqrt{3}/3 & 0 \\
0 & \sqrt{3}/3 & \sqrt{6}/3 & 0 & 0 & 0 \\
1 & 0 & 0 & 0 & 0 & 0
\end{pmatrix}.
$$

Before we proceed further, we have to check whether this matrix is unitary, as requested by Eq. (B.1). A unitary matrix S is by definition a matrix whose Hermitian conjugate S^+ is equal to its inverted matrix S^{-1} :

$$
S^+ = S^{-1}. \tag{B.5}
$$

Remembering the definition of a Hermitian conjugate matrix $S^+_{ij} = S^*_{ji}$ (i.e. a simple interchange of columns for rows in our case) we obtain

$$
S^+ = \begin{pmatrix}
0 & 0 & 0 & 0 & 0 & 1 \\
0 & -\sqrt{6}/3 & 0 & 0 & \sqrt{3}/3 & 0 \\
0 & \sqrt{3}/3 & 0 & 0 & \sqrt{6}/3 & 0 \\
-\sqrt{3}/3 & 0 & 0 & \sqrt{6}/3 & 0 & 0 \\
\sqrt{6}/3 & 0 & 0 & \sqrt{3}/3 & 0 & 0 \\
0 & 0 & 1 & 0 & 0 & 0
\end{pmatrix} \tag{B.6}
$$

and then we can easily calculate the product SS^+:

$$SS^{+}=S^{+}S=\begin{pmatrix}1&0&0&0&0&0\\0&1&0&0&0&0\\0&0&1&0&0&0\\0&0&0&1&0&0\\0&0&0&0&1&0\\0&0&0&0&0&1\end{pmatrix}\equiv I.$$

It means that S^{+} possesses the property of an inverted matrix, $S^{+}=S^{-1}$, indeed. Consequently, the S matrix is unitary and we may proceed to calculate the product defined by Eq. (B.2), namely, $A=SA'S^{-1}$. At first, applying a bit tedious but straightforward algebra, we calculate using Eq. (B.4), (B.5), and (B.6):

$$A'S^{-1}=P_{34}\text{''}S^{-1}=\begin{pmatrix}0&-1&0&0&0&3\\-1&0&0&0&3&0\\0&0&0&2&0&0\\0&0&2&0&0&0\\0&3&0&0&0&-1\\3&0&0&0&-1&0\end{pmatrix}\begin{pmatrix}0&0&0&0&0&1\\0&-\sqrt{6}/3&0&0&\sqrt{3}/3&0\\0&\sqrt{3}/3&0&0&\sqrt{6}/3&0\\-\sqrt{3}/3&0&0&\sqrt{6}/3&0&0\\\sqrt{6}/3&0&0&\sqrt{3}/3&0&0\\0&0&1&0&0&0\end{pmatrix}=$$

$$=\begin{pmatrix}0&\sqrt{6}/3&3&0&-\sqrt{3}/3&0\\\sqrt{6}&0&0&\sqrt{3}&0&-1\\-2\sqrt{3}/3&0&0&2\sqrt{6}/3&0&0\\0&2\sqrt{3}/3&0&0&2\sqrt{6}/3&0\\0&-\sqrt{6}&-1&0&\sqrt{3}&0\\-\sqrt{6}/3&0&0&-\sqrt{3}/3&0&3\end{pmatrix}\equiv Q.$$

Finally

$$A=SQ=\begin{pmatrix}0&0&0&-\sqrt{3}/3&\sqrt{6}/3&0\\0&-\sqrt{6}/3&\sqrt{3}/3&0&0&0\\0&0&0&0&0&1\\0&0&0&\sqrt{6}/3&\sqrt{3}/3&0\\0&\sqrt{3}/3&\sqrt{6}/3&0&0&0\\1&0&0&0&0&0\end{pmatrix}\begin{pmatrix}0&\sqrt{6}/3&3&0&-\sqrt{3}/3&0\\\sqrt{6}&0&0&\sqrt{3}&0&-1\\-2\sqrt{3}/3&0&0&2\sqrt{6}/3&0&0\\0&2\sqrt{3}/3&0&0&2\sqrt{6}/3&0\\0&-\sqrt{6}&-1&0&\sqrt{3}&0\\-\sqrt{6}/3&0&0&-\sqrt{3}/3&0&3\end{pmatrix}=$$

$$=\begin{pmatrix}0&-8/3&-\sqrt{6}/3&0&\sqrt{2}/3&0\\-8/3&0&0&-\sqrt{2}/3&0&\sqrt{6}/3\\-\sqrt{6}/3&0&0&-\sqrt{3}/3&0&3\\0&-\sqrt{2}/3&-\sqrt{3}/3&0&7/3&0\\\sqrt{2}/3&0&0&7/3&0&-\sqrt{3}/3\\0&\sqrt{6}/3&3&0&-\sqrt{3}/3&0\end{pmatrix}\equiv P_{34so}\text{''}$$

in full agreement with Eq. (3.2.20) of the main text.

We encourage the reader to transform in a quite similar way (and using the same matrix S) Z_{51}" from Eq. (3.2.18) to obtain Z_{51so}" as given by Eq. (3.2.21) and also to transform Z_{44}" from Eq. (3.2.19) to Z_{44so}" in Eq. (3.2.22).

Appendix C
Matrix Direct Product

A well-known matrix product occurring quite often both in linear algebra and in physics is that of two $(q \times q)$ square matrices of the same order q,

$$A = (a_{mn}) \text{ and } B = (b_{mn}),$$

which is defined as

$$C = A \cdot B = (a_{mn})(b_{mn}) = \sum_{p=1}^{q} a_{mp} b_{pn} = (c_{mn}) \text{ with } (m, n) = 1, \ldots, q.$$

Or, explicitly, an element c_{mn} standing on the point of intersection of the m-th row and the n-th column of the resulting matrix C is equal to a scalar product of the m-th row of matrix A with the n-th column of matrix B, or

$$c_{mn} = a_{m1} b_{1n} + a_{m2} b_{2n} + \cdots + a_{mq} b_{qn}.$$

The matrix C remains to be a $(q \times q)$ square matrix.

However, apart from this common matrix product, another product called "Matrix direct product" or "Kronecker product" can be defined. This kind of matrix product is of particular importance in physics. Let us have a $(m \times n)$ matrix A and a $(p \times q)$ matrix B. (In general, since $m \neq n$ and $p \neq q$, neither A nor B thus need to be a square matrix.) The Kronecker product

$$C = A \otimes B$$

is defined as

$$c_{\alpha\beta} = a_{ij} b_{kl}, \tag{C.1}$$

where

© Springer International Publishing AG, part of Springer Nature 2018

B. Hönerlage and I. Pelant, *Symmetry and Symmetry-Breaking in Semiconductors*, Springer Tracts in Modern Physics 279,
https://doi.org/10.1007/978-3-319-94235-3

$$\begin{aligned} \alpha &= p(i-1)+k, \alpha = 1, \ldots, mp \\ \beta &= q(j-1)+l, \beta = 1, \ldots, nq. \end{aligned} \tag{C.2}$$

To understand better this definition, it might be useful to discuss a simple example. Let

$$A = \begin{pmatrix} a_{11} & a_{12} & a_{13} \\ a_{21} & a_{22} & a_{23} \\ a_{31} & a_{32} & a_{33} \end{pmatrix} \text{ with } (m = n = 3) \text{ and } B = \begin{pmatrix} b_{11} & b_{12} \\ b_{21} & b_{22} \end{pmatrix} \text{ with } (p = q = 2).$$

In this case, when applying Eqs. (C.1) and (C.2), we obtain

$$C = A \otimes B = \begin{pmatrix} a_{11} & a_{12} & a_{13} \\ a_{21} & a_{22} & a_{23} \\ a_{31} & a_{32} & a_{33} \end{pmatrix} \otimes \begin{pmatrix} b_{11} & b_{12} \\ b_{21} & b_{22} \end{pmatrix} = \\ = \begin{pmatrix} c_{11} & c_{12} & c_{13} & c_{14} & c_{15} & c_{16} \\ c_{21} & c_{22} & c_{23} & c_{24} & c_{25} & c_{26} \\ c_{31} & c_{32} & c_{33} & c_{34} & c_{35} & c_{36} \\ c_{41} & c_{42} & c_{43} & c_{44} & c_{45} & c_{46} \\ c_{51} & c_{52} & c_{53} & c_{54} & c_{55} & c_{56} \\ c_{61} & c_{62} & c_{63} & c_{64} & c_{65} & c_{66} \end{pmatrix},$$

where

$$\begin{aligned} c_{11} &= a_{11}b_{11}, c_{12} = a_{11}b_{12}, c_{13} = a_{12}b_{11}, c_{14} = a_{12}b_{12}, c_{15} = a_{13}b_{11}, c_{16} = a_{13}b_{12} \\ c_{21} &= a_{11}b_{21}, c_{22} = a_{11}b_{22}, c_{23} = a_{12}b_{21}, c_{24} = a_{12}b_{22}, c_{25} = a_{13}b_{21}, c_{26} = a_{13}b_{22} \\ &\ldots \\ &\ldots \\ &\ldots \\ &\ldots, c_{64} = a_{32}b_{22}, c_{65} = a_{33}b_{21}, c_{66} = a_{33}b_{22}. \end{aligned} \tag{C.3}$$

It means that the first row in the $C = A \otimes B$ matrix is formed by an ordered sequence of three pairs of numbers: the first pair is created by multiplying a_{11} at first with b_{11} and then with b_{12}, the second pair originates when multiplying successively a_{12} with b_{11} and then with b_{12} etc. The fifth row is formed e.g. by multiplying in an analogous way the elements a_{3i} (the 3rd row of A) with the elements b_{1j} (the 1st row of B).

As a numerical example, let us calculate the Kronecker product P_{34}" quoted in the main text with the aid of Eq. (3.2.7):

$$P_{34}" = \left(3\begin{pmatrix}0&0&1\\0&0&0\\1&0&0\end{pmatrix} - \begin{pmatrix}1&0&0\\0&-2&0\\0&0&1\end{pmatrix}\right) \otimes \begin{pmatrix}0&1\\1&0\end{pmatrix} =$$

$$= \begin{pmatrix}-1&0&3\\0&2&0\\3&0&-1\end{pmatrix} \otimes \begin{pmatrix}0&1\\1&0\end{pmatrix}.$$

Simple application of Eq. (C.3) yields immediately

$$P_{34}" = \begin{pmatrix}0&-1&0&0&0&3\\-1&0&0&0&3&0\\0&0&0&2&0&0\\0&0&2&0&0&0\\0&3&0&0&0&-1\\3&0&0&0&-1&0\end{pmatrix}$$

in compliance with the right hand side of Eq. (3.2.7).

Note: The Kronecker product of two unit matrices results again in a unit matrix (of a higher order), for example:

$$\begin{pmatrix}1&0\\0&1\end{pmatrix} \otimes \begin{pmatrix}1&0&0\\0&1&0\\0&0&1\end{pmatrix} = \begin{pmatrix}1&0&0\\0&1&0\\0&0&1\end{pmatrix} \otimes \begin{pmatrix}1&0\\0&1\end{pmatrix} =$$

$$= \begin{pmatrix}1&0&0&0&0&0\\0&1&0&0&0&0\\0&0&1&0&0&0\\0&0&0&1&0&0\\0&0&0&0&1&0\\0&0&0&0&0&1\end{pmatrix}.$$

Appendix D
Some Elements of Group Theory Applied to Crystalline Solids

As discussed in Sect. 1.4, a crystal is characterized by the "Bravais lattice", which consists of a periodical arrangement of atoms in one, two or three dimensions (dimension n =1, 2, or 3). The Bravais lattice is characterized by the fact that, if some well defined discrete spatial translations are applied to the system, the environment of a point remains invariant under these translations.

If we neglect translations, there are five other symmetry operations, which may leave an object invariant. To each operation belong symmetry elements, which specify the fixed points. The symmetry operations and symmetry elements are in general denoted by:

- E: The identity operation in which no action is applied to the object
- C_n: An n-fold rotation, i.e. the symmetry element is a rotation by $2\pi/n$ around an axis, which is called the "symmetry axis" of the system
- σ: A reflection on a mirror plane which has to be defined
- i: The inversion with respect to one point, which is called the "center of inversion". Taken as the origin, the coordinates $\boldsymbol{r}$ of a point are changed into $-\boldsymbol{r}$
- S_n: An n-fold improper rotation, i.e. the symmetry element is a rotation C_n by $2\pi/n$ around an axis, followed by a reflection on a plane perpendicular to this rotation axis.

In S_n the axis is called an "n-fold rotation-reflection axis". Similarly, the rotation may be followed by an inversion in a point lying on the rotation axis. The axis is then called an "n-fold rotation-inversion axis". Although neither of the operations alone are symmetry elements, the combined actions are. For example, the operation S_1 corresponds to a reflection, S_2 to an inversion.

Concerning the n-fold rotation C_n, n may take every value in centro-symmetric atoms. In crystals, however, n can only take the discrete values n = (2, 3, 4, or 6) because of the translational invariance of the system. If a system has several axis of rotation, the one with the largest value of n (if it is unequivocal) is called the "principal axis" of rotation. If $n \geq 3$ the sense of rotation becomes important. Then

© Springer International Publishing AG, part of Springer Nature 2018

B. Hönerlage and I. Pelant, *Symmetry and Symmetry-Breaking in Semiconductors*, Springer Tracts in Modern Physics 279,
https://doi.org/10.1007/978-3-319-94235-3

a clockwise rotation (seen from above) is noted C_n^- and a counter-clockwise rotation (positive in the mathematical sense) C_n^+.

One remarks that C_n^- and C_n^+ applied one after the other to the system annihilate the operations performed by the two symmetry elements. (This corresponds to the identity operation "E".) We call this successive application of elements to a system a "multiplication". Let us call $R = C_n^-$ and $S = C_n^+$. If we apply first R and then S to the system, we write it as the product SR. As discussed above, we obtain $SR = E$. So, in the language of mathematics, S is the inverse element of R or $S = R^{-1}$. In addition, we find that $RS = SR = E$, i.e. $RR^{-1} = R^{-1}R = E$.

Concerning a reflection on a mirror plane σ, this reflection plane is called a "vertical mirror plane" if it contains the principal axis of rotation. This is indicated by σ_v. If the plane is perpendicular to the principal axis, it is called a "horizontal mirror plane" indicated by σ_h. A "diagonal mirror plane" is a vertical plane (i.e. a plane containing the principal axis), which bisects the angle between two C_2 axis, which are orientated perpendicular to the principal axis. The symmetry element is indicated by σ_d.

In general, if R is an element of the set of operations its inverse element R^{-1} is also an element of this set. In addition, if $[R, S]$ are two elements of the set, their product $(RS) = T$ is also an element of the set. Furthermore, the multiplication, which we have defined above, is associative, i.e. if $[R, S, U]$ are elements of the set of operations, which we have considered, one obtains that $(RS)U = R(SU)$.

One sees that the above mentioned set of symmetry elements form a group in the language of mathematics. This means that they obey the following rules:

1. The identity is an element of the set
2. The multiplication of elements is associative
3. If two elements R and S are members of the set, their product RS is also a member of the set
4. The inverse R^{-1} of an element R is also a member of the set.

Using the symmetry operations discussed above together with the condition that a crystal forms a Bravais lattice, thirty-two distinct crystallographic point groups can be constructed. Using the Schoenflies' notation for their classification, one obtains:

The five cubic point groups O_h, O, T_d, T, T_h :

- O_h is the full symmetry group of a cube
- O the cubic group without improper rotations
- T_d is the full symmetry group of a regular tetrahedron
- T is the full symmetry group of the regular tetrahedron without improper operations (i.e. operations that transform right-handed objects into left-handed ones as reflections or inversions)
- T_h is the result when inversion symmetry is added to the elements of T.

Using Schoenflies' notation for their classification, the other crystallographic point groups in the different crystal classes are given by:

- C_n: groups containing only an n-fold rotations

- C_{nv}: groups containing an n-fold rotation axis as well as vertical mirror planes that contain the rotation axis. Their number is determined by the symmetry operation C_n
- C_{nh}: groups containing an n-fold rotation axis as well as a horizontal mirror plane perpendicular to the rotation axis
- S_n: groups containing only an n-fold rotation-reflection axis
- D_n: groups containing an n-fold rotation axis as well as two-fold rotation axes that are perpendicular to the n-fold rotation axis. Their number is determined by the symmetry operation C_n
- D_{nh}: groups containing the elements of D_n as well as a horizontal mirror plane perpendicular to the n-fold rotation axis
- D_{nd}: groups containing the elements of D_n as well as mirror planes, containing the n-fold rotation axis and bisecting the angles between the two-fold rotation axes. Their number is determined by the symmetry operation C_n.

In order to summarize we give the symmetries of the thirty-two distinct crystallographic point groups in the different three-dimensional crystal classes which are:

- 5 Cubic point groups: O_h, O, T_d, T, T_h
- 7 Tetragonal point groups: C_4, C_{4v}, C_{4h}, S_4, D_4, D_{4h}, D_{2d}
- 3 Orthorhombic point groups: C_{2v}, D_2, D_{2h}
- 3 Monoclinic point groups: C_2, C_{2h}, C_{1h}
- 2 Triclinic point groups: C_1, S_2
- 5 Trigonal point groups: C_3, C_{3v}, S_6, D_3, D_{3d}
- 7 Hexagonal point groups: C_6, C_{6v}, C_{6h}, C_{3h}, D_6, D_{6h}, D_{3h}.

We are mostly interested in the simple cubic semiconductors with T_d and O_h symmetry and the hexagonal ones with C_{6v} symmetry. The symmetry elements of the cubic T_d and O_h structures are:

- T_d: $E, 8C_3, 3C_2, 6\sigma_d, 6S_4$
- O_h: In addition to the elements of the T_d structure the inversion i.

For C_{6v} symmetry one obtains:

- C_{6v}: $E, C_2, 2C_3, 2C_6, 3\sigma_d, 3\sigma_v$.

When considering T_d point-group symmetry, we see that there are five symmetry classes (E, C_3, C_2, σ_d, and S_4) into which the twenty-four symmetry elements are divided. One can state that rotations by the same angle around equivalent axes or reflections by equivalent planes belong to the same class. Each class gives rise to one "irreducible representation" for which one or several wave-functions are defined. Since the electron-wave functions have to be compatible with the symmetry operations of the crystal (i.e. have to be invariant under the symmetry operations) they can be characterized by their symmetry properties, and the functions are then called to transform as the irreducible representation. Each crystal-wave function can then be decomposed into a linear combination of electron-wave functions transforming as an irreducible representation of a definite symmetry.

Reference

1. Yariv, A.: Quantum Electronics, 2nd edn. J. Wiley and Sons, New York (1975)

Index

© Springer International Publishing AG, part of Springer Nature 2018

B. Hönerlage and I. Pelant, *Symmetry and Symmetry-Breaking in Semiconductors*, Springer Tracts in Modern Physics 279,
https://doi.org/10.1007/978-3-319-94235-3

D

E

The manufacturer's authorised representative in the EU is Springer Nature Customer Service Centre GmbH, Europaplatz 3, 69115 Heidelberg, Germany. If you have any concerns regarding our products, please contact ProductSafety@springernature.com

Printed and bound by CPI Group (UK) Ltd, Croydon, CR0 4YY
05/10/2026
02228135-0006